NON-CHROMATOGRAPHIC CONTINUOUS SEPARATION TECHNIQUES

Non-chromatographic Continuous Separation Techniques

M. Valcárcel and M.D. Luque de Castro
Department of Analytical Chemistry
University of Cordoba, Spain

THE ROYAL SOCIETY OF CHEMISTRY
1841–1991

A catalogue record for this book is available from the British Library.

ISBN 0-85186-986-6

Published by The Royal Society of Chemistry,
Thomas Graham House, The Science Park, Cambridge CB4 4WF

Typeset by AsTec, Newport, Saffron Walden, Essex
Printed and bound in Great Britain by Bookcraft (Bath) Ltd.

Preface

Separation techniques play a crucial role in today's analytical processes as applied to real samples. Chromatographic methods have gradually superseded their batch manual counterparts, which they clearly surpass in analyte discrimination and interference removal capabilities. With the late boom of automatic continuous analysers—undoubtedly promoted by the elimination of the traditional air bubbles of classical air-segmented methodologies—non-chromatographic continuous separation techniques have aroused increasing interest and gained widespread popularity as intermediate alternatives to manual and chromatographic techniques on account of their relevance to laboratories concerned with routine analyses or R&D strategies. The intrinsic dynamic character of non-chromatographic continuous systems allows for on-line coupling not only to chromatographs (GC, HPLC, SFC), but also to optical and electroanalytical detectors, which lend themselves readily to automation and miniaturization.

Our ultimate aim in writing this book was to provide a systematic overview of non-chromatographic continuous separation techniques, with special emphasis on those already firmly established or about to become serious alternatives, and those already available for implementation on commercial equipment. Rather than dealing with them in isolation, we opted for grouping them according to the type of interface across which mass transfer takes place in each instance. Thus, Chapter 2 deals with gas–liquid systems (direct introduction of gas samples, gas diffusion, hydride generation, vaporization), while Chapter 3 discusses the two most common liquid–liquid systems, namely dialysis and extraction, and Chapter 4 is devoted to the wide variety of non-chromatographic separation techniques involving solid–liquid interfaces (sorption, precipitation–dissolution, electrochemical stripping, leaching). Finally, Chapter 1 introduces the reader to the subject matter of the book and Chapter 5 provides an overall evaluation of the techniques dealt with in the previous three chapters and a brief description of two late techniques of increasing interest, *viz* field flow fractionation and capillary electrophoresis.

The wide variety of analytical problems, species (inorganic, organic, biochemical) and samples (clinical, pharmaceutical, environmental, industrial, nutritional) that can be managed by non-chromatographic continuous separation techniques make them equally relevant to analytical chemists and professionals concerned with methods of analysis.

This book is intended for use neither as a monograph nor as a textbook, but as something in between. Each technique is dealt with by describing not only its fundamentals, but also its different variations and applications wherever applicable. As far as possible, each is also critically compared with some other separation technique and outstanding advantages or disadvantages in this respect are underlined.

We should like to thank all those who have helped this book to materialize. Thus, we are grateful to Antonio Losada for translating the original Spanish manuscript and to Pablo Ruiz for drawing the profuse artwork. The warm welcome of the project by the Royal Society of Chemistry also deserves our grateful acknowledgement, as does financial support granted by the Spanish *Comisión Interministerial de Ciencia y Tecnología* (CICyT) for the realization of work relevant to the contents of this book.

THE AUTHORS
Cordoba, November 1990

Contents

CHAPTER 1

Introduction to Continuous Separation Techniques

1.1 Analytical Chemistry Today

Like many other branches of science and technology, analytical chemistry has experienced an enormous expansion in the last few decades of this century, promoted by scientific advances in general and by the occasionally pressing demands of a permanently, almost frantically changing society. As can be seen from Figure 1.1, the present challenges of analytical chemistry are basically the obtainment of more qualitative, quantitative, and structural information of greater quality on the matter or system concerned by expending less material, time, effort, and economic resources. The pursuit of these goals currently relies on a number of analytical chemical strategies based on the development of novel instrumentation, automation, hyphenated techniques, work outside the laboratory environment (*e.g.* on-line process control, relations with other scientific and technical areas), application of chemometric techniques, immobilization of analytes, reagents, catalysts, and reaction products, development of physical and (bio)chemical sensors, use of biological systems, implementation of multi-determinations, increased accuracy and precision, and higher safety and quality levels at control laboratories, among others.[1]

All these trends have an essential common aim, namely improving on the existing levels of such basic analytical features as sensitivity, selectivity, precision (accuracy), rapidity, reliability, scope of application, analytical costs, versatility, and sample throughput, in addition to other, more specific properties. To fulfil these objectives, analytical chemists have developed a host of methods and techniques to cope with the growing demands of society.

Research and development in analytical chemistry can be approached in various ways. Addressing a given analytical problem (*e.g.* the determi-

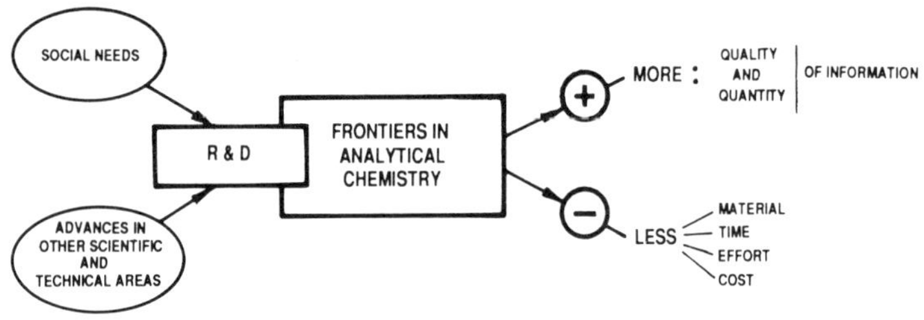

Figure 1.1 *General frontiers of R&D in analytical chemistry*

nation of the proportion in which enantiomers occur in pharmaceutical preparations) stimulates basic research, the results of which materialize in the development of new analytical methods to solve the problem in question. On the other hand, innovations occasionally result from advances in other areas of science and technology that allow existing analytical methods to be superseded by others improving on previous analytical performance. In any case, research exclusively aimed at publication of results—still not a rare occurrence, unfortunately—should be discouraged.

No doubt the word 'interface' is rather common in today's scientific jargon including that of analytical chemistry. Figure 1.2 illustrates four of the most significant accepted uses of this word in the context of this subject. Mass transfer between two phases previously brought into contact is the universal foundation of separation processes used for centuries, processes that still occupy a prominent place in analytical chemistry and are the subject matter of this book.

Hyphenated techniques resulting from the combination of two or three chromatographic (HPLC, GC, SFC) and spectroscopic techniques (MS, FTIR, NMR, AAS, AES) are rather an apt development as the shortcomings of chromatography (*e.g.* small identification power) and spectroscopy (the need for analytes to be previously isolated) are offset by their complementary advantages (*e.g.* chromatographic resolution and the high identification capacity of spectroscopy). In hyphenated methodologies the interface between the two techniques that are coupled on-line is one of the cornerstones on which the efficiency of the resultant hyphenated technique rests.[2, 3]

Connecting a microcomputer to an analytical instrument also calls for an appropriate interface, whether *passive* (devoted to acquiring analytical signals exclusively) or *active* (allowing instrumental parameters to be controlled via the computer keyboard). Many of the problems with which modern laboratories are currently confronted arise from the communication between instruments and computers, which can only be accomplished efficiently via an appropriate interface.[4, 5]

Real scientific and technological advances rely increasingly heavily on integration of the efforts of professionals from various fields. The future

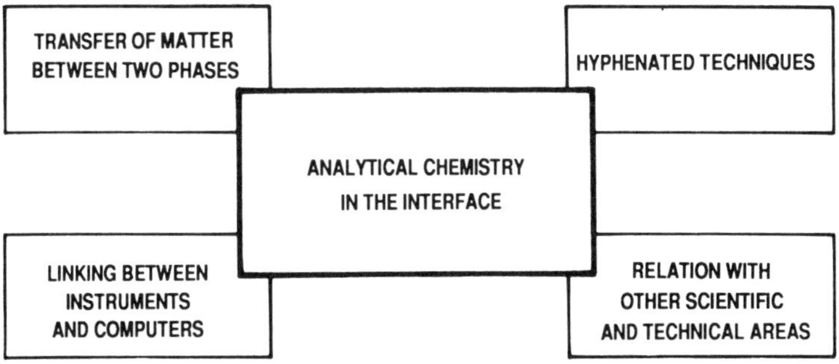

Figure 1.2 *Most commonly accepted uses of the word 'interface' in the context of analytical chemistry*

of science and technology rests on the interfaces resulting from the combined work of biologists, chemists, physicists, engineers, *etc.* Confining analytical chemistry to the laboratory is rather a serious mistake, occasionally fostered by those who put unconfessable personal interests before universal interests. Society poses problems (*e.g.* river pollution) that must be addressed not only by analytical chemists, but also by other professionals concerned (see Figure 1.3). From the problem becomes apparent the subject (the river, with all its features at different times of the year), which in turn dictates the type of sampling (different places at different times). Development of the analytical process gives rise to results that call for interpretation and processing by other professionals. This

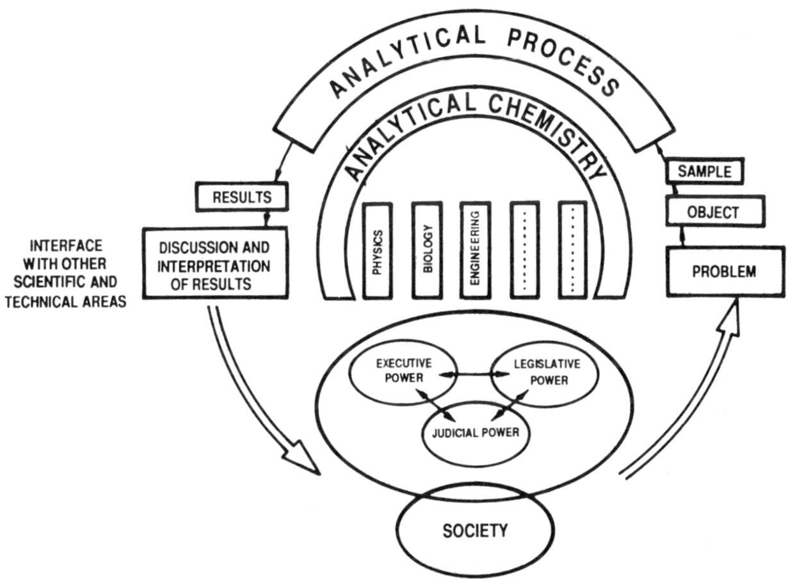

Figure 1.3 *Action of analytical chemistry in co-operation with other areas of science and technology to solve problems posed by society*

closes the circle by returning the solved problem—or proposals for a solution through the legislative, judicial, or executive powers—to society. Thus analytical chemistry must establish co-operative interfaces as regards both preliminary approaches and the end of the circle in order to ensure as efficient a performance as possible.

1.2 The Analytical Process

The analytical process can be defined as the set of operations separating the untreated, unmeasured sample from the results expressed as required in accordance with the 'analytical black box' as defined by Bailescu and Chalmers.[6] Figure 1.4 shows a scheme of the three basic stages of this process.

The so-called 'preliminary operations' comprise a series of steps such as sampling, sample preservation and treatment (*e.g.* dissolution, disaggregation), application of separation techniques, development of analytical reactions, and transfer of an adequate portion of the treated sample to the detector.

The second stage of the analytical process requires the use of one or several instruments to generate the pertinent information. After transducing, the analytical signal (optical, electroanalytical, thermal, *etc.*) should be unequivocally related with the presence, amount, or structure of one or several analytes.

Finally, a data acquisition and processing system—usually a microcomputer—performs the mathematical and statistical computations required to obtain the best possible analytical results from the data obtained. Chemometrics obviously plays a major role in this context.[9]

Notwithstanding the fact that each stage of the analytical process might determine its overall efficiency in every specific situation, the basic features of virtually every analytical methodology (sensitivity, selectivity, precision, rapidity) are affected by all three stages. Traditionally—and also mistakenly—these properties are exclusively ascribed to some operations. Thus, sensitivity is thought to be essentially reliant on the instrumental measurement, selectivity on the analytical reaction and type of instrument used, precision on the instrument and data processing method, and rapidity on the first and second stages.

The significance of preliminary operations is often underestimated, despite the fact that they are occasionally the source of major errors, that they enable analyte preconcentration and elimination of matrix effects, and that they can be rather complex and time-consuming and thus require ample dedication from the operator. Thus, the first stage of the analytical process decisively influences its precision, sensitivity, selectivity, rapidity, and cost. Analytical separation techniques, the subject matter of this book, are among the sub-stages most frequently included in preliminary operations and thus significantly contribute to the overall features of the analytical process.

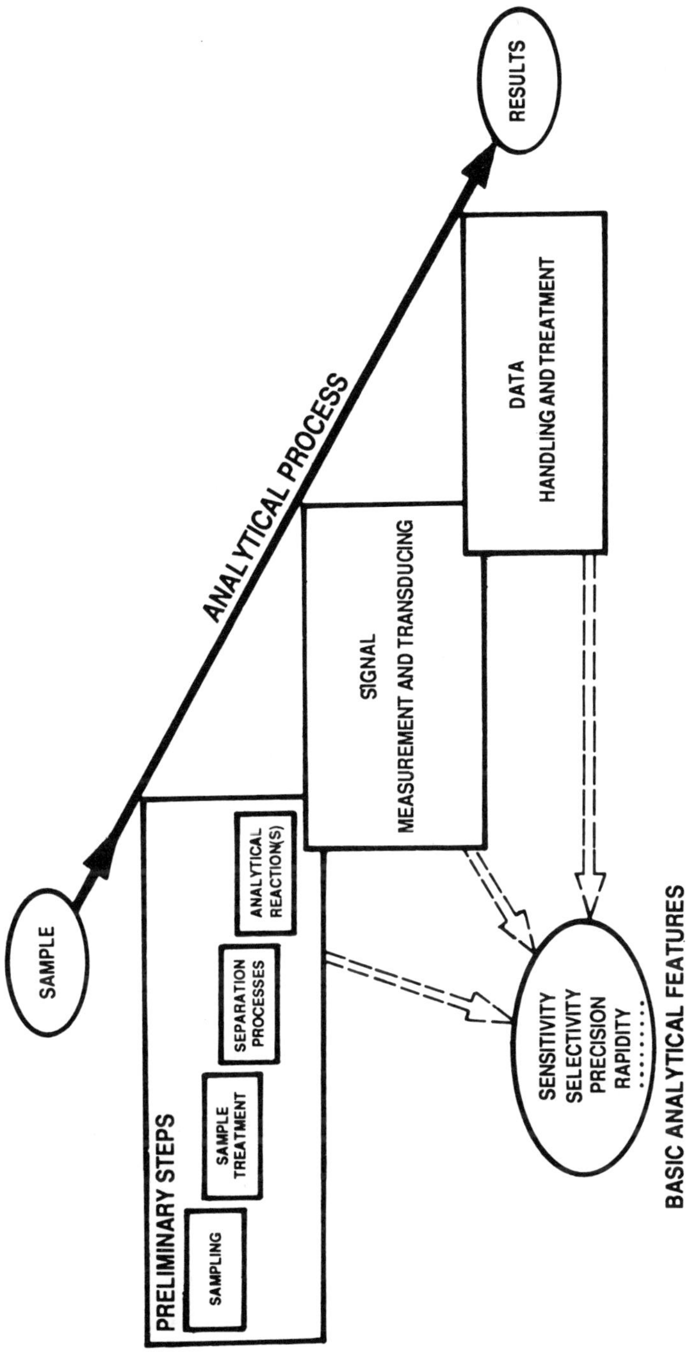

Figure 1.4 *General stages of the analytical process*

1.3 Automation of the Analytical Process

As in other scientific, technical, and social fields, reducing the need for human participation is one of the primary goals of today's analytical chemistry, a goal to which major endeavours[7] and several specialized journals (*e.g. Journal of Automatic Chemistry, Laboratory Robotics and Automation*) are currently devoted. The significance of the process laboratory for automation was recently acknowledged by the National Institute for Science and Technology (formerly the National Bureau of Standards) of the USA in creating a Consortium on Automated Analytical Laboratory Systems with the chief goal of fostering automation in analytical control laboratories.[8]

Despite the spectacular advances in automation in analytical chemistry over the past ten years or so, the degree of automation reached by each of the three stages making up the analytical process is rather unequal. The last stage, which involves the acquisition of analytical signals and processing of the data obtained from them, is currently implemented with a high degree of automation thanks to the use of increasingly powerful computers and the outstanding development of chemometrics.[9] Analytical instrumentation, typically associated with the second stage of the analytical process, has also attained quite a high degree of automation by now. The earliest generation of measuring instruments abounded with switches, selectors, dials, *etc.* The instruments of the second generation, launched in the 1970s, used built-in microprocessors that enabled control of the instrument via a straightforward keyboard. The past few years have seen the emergence of a third generation of instruments characterized by the absence of built-in control systems; this task is entrusted to microcomputers connected on-line to the instrument. Automation has changed the external appearance of instruments, facilitated their use, and expanded their potential.

Attempts at the reduction of human participation in preliminary operations have not met with such a great success for obvious reasons: on the one hand, these operations are rather varied in nature, so generalizing in this context is rather difficult; on the other, they are complex operations and hence difficult to automate. Many of the so-called 'automatic analysers', whether continuous or discrete,[7] only allow human participation in the preliminary operations to be partly reduced. Only robot stations[10] enable automation of operations inaccessible to other types of analyser (*e.g.* weighing, dissolution, disaggregation, evaporation, solid–liquid extraction, *etc.*) as they normally required intensive human participation.

Therefore, the automation of the large variety of operations that make up the first stage of the analytical process is one of the most daring challenges facing the analytical chemistry of the late 20th century. This is the setting for this book. Analytical separation techniques carried out in a continuous fashion and based on mass transfer through a solid–liquid, solid–gas, liquid–liquid, or liquid–gas interface make a major contribution to the automation of these preliminary steps by contrast with their batchwise counterparts that involve intensive participation of the operator (*e.g.* compare liquid–liquid extraction as done in separatory funnels with its continuous counterpart, described in Chapter 3, which requires no measuring or transfer of solvents, samples, or reagents, and no shaking, emulsion breaking, phase separation, or introduction of an aliquot into the detector).

1.4 Separation Techniques

1.4.1 Fundamentals

An analytical separation can be regarded as an operation involving splitting a mixture (sample) into at least two parts of different composition in order to enrich one of the fractions with one component in relation to the rest. The act of separation requires that different components be finally pulled apart, so there must always be a physical displacement through space. Therefore, transport phenomena, which should logically be differential by nature, are essential to separations. Simplistically, displacements can be classified into: (*a*) bulk or flow displacements, in which the components are carried along in their medium, which is optional; and (*b*) relative displacements, which take place through the surrounding medium, which is essential to accomplish the separation.

According to Giddings,[11] the relative displacement of a component that underlies every separation is given by the transport rate (J) per unit area along axis x

$$J = -(c\mathrm{d}\mu^*/f\mathrm{d}x) - (RT\mathrm{d}c/f\mathrm{d}x)$$

where c is the concentration, μ^* the chemical potential driving transport—at least in isothermal systems—, f the friction coefficient, which measures the drag force of the components undergoing relative transports, R the gas constant, and T the absolute temperature. From the above expression it follows that relative transport rates are inversely proportional to the friction coefficient and that the time scale of separation is inversely related to the quantity f. On the other hand, the separation speed is inversely proportional to f. According to Stokes's law, f is directly proportional to viscosity (φ) and the particle radius (a):

$$f = 6\pi\varphi a$$

Decreased carrier viscosity (medium changeover or increased temperature) will thus hasten separation. The friction coefficient, f, and the diffusion coefficient, D, are inversely related through the Planck–Einstein equation:

$$f = RT/D$$

so the time scale of separation, directly related to *f*, will be inversely proportional to *D*. The separation speed almost always increases with *D*.

The occurrence of a chemical potential (μ^*) is essential to a host of analytical separations. The μ^* profile can take three shapes depending on the components involved, namely: (1) a continuous profile resulting from application of an external field; (2) a discrete profile arising from the abrupt partitioning forces acting across an interface; and (3) a combined (continuous and discrete) profile.

There are thus two broad types of separation systems, static and flow systems, since, as noted earlier, the flow is optional. In addition, flow (dynamic) systems can be physically arranged in such a way that the flow displacement is essentially perpendicular or parallel to the relative displacement.

1.4.2 Objectives

Analytical separation techniques play a transcendental role in today's analytical chemistry. Few laboratories, however modest, do not include one or several separation techniques in their daily work, both in routine control and in R&D operations. This is chiefly a result of the improved sensitivity and selectivity afforded by these techniques. Many analytical methods lack adequate levels of these two essential features, which are significantly—indirectly—enhanced by use of one such technique.

Figure 1.5.1 illustrates the contribution of analytical separation techniques to selectivity. Direct application of the analytical determinative technique to the sample results in interferences from other species with physico-chemical properties similar to those of the analytes or from disturbances to signal measurement. Isolating the analyte of interest from

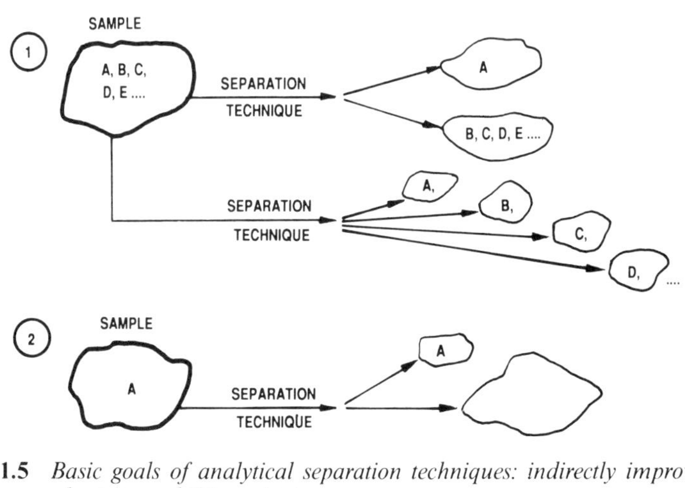

Figure 1.5 *Basic goals of analytical separation techniques: indirectly improved* (1) *selectivity and* (2) *sensitivity*

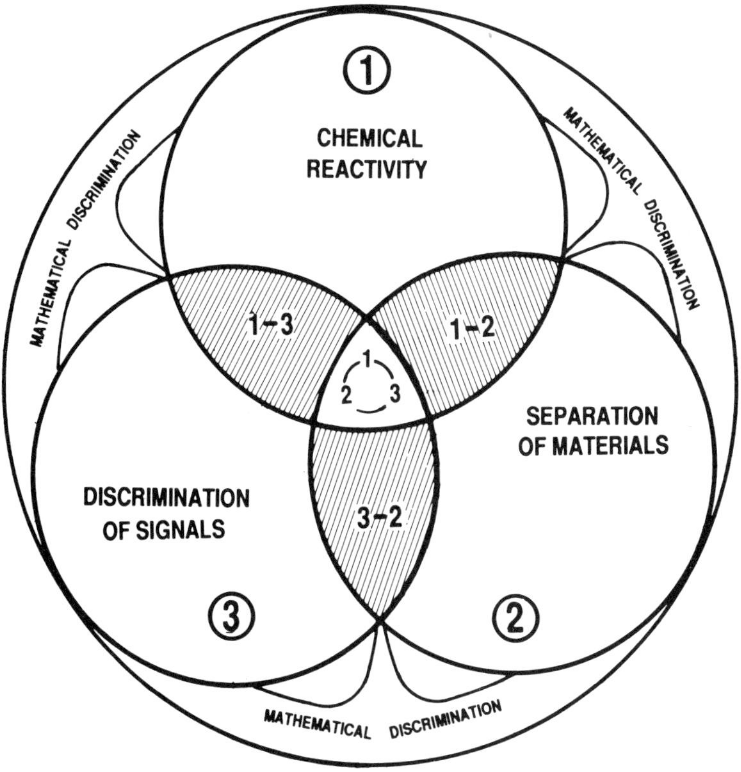

Figure 1.6 *Different alternatives to accomplishing or enhancing selectivity in analytical processes*

the rest logically avoids these problems. However, the goal is sometimes more ambitious (*e.g.* multi-determinations, which require individual isolation of each analyte to be determined).

Separation techniques are one of the three primary ways—the other two are chemical reactivity and signal discrimination—of achieving or enhancing selectivity (see Figure 1.6).[12] Obviously, a binary or ternary combination of the three ways will result in high selectivity levels. Thus, a combination of liquid–liquid extraction and the use of masking agents enables the selective separation of metal ions by formation of complexes. This is also the case with the combination of pre- or post-column derivatization and column chromatography. Combinations of instrumental techniques with analyte discrimination capabilities and separation techniques compensates for the incomplete fulfilment of these objectives—such is the case with the use of a diode array detector in liquid chromatography or with hyphenated methods. A triple combination (chemical reactivity/separation/signal discrimination) allows complex analytical problems arising from great similarities in physico-chemical behaviour between the analytes involved and with potential interferents to be readily solved. In addition, application of chemometric techniques[9] further enhances selectivity in most instances.

Another basic objective of analytical separations is the indirect enhancement of sensitivity. With very low analyte concentrations (*e.g.* in trace analysis), direct application of the analytical method concerned usually results in the obtainment of no signal or of rather a small one that cannot be distinguished from background noise and hence in working below the detection or determination limit. The use of an analytical separation technique by transferring the analyte(s) of interest to a new phase with a markedly (2 to 10^5 times) smaller volume enables satisfactory application of the analytical determinative method thanks to the preconcentration thus achieved (see Figure 1.5.2).

In addition to the above two objectives, analytical separation techniques provide the analytical process with some otherwise unavailable possibilities, namely:

(*a*) The analytical reaction can take place simultaneously with mass transfer, thereby avoiding some steps of the process.
(*b*) Transport of the analyte(s) or reaction product(s) to the detector can be facilitated by using dynamic (flow) separation techniques.
(*c*) The analyte(s) can be appropriately stored in the event that sampling is done outside the laboratory.
(*d*) The analytical instrumentation can be protected from the potential aggression of species or particles that may give rise to reversible or irreversible problems with adverse economic repercussions.
(*e*) The phase carrying the analytes can be made compatible with the detection system.

1.4.3 Classification

The wide and still growing variety of analytical separation techniques, their interrelations, theoretical studies, and the host of applications reported so far justify the fact that scientific journals, treatises,[14] and textbooks[15,16] all speak of a Separation Science. Any attempt at an illustrative description, however brief, of all the separation techniques is obviously beyond the scope of this introductory chapter. Therefore, only the most relevant are dealt with here, grouped into different categories according to various criteria.

Analytical separation techniques can be roughly classified according to three basic criteria. One such criterion is the type of interface across which mass transfer (solid–liquid, solid–gas, liquid–liquid, or gas–liquid) takes place. If the phase containing the sample is considered to be the first phase, then the second phase can be generated *in situ* (*e.g.* by distillation or precipitation) or be an external phase (*e.g.* in extraction, chromatography, or dialysis). The forces involved in the process can be purely mechanical (*e.g.* filtration, centrifugation), physical, based on state changes (*e.g.* distillation, zone fusion), solubility-based partitioning processes (*e.g.* liquid–liquid and gas–liquid chromatography), based on the behaviour towards an electric field (*e.g.* electrophoresis), or chemical (*e.g.* precipitation, ion exchange). However, most separation techniques involve several types of force.

Depending on their foundation, analytical separation techniques can be thermodynamically or kinetically controlled, or both. To separations based on partitioning equilibria involving two phases can be applied such classical thermodynamic relations as the phase rule, partition constants, and the Gibbs and Clausius–Clapeyron equations; they are referred to as 'equilibrium techniques' and are represented by liquid–liquid extraction, precipitation, distillation, and sublimation, among others. In the so-called 'kinetic separation techniques' it is the different rate at which the mixture components travel that facilitates their separation. These techniques rely on the occurrence of concentration, electric potential, temperature, and gravity gradients, which are the foundation of dialysis, electrophoresis, thermal diffusion and centrifugation, respectively. These migration processes are governed by diffusion laws. However, such widely used separation techniques as chromatography and ion exchange are subject to both types of control. Chromatography is traditionally regarded as an equilibrium technique governed by a partition coefficient; however, chromatography is a dynamic process that actually involves gradients, so longitudinal and whirl diffusion will be kinetically controlled. Similar considerations can be made on ion exchange, where both particle and film diffusion are the elements controlling the separation, even though predictions in this respect are based on the selectivity coefficient.

Other classifications of analytical separation techniques are based on dynamic or operational aspects. A first distinction is made on the basis of whether mass transfer takes place across a well defined interface (surface or interfacial techniques primarily based on adsorption phenomena) or in the bulk phase (*e.g.* liquid–liquid extraction under agitation, precipitation, distillation). According to the number of times the partition equilibrium is reached, separation techniques can be classified into simple (once), repetitive (several times), and multiple (many times); the separation efficiency obviously increases in the same order. Thus, liquid–liquid extraction can be done in a single operation in a separatory funnel, by circulation, and by the partition mode in liquid chromatography.

As noted earlier, separation techniques can be classified into two broad groups according to whether or not a flow of one or of the two phases is used to facilitate the differential displacement of the mixture (sample) components. This distinction allows separation into the following two major groups (see Figure 1.7):

(*a*) *Static separation techniques*, also referred to as 'discrete' or 'batch' techniques. They are characterized by the absence of a flow of the phases; in addition, kinetics usually plays no relevant role. In most cases, detection is carried out in isolation (off-line) from the separation technique.

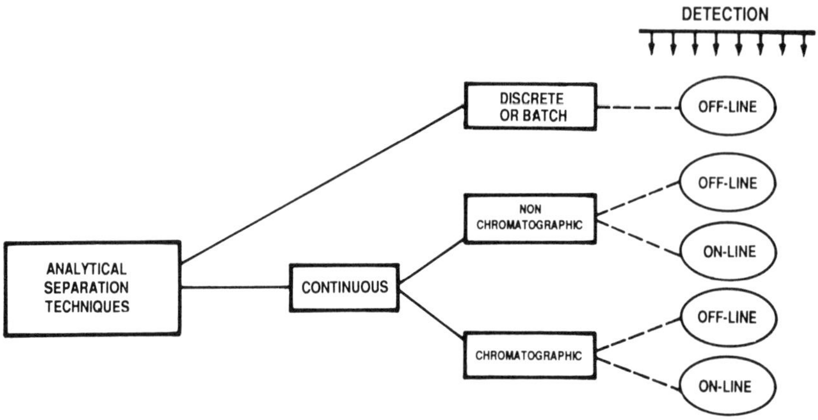

Figure 1.7 *Classification of analytical separation techniques according to operational dynamics, efficiency, and the way in which detection is performed*

Depending on the profile of the chemical potential (μ*), static techniques can be further divided according to whether such a profile is continuously increasing or decreasing (*e.g.* electrophoresis, isoelectric focusing), or discontinuous (*e.g.* batch extraction, batch crystallization, batch ion exchange), or a combination of the two (*e.g.* electrodeposition).

(*b*) *Dynamic separation techniques*, also known as 'continuous' techniques, which are characterized by the presence of a flow of one or the two phases (concurrent or countercurrent) that facilitates the separation. As a rule, these are more efficient techniques than their static counterparts. They can be classified according to the flow direction in relation to that of the relative displacement of the components to be separated, in addition to the type of profile of the chemical potential imposed. Thus, filtration and reverse osmosis are two dynamic separation techniques involving parallel flow and rely on a discontinuous profile of the applied chemical potential, while chromatography and field-flow fractionation are dynamic techniques involving perpendicular flow and a discontinuous, combined profile of the chemical potential.

From a more practical point of view, separation techniques can be classified according to operativeness and efficiency into chromatographic and non-chromatographic (see Figure 1.7), the latter being much more efficient in separating a large number of analytes as partition equilibria are reached many times (10^2–10^7). Also, it is worth considering whether the isolated analytes (or their reactions products) are subsequently detected on-line or off-line. Continuous non-chromatographic separation techniques using a detector—usually optical or electroanalytical—connected on-line to the separation system are the subject matter of this book. This last group also includes separation techniques of growing interest such as field-flow fractionation and capillary electrophoresis (isotachophoresis), that are commented on in Chapter 5. Continuous chromatographic separations effected in columns and involving a gas (GC), liquid (HPLC), or supercritical fluid (SFC) always entail the use of an on-line detector, so they are implemented on systems categorized as instruments insofar as they not only enable the separation, but also provide analytical information. Continuous separation techniques involving off-line detection (*e.g.* classical liquid chromatography with eluate fraction collection) are gradually falling into disuse, with the exception of modern thin-layer chromatographic variants (HPTLC).

1.5 Non-chromatographic Continuous Separation Techniques

As stated above, this type of separation is characterized by the occurrence of at least one flow (generally liquid, though also occasionally gaseous) of one of the phases involved in the mass transfer. It bears more than a passing resemblance to column chromatographies as it uses modules such as the propulsion, sample introduction, separation, and continuous detection systems. However, there are major differences in the foundation of the separation process in relation to the number of times equilibrium is reached (the number of theoretical plates), which in turn determines their efficiency. This makes these two groups of analytical techniques appropriate for unlike purposes. Thus, non-chromatographic techniques usually pursue the determination of one or more analytes, the separation of components into groups, or interference removal, among others, while chromatographic techniques are chiefly intended to separate analytes from one another.

The implementation of non-chromatographic continuous separation techniques relies on the principles behind, and developments in, the field of automatic continuous analysers, both in their air-segmented[17] and unsegmented variants—of which flow injection analysis (FIA) is the most representative example.[18, 19]

The dynamic character of this type of continuous separation arises from the use of a propulsion system to set a liquid or gas (or the two) in motion through a continuous separation assembly. Of the variety of possibilities available in this respect, the following two are the most frequently resorted to: (*a*) the use of peristaltic pumps based on the regular squeezing of flexible tubing or of piston pumps through which the liquid phase is circulated—the former can establish one or more flows; and (*b*) the use of gas-pressure devices to transport liquids or gases.

A continuous separation system can give rise to various types of interface across which mass transfer can take place. Thus, there can be gas–liquid (introduction of gas-samples into hydrodynamic systems, distillation, gas diffusion, hydride generation), liquid–liquid (dialysis, extraction) and liquid–solid systems (extraction, sorption, precipitation, electrochemical stripping). The chapters of this book have been written according to this classification.

Initially, samples can be solid, liquid, or gaseous. Liquid and gas samples can be introduced into the continuous separation system either by insertion or injection into a liquid or gas carrier phase or by continuous aspiration into the system. Solid samples are previously dissolved or disaggregated or, occasionally, can also be directly incorporated into the continuous separation system (*e.g.* in automatic leaching).

The second phase, which collects the analytes (or their reaction products) or the interferents, can be introduced into the system in a direct fashion (*e.g.* liquid–liquid extraction, dialysis, gas diffusion) or be a part of the separation system proper (*e.g.* the active material in a sorption microcolumn). Also, this second phase can be generated *in situ* by a physical change (distillation), a chemical reaction (precipitation), or a combination of the two.

Dynamic separation systems typically use continuous on-line detection systems to which the stream of the phase carrying the analyte(s) or reaction product(s) are driven after the separation. This is usually entrusted to the flow-cell of a conventional optical (photometer, fluorimeter) or electroanalytical instrument (potentiometer, voltammeter). Continuous separation systems with integrated separation and detection are a class of their own; the two operations can be carried out sequentially (*e.g.* cathodic or anodic stripping) or simultaneously (*e.g.* use of continuous-flow chemical sensors constructed by packing a photometric or fluorimetric flow-cell with an exchanging material).

Analytical or complementary reactions can take place *before* (*e.g.* formation of volatile species in gas diffusion), *during* (*e.g.* precipitation and formation of coloured or fluorescent chelates or ion-pairs in liquid-liquid extraction), or *after* the separation process (*e.g.* formation of detectable derivatives). In most instances, additional channels are required to incorporate reagents at the appropriate place into the continuous separation system.

Most of the continuous separations described in this book can be done discretely or batchwise. In some instances (*e.g.* dialysis, gas diffusion, ion exchange), the static or batch alternative is scarcely practical for analytical purposes, though usable for basic studies.

In addition to the general assets provided by the inclusion of an analytical separation in the analytical process, the use of non-chromatographic continuous separation systems offers the following advantages over the corresponding batch counterparts: (*a*) ease of incorporation into automatic analytical systems, (*b*) reduced hazards (*e.g.* those arising from handling toxic solvents) for manual operators, (*c*) increased precision resulting from elimination of the typical errors due to the 'human factor', (*d*) markedly increased sample throughput, (*e*) drastically reduced sample and reagent consumption, and (*f*) reduced analytical costs. All these advantages justify the present significance of non-chromatographic continuous separation techniques in the context of analytical chemistry.

References

1. M. Valcárcel, *Quím. Anal.*, 1990, **9**, 225.
2. T. Hirschfeld, *Anal. Chem.*, 1980, **52**, 297A.
3. F. Karasek and R. Clement, 'Basic GC/MS. Principles and Techniques', Elsevier, Amsterdam, 1986.
4. D.J. Malcome-Lawes, 'Microcomputers and Laboratory Instrumentation', Plenum Press, New York, 1984.
5. 'Computers in the Laboratory', ed. J.G. Liscouski, ACS Symposium Series, Washington DC, 1984.
6. G.E. Bailescu, C. Patrocescu, and R.A. Chalmers, 'Education and Training in Analytical Chemistry', Ellis Horwood, Chichester, 1982.
7. M. Valcárcel and M.D. Luque de Castro, 'Automatic Methods of Analysis', Elsevier, Amsterdam, 1988.
8. H.M. Kingston, *Anal. Chem.*, 1989, **61**, 1381A.
9. D.L. Massart, B.G.M. Vandeginste, S.N. Deming, Y. Michotte, and L. Kaufman, 'Chemometrics: A textbook', Elsevier, Amsterdam, 1988.
10. W.J. Hurst and J.W. Mortiner, 'Laboratory Robotics', VCH Pu., New York, 1987.
11. J.C. Giddings, *Anal. Chem.*, 1981, **53**, 945A.
12. M. Valcárcel and A. Ríos, *Analusis*, 1990, **18**, 469.
13. J.M. Miller, 'Separation Methods in Chemical Analysis', Wiley, New York, 1975.
14. J.C. Giddings, 'Principles of Chemical Separations', in 'Treatise on Analytical Chemistry', Part 1, Vol. 5 (2nd Edn.), eds. P.J. Elving, E. Grusuka, and I.M. Kolthoff, Wiley, New York, 1982.
15. B.L. Kargen, L.R. Snyder, and C. Horvath, 'An Introduction to Separation Sciences', Wiley, New York, 1973.
16. M. Valcárcel and A. Gómez-Hens, 'Técnicas Analíticas de Separación', Reverté, Barcelona, 1988.
17. W.B. Furtman, 'Continuous Flow Analysis: Theory and Practice', Marcel Dekker, New York, 1976.
18. M. Valcárcel and M.D. Luque de Castro, 'Flow Injection Analysis: Principles and Applications', Ellis Horwood, Chichester, 1987.
19. J. Ruzicka and E. Hansen, 'Flow Injection Analysis', (2nd Edn.), Wiley, New York, 1988.
20. M. Valcárcel and M.D. Luque de Castro, *J. Chromatogr.*, 1987, **393**, 3.
21. M. Valcárcel and M.D. Luque de Castro, 'Continuous Separation Techniques in FIA', Chapter VII of 'Selective Sample Handling and Detection in HPLC', eds. K. Zech and R.W. Frei, Elsevier, Amsterdam, 1989.

CHAPTER 2

Gas–Liquid Systems

2.1 Introduction

Mass transfer between a gas phase and a liquid phase in continuous automatic analytical systems offers some major advantages over other types of interface; however, the co-occurrence of the two phases in a hydrodynamic system also gives rise to interesting specific problems which call for ingenious technical solutions.[1, 2]

Liquid phase streams may contain several reactants, the liquid sample (either injected or aspirated), or both. The gas phase, which may be the sample itself, can be generated *in situ* in the continuous system or can be introduced either directly or intermittently into the system to act as a carrier.

The formation and isolation of a gas from a liquid phase can be accomplished in one or a combination of the following ways: (*a*) by a chemical reaction yielding volatile products (releasing techniques); (*b*) by increasing the temperature in a controlled fashion; (*c*) by using an inert gas stream (sweeping techniques); and (*d*) by using a special gas-permeable material (usually a thin membrane).

As a rule, the gas phase contains the analyte(s) at some time during the process, with the exception of continuous evaporation systems, where only the solvent is volatilized.

The mass transfer on which the separation is based usually takes place from the liquid to the gas phase, although it can also proceed in the opposite direction if the sample is a gas. A second gas-to-liquid phase transfer is also customarily employed to adapt the continuous separation system to the detector used.

The analyte(s) can be transferred to the continuous detection system in various ways, namely: (*a*) by direct introduction of the gas phase containing the analyte, which is the alternative of choice when an atomic spectroscopic detector is used to sense analytes that have been derivatized as volatile hydrides; (*b*) by absorption into a collecting liquid stream that drives the analytes to an optical or electroanalytical detector with or without an intervening derivatization reaction intended to suit the analytes to the detector features; (*c*) by dissolving the (solid) analytes after the solvent has been evaporated.

There is a large variety of continuous separation systems relying on the use of gas–liquid interfaces and where the samples (gas or liquid) may be aspirated or injected, the liquid streams may serve different functions apart from that of transport (*e.g.* contain reactants, buffers and/or masking agents, act as carriers or collectors), special units such as distillers, gas–liquid separators or gas diffusers may be needed, and an inert gas may be used as carrier–collector.

The state of the art in each of these alternatives will be discussed separately in detail below as no generalization can be made in this respect. They are dealt with in order of increasing complexity, beginning with hydrodynamic analytical systems based on the direct introduction of gas samples. These are followed by some systems requiring no external energy contribution or additional gas stream, *i.e.* those based on gas diffusion. Systems based on hydride generation commonly use a stream to carry/collect the gas. Finally, due mention is made of those techniques requiring some type of external energy, namely distillation and evaporation.

2.2 Direct Introduction of Gas Samples

Over 99% of all continuous automatic methodologies, whether based on the use of unsegmented[3-5] or air-segmented streams,[6,7] rely on the introduction of liquid samples by aspiration or injection. Solid, heterogeneous, and gas samples must be pretreated to dissolve the analytes. A number of approaches to the direct introduction of this type of sample have been developed in the past few years with the aim of reducing human participation in the preliminary stages of the analytical process. Such is the case with various continuous automatic methodologies based on the direct introduction of gas samples.

This type of system usually involves transferring the analytes from the gas to the aqueous phase through interfaces established in the continuous system. A suitable chemical reaction between reactants dissolved in the aqueous phase not only facilitates the transfer, but also enables detection. There are two possibilities in this respect according to whether the configuration used includes a continuous degassing system to avoid the arrival of parasitic signals at the detector.

Figure 2.1 depicts various continuous automatic configurations with direct introduction of gas samples.[8-10] They are closed-loop systems involving the reversible redox couple formed by the mononuclear complex of Fe^{II} (red) and the binuclear complex of Fe^{III} (pale yellow) with 1,10-phenanthroline dissolved in the carrier. The binuclear Fe^{III} complex acts as the main reagent and is reduced to the mononuclear coloured ferroin by the gas-analyte (sulphur dioxide[8,10] which has a reducing character) at the gas–liquid interface formed by the intercalated sample plug and the carrier solution during transport to the detection unit according to the following indicator reaction:

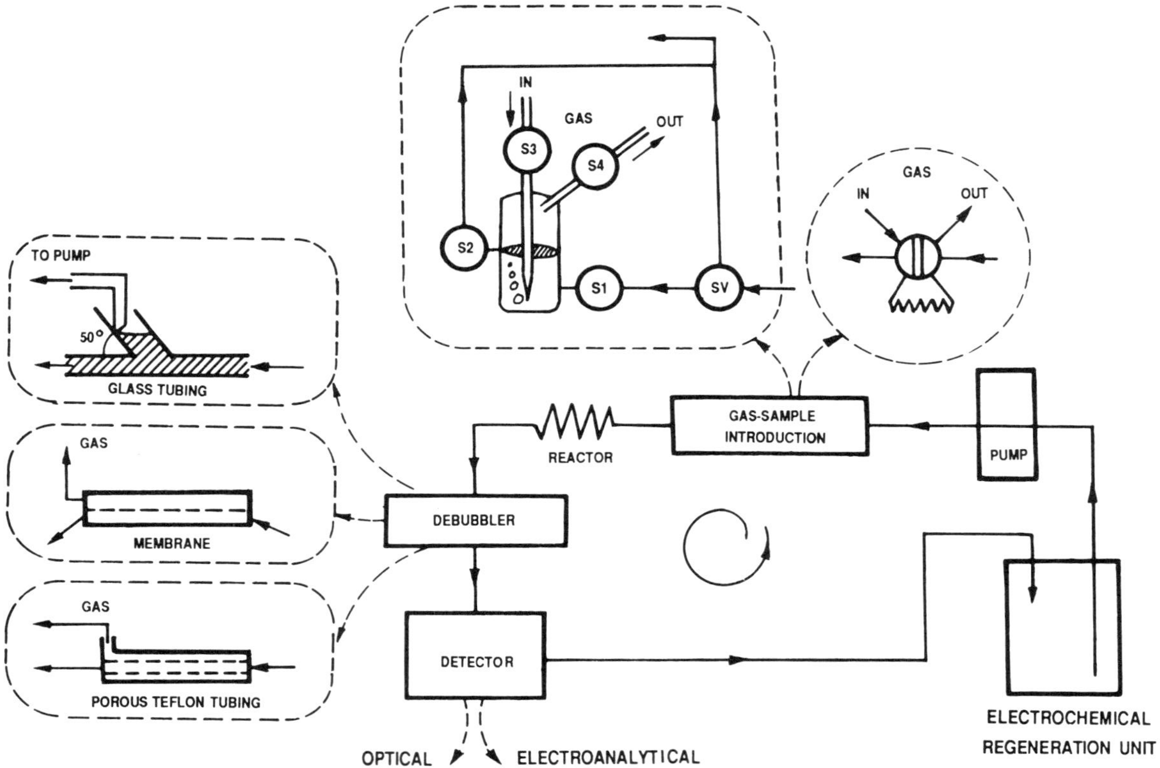

Figure 2.1 *General scheme of the different alternatives to the direct introduction of gas samples into automatic continuous analysers based on the closed-loop approach. For details, see text*

$$[Fe_2(phen)_4(OH)_2]^{4+} + 2\ phen + SO_2(g) \leftrightarrow 2\ [Fe(phen)_3]^{2+} + 2\ H^+SO_4^{2-}$$

When the analyte possesses oxidizing properties (*e.g.* nitrogen dioxide[11]), the indicator reaction takes place in the opposite direction, so the main reagent is the mononuclear Fe^{II}–phen complex.

The regeneration unit used is intended to maintain a constant level of the main reagent in the carrier, so it will involve an electrochemical oxidation or reduction depending on the nature of the indicator reaction:

$$2\ [Fe(phen)_3]^{2+} + 2\ H_2O \rightleftharpoons [Fe_2(phen)_4(OH)_2]^{4+} + 2\ phen + 2\ e^-$$

Thus, the system can work with no reagent addition over a long period.

The gas (usually air) containing the analyte can be directly intercalated into the system by using a conventional rotary injection valve (IV) to fill the loop under pressure.[8-10] As the gas plug travels through the coil, the reaction product builds up at the rear because the largest contact zone between the two phases is in the solution film adhering to the tube walls.

Samples can also be introduced by means of a customized midget impinger (see Figure 2.1).[8] In the sampling position, stopcocks S_1 and S_2 are closed, while S_3 and S_4 are open, so that the gas sample passes through the absorbing solution together with the main reagent held in the impinger. After this stage, the flow is diverted through the by-pass mounted on the three-way selecting valve (SV), stopcocks S_3 and S_4 are closed as S_1 and S_2 are opened, and the absorbing solution is driven to the detector. This alternative, of greater technical complexity, has the advantage that it allows preconcentration by aspiration of large volumes of gas; however, it features a substantially decreased sampling rate.

These closed-loop systems inevitably require a gas–liquid separation unit (a debubbler such as that in Figure 2.1) in order to prevent the gas from reaching the detector and hence give rise to parasitic signals. This operation should be carried out in such a way as to avoid dispersion or dilution of the reaction product, which requires using units with very low inner volumes. On the other hand, it is indispensable to ensure smooth operation in order to avoid the irreproducibility typically arising from this type of unit, particularly in relation to continuous automatic air-segmented analysers.[3,4] Figure 2.1 illustrates several alternatives, namely: (*a*) an open glass outlet with tightly designed dimensions to ensure smooth operation, which is accomplished by using an aspiration tube that maintains a constant liquid level at the opening to the outside; (*b*) a gas-diffusion unit furnished with a gas-permeable membrane; (*c*) a microporous Teflon tube allowing the gas to be expelled with no technical complication whatsoever.

The detection system can be a straightforward photometer monitoring the reduced form of the redox couple (ferroin complex) at 510 nm,[8] which can also be measured amperometrically at 965 mV against an Ag/AgCl reference electrode.[9,10]

The problems arising from the functioning of the degassing unit can be circumvented by applying the novel approach illustrated in Figure 2.2,[11] which is based on the iterative change of the flow direction by means of an electronic device or a microcomputer that acts on the propulsion system.[12] The single channel used (A.1) carries the reagent solution. The loop of the injection valve (IV) is filled with the gas sample by means of a syringe. The photometric detector employed is located in an unusual position, *viz.* between the pump and IV. Injection of the sample yields two gas–liquid interfaces (Figure 2.2A.2) at which the analyte transfer and the indicator reaction take place. Dispersion of the reaction product into the liquid phase is favoured by the iterative change of the flow direction, which is started simultaneously with the sample injection. The repeated passage of the liquid zone close to the gas–liquid phase nearest to the detector through the flow-cell yields a multipeak absorbance–time recording (normal and derivative, Figure 2.2A.3) which is the basis for the determination. Once the recording process is finished, the sample is driven to waste in a drain cycle (flow direction from left to right in Figure 2.2A.1), so the gas phase never reaches the detector flow-cell.

The direct determination of very dilute analytes in gas samples is often confounded by inadequate sensitivity from the analytical methodology to be applied, a problem which is frequently addressed by implementing a preconcentration stage, as the sample is generally available in sufficiently large volumes. Figure 2.2B depicts a configuration resulting from replacing the conventional injection valve in Figure 2.2A with an internally coupled valve system. In the filling position, the loops of the two valves are interconnected and filled with the gas sample. The two valves are switched simultaneously as the program enacting the iterative change in the flow direction is started. After 1.5 cycles, valve IV_2 is switched to its filling position, the loop is refilled with a fresh portion of sample and the valve is switched back to its emptying position when two cycles have been completed. In this way, most of the initially injected sample is replaced with a fresh portion of sample. This step can be repeated as many times as required. Figure 2.2B.2 shows three recordings given by three samples with different analyte concentrations (number of steps = 10).

These novel approaches have been applied to the determination of nitrogen dioxide over the concentration range 0.1–100 μl^{-1} with an r.s.d. less than 3% by using Shinn's modified version of the classical Griess reaction.[11]

2.3 Gas Diffusion

2.3.1 General Considerations

Mass transfer across membranes is of great relevance to biological and physico-chemical processes alike. Research aimed at elucidating the intricate mechanisms involved may shed some light on this subject and

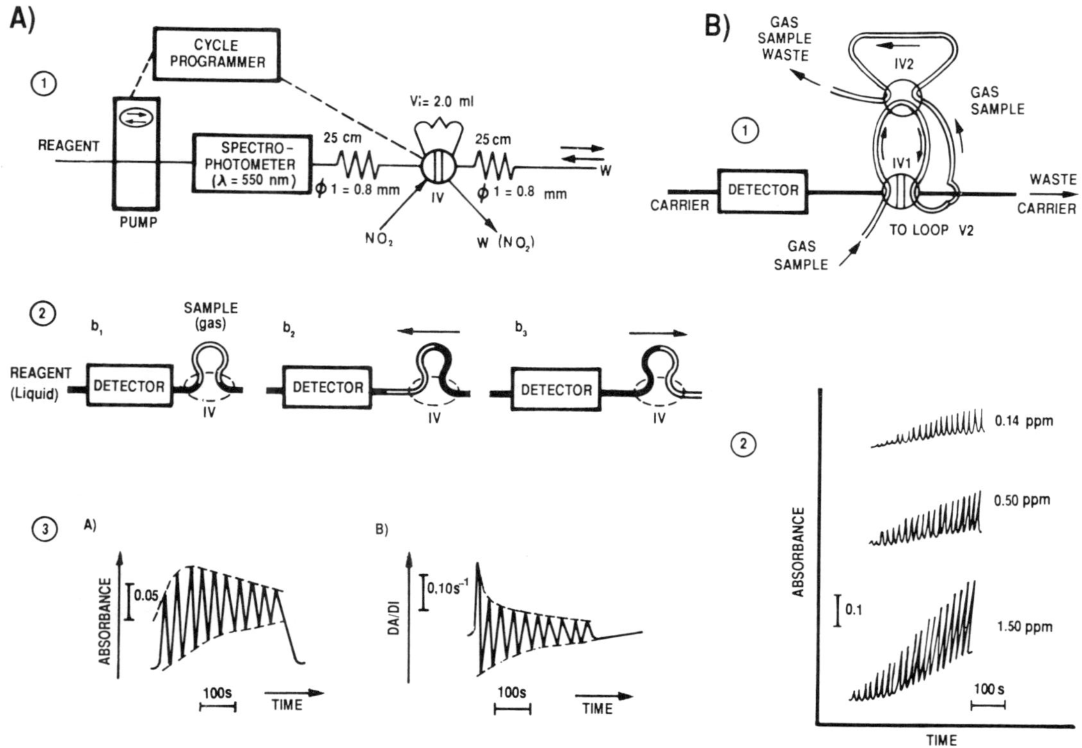

Figure 2.2 *Direct introduction of gas samples into a continuous unsegmented analyser based on the iterative change of the flow direction.* (A) *Manifold* (1), *operational sequence* (2) *and recordings obtained* (3) *with a simple injection valve.* (B) *Manifold with two internally coupled valves* (1) *and recordings obtained* (2).
(Adapted from reference 11, with permission of Elsevier Science Publishers.)

allow the development of new biochemical (clinical and non-clinical) and technical approaches, one of the priority objectives of many a research and development strategy in developed countries.

As a rule, membranes separate two usually miscible liquids which are kept from mixing; however, some solvent may be transferred across the membrane by osmosis, electro-osmosis, or thermal diffusion. The foundation of the selective separation of species (gases, ions, molecules) passing from one carrier solution to another varies widely and depends on the nature of such species. The membranes can be physico-chemically or biologically active and thus selectively permeable to some species (*e.g.* ion-exchange membranes or supported thin films of an immiscible solvent acting through solubility phenomena). Other membranes rely on the use of a film of microporous polymer (pore size 5–1000 Å) which allows the selective passage of dissolved ions or molecules, or even gases through a sifting effect. In most cases, chemical and/or physico-chemical kinetics plays a major role in the efficiency and selectivity of these separations.

In particular, separations based on gas diffusion are based on the permeability of microporous membranes to gases. The process is typically kinetically controlled; each gas has its own diffusion constant (D) which is related to the so-called 'coefficient of permeability' (P) through the solubility constant in Henry's law, according to the expression

$$D = 1/RP$$

If both phases are stationary, then the flow velocity of a gas across a membrane per unit area, J, will be

$$J = P(P_1 - P_2)/L$$

where P_1 and P_2 denote the partial pressure of the gas in the initial and collecting phase, respectively, and L the membrane thickness.

Nevertheless, the separation of gases by gas diffusion (gas–gas separations) still poses major practical difficulties despite its intrinsic interest and the technical endeavours devoted to improving the process. On the one hand, polymer membranes are not selective enough; on the other, their rate of transfer is quite low, so efficient separations can only be accomplished by using very large transfer areas, which makes the process even more complex from a technical point of view.

Only those continuous systems using a gas diffuser including a membrane to separate two liquids, one acting as the sample carrier and the other as collector of the transferred gas (usually the analyte or its reaction product), are of practical analytical interest.[40] Obviously, static gas diffusion systems, *i.e.* those where both liquids remain steady, are of only theoretical interest.

Described below are various continuous automatic systems involving the use of a gas diffuser and an on-line coupled detector for quantifying the concentration of the analyte(s) separated in one or more derivatizing reagents. These are followed by the different types of gas diffuser reported so far and their most salient features, as well as by various experimental set-ups used in this context. Finally, the most significant applications of these systems are gathered in a common section.

2.3.2 Chemical Reactions Involved

Continuous gas-diffusion systems can be classified according to whether or not they involve a chemical reaction, and to whether this takes place outside the continuous system, in the carrier or acceptor stream, or in both.

Volatile analytes require no derivatizing reaction to be implemented outside the system or in the carrier stream. Such is the case with the determination of ozone,[13] nitrogen oxides,[41] chlorine dioxide,[14] and ethanol,[15] among other species.

Derivatizing reactions carried out prior to the gas–liquid separation in the gas-diffusion cell are basically intended to generate volatile products from the analyte. Acid–base reactions are by far the most frequently used in this respect: an acid is incorporated into the carrier stream to cause the release of gases such as CO_2, SO_2, HCN, H_2S, or NO_2 from their respective analytes (HCO_3^- or CO_3^{2-},[16,17] HSO_3^- or SO_3^{2-},[17-19] CN^-,[20,21] S^{2-},[22] and NO_2^-[10,39]) or alternatively, a basic carrier stream, induces the formation of ammonium gas from samples containing ammonium ion.[23-31] The possibility of using enzymatic derivatizing reactions yielding ammonium ions considerably expands the number of analytes that can be determined in this fashion.[32,33] Other types of reaction, the most significant of which are those based on the generation of volatile hydrides[34-37] and cold mercury vapour[49-52] have also been used as prior reactions in continuous systems and are commented on below. Occasionally, the derivatizing reaction is implemented in a conventional manner (off-line) in the continuous system; such is the case with the determination of oxidized ketone bodies in milk on the basis of the permeation of the ketone through a microporous membrane.[38]

Chemical reactions implemented in the acceptor stream have a twofold purpose. Firstly, in taking place at the interface, they facilitate transfer of the gas-analyte by preserving a favourable concentration gradient and considerably increasing the kinetics of the separation process. Secondly, the derivatizing reaction can also adapt the transferred analyte to the detector used, thereby increasing the selectivity and sensitivity of continuous analytical methodologies in many cases. Absorption of the transferred gas-analyte by the acceptor stream is substantially increased if the analyte has acid–base properties. Thus, gases such as CO_2, SO_2, HCN, and NO_x can be readily dissolved in a basic stream, and NH_3 in an acid stream.

The separation process can be followed by a variety of analytical indicator reactions. Analytes with acid–base properties can be determined by changing the pH of the medium to absorb the gas and using a direct potentiometric sensor (pH electrode) or an indirect detector (*e.g.* an acid-base photometric indicator). Other types of analyte call for different derivatizing reactions of varied nature: redox (*e.g.* use of an irreversible coloured indicator for ozone or ferrous dichromate for ethanol), coupling (hydroxylamine with acetone), chemiluminescence (use of luminol as reagent for chlorine dioxide or luminol plus hypochlorite ion for ammonium).

However, no derivatization reaction is used when the physico-chemical properties of the analyte enable direct detection. The gas-analyte can be directly driven to the detector by means of an acceptor gas stream (*e.g.* in hydride and cold mercury vapour-generation methods). Alternatively, the analyte can be absorbed into an appropriate liquid stream for transport to the detector—usually potentiometric (ISE), conductimetric or optical (UV).

2.3.3 Gas-diffusion Cells

Despite the fact that some manufacturers (*e.g.* Tecator) market a few gas-diffusion cells, most of them are laboratory-made as their construction is technically affordable. Such ease of construction has resulted in the occurrence of a large variety of models which, however, bear some similarities in their elements and operation. Figure 2.3 depicts the two generic types of gas-diffusion design, namely sandwich and tubular. Both use a low-volume mini-chamber as central zone, which is split into two by a flat or cylindrical (tubular) membrane, the resulting half-chambers allowing passage of the sample and the carrier stream (with or without a reagent), respectively, although a stream of acceptor gas is occasionally used for special applications. The gas-analyte obviously diffuses across the membrane.

A typical sandwich gas-diffusion cell consists of two blocks (Perspex, Teflon, aluminium) with identical internally engraved conduits that make up the inner chamber, the geometry of which normally varies from model to model (Figure 2.3.A.2).[31] The membrane is placed between the two blocks, which must be joined tightly in order to avoid leaks. Each engraved microconduit has two holes on its ends that connect it with the manifold tubing. The dimensions of the inner parallelepiped—the most common shape—usually vary over the following ranges: 0.1–0.3 mm deep, 1–3 mm wide and 3–10 cm long.

Tubular gas-diffusion cells use built-in Teflon tubes (2–5 mm OD, 2–4 mm ID, 8–20 cm long) acting as membranes. The acceptor stream (sample) is circulated internally, while the acceptor stream—containing a reagent or not—is circulated externally (Teflon or Pyrex tubing of 5–10 mm ID).

Any membrane to be used as a gas-diffuser must meet the following basic requirements: (*a*) it should be permeable to gases and to differing

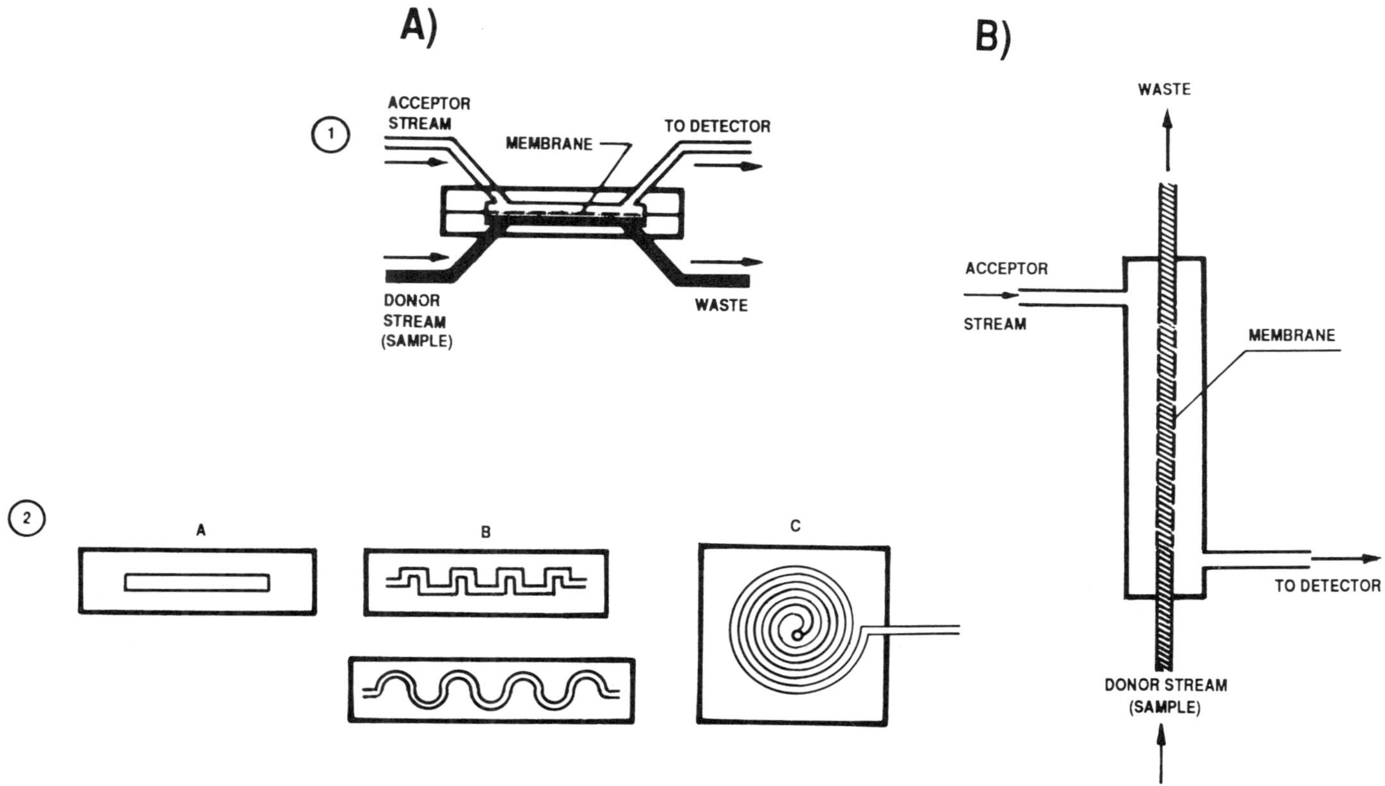

Figure 2.3 *Different types of gas-diffusion cell.* (A) *Sandwich-type, with different minichamber designs* (2): (a) *parallelepipedal,* (b) *winding,* and (c) *spiral.* (B) *Tubular-type*

extents in order to allow for kinetic discrimination; (*b*) it should rapidly and efficiently effect the gas–liquid separation; (*c*) it should be hydrophobic in order to avoid dialysis phenomena (*i.e.* the passage of dissolved ions or molecules) and their potential interference; (*d*) it should be of high mechanical resistance; (*e*) it should not be easily clogged by suspended particles occasionally present in the sample plug; and (*f*) it should have a reasonably long lifetime to avoid frequent replacement in the diffusion cell.

Depending on the type of membrane used, gas species diffuse according to two types of mechanism. Non-porous membranes (*e.g.* silicone rubber) rely on the solubility difference between them and the acceptor and donor fluids. Porous membranes (*e.g.* Teflon, polypropylene), are also referred to as 'heterogeneous' or 'air-gap' membranes because the gas diffuses across a gas layer, the thickness of which is determined by that of the membrane and its pore size. As a rule, microporous membranes of controlled pore size (0.05–10 nm) and thickness (0.01–0.05 mm) provide much better results and are commercially available under a variety of names. More specifically, polypropylene membranes have been shown to offer increased mechanical resistance in some cases.[30]

Several authors[30, 41–43] have studied the influence of different variables affecting gas transfer in sandwich membrane diffusers. Their conclusions in this respect can be summed up as follows:

(1) Membrane permeability and the characteristic membrane parameters are qualitatively correlated. Thus, the larger its pore size and the thinner it is, the more efficient—though also the less mechanically resistant and the less discriminating—a membrane will be.

(2) The extent of analyte transfer is essentially a function of the analyte solubility. Thus, carbon dioxide is much more permeable than ammonia—their water solubility is 0.515 $g\,g^{-1}$ and 1.72×10^{-3} $g\,g^{-1}$ for CO_2 and NH_3, respectively.

(3) In the absence of chemical reaction, the highest gas-diffusion efficiency is accomplished at low flow-rates. Changes in the flow-rate ratio result in no significant improvement in this respect. If the analyte reacts with the acceptor stream, the flow-rate of the latter should be kept as low as possible. In addition, halting the flow of the acceptor stream enables preconcentration steps to be efficiently implemented.

(4) The use of counter-current instead of concurrent flows does not significantly increase separation efficiency.

(5) The pressure should be the same on both sides of the membrane.

(6) Large effective contact areas result in increased separation efficiency and even allow the mixed gases to be discriminated. However, too large a surface area may detract from the mechanical resistance of parallelepipedal cells. Thus, the winding or spiral track of tubular cells ensures better results than those provided by

parallelepipedal cells of the same external dimensions, to the slight detriment of the analyte dispersion or dilution.[39]

(7) Conduit depth in tubular cells should be as shallow as possible in order to avoid problems arising from diffusional transport by the two liquids, which in turn may give rise to kinetic complications.

(8) The efficiency with which scarcely volatile species are separated can be increased by 10–30% by using high temperatures (50–70 °C)[31,41] or irradiating the diffusion cell with ultrasonic energy.[31] Even more substantial benefits are occasionally achieved by increasing the ionic strength of the donor stream (*e.g.* in the case of ammonia[30]).

Van der Linden[41] used the tanks-in-series model to derive mathematical expressions relating the analyte concentration on both sides of the membrane to time, the transfer coefficient and the flow-rate. Other theoretical contributions from this author deal with the optimal composition of the acceptor stream.[42] The transport of the gas-analyte through microporous hydrophobic membranes involves diffusion of the solute to the solution–membrane interface, partitioning between the membrane and the acceptor solution and diffusion from the surface into the bulk acceptor phase. This requires the occurrence of a solution–membrane interface along the pore line to the inside of the pore. Such an interface is made up of an aqueous layer fixed to the membrane frame that forms a path connecting the two sides of the membrane.

2.3.4 Continuous Configurations

The use of automatic continuous configurations coupled to gas-diffusion cells has opened a host of prospects thanks to their versatility and the vast development of FIA over the last few years.[5,6] Thus, injection valves can be used to insert the sample (Figure 2.4A), while selecting valves (Figures 2.4B and C) can be devoted to choosing the appropriate stream at each operational stage. Gas-diffusion cells are usually located in the transport–reaction zone (Figures 2.4A and C), although they are also occasionally placed within the loop of an injection valve (Figure 2.4B). The sample can be injected into the channel through which the carrier acting as donor solution is circulated (Figure 2.4A) or, alternatively, it can be continuously inserted by aspiration (Figures 2.4B and C), in which case the inserted volume is controlled through the flow-rate and the aspiration time—an additional reagent stream, R_1, is required in those cases where the analyte must be volatilized. Continuous introduction of the sample often calls for an additional carrier in order to condition the diffusion cell between successively introduced samples. As a rule, the acceptor stream is directly driven to the gas-diffusion cell; an additional stream, containing another reagent (R_2), allows a derivatizing-indicator reaction to be implemented in a coil located after the merging point for the two streams wherever necessary.

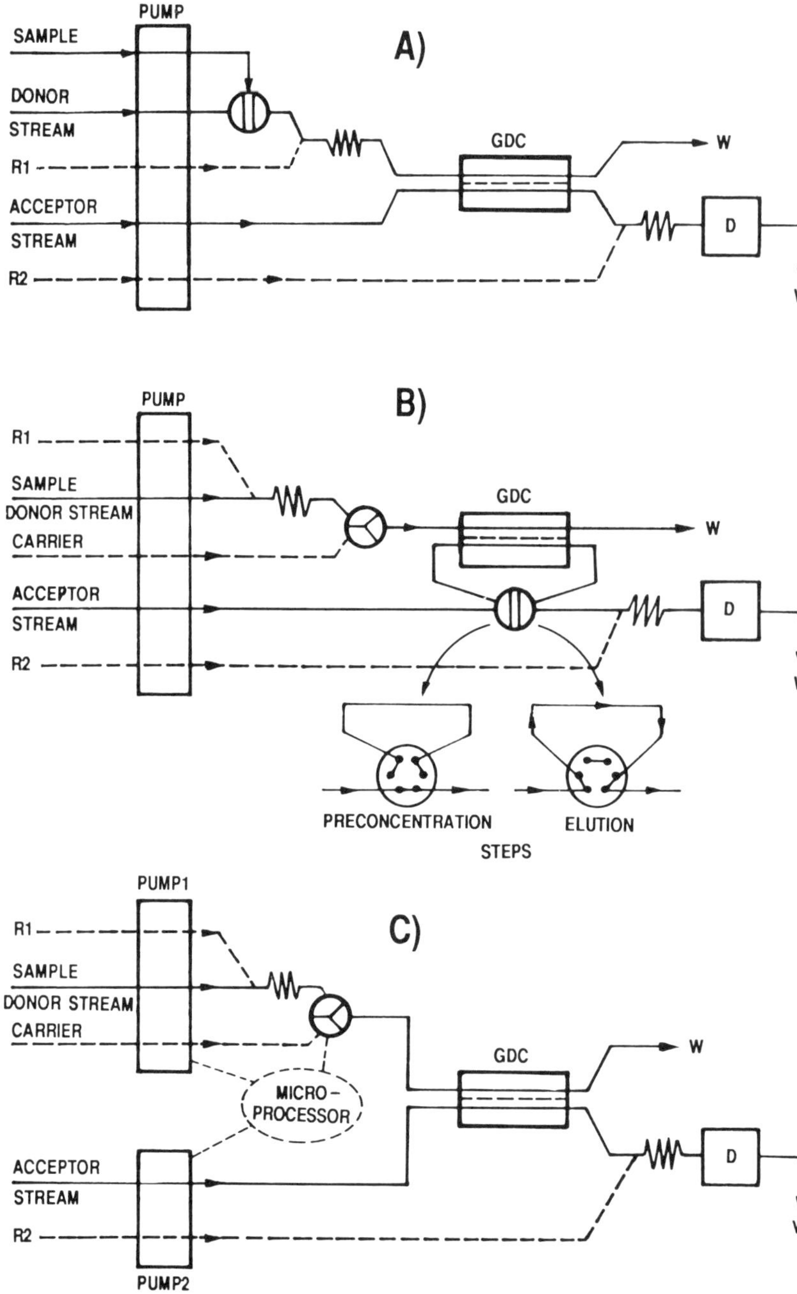

Figure 2.4 *Automatic continuous configurations using a gas-diffusion cell (GDC) and, optionally, auxiliary reagents* (R_1, R_2). (A) *Continuous operational mode with injection of the sample into the donor stream.* (B) *Intermittent functioning for preconcentration (the GDC is located in the injection valve).* (C) *Stopped-flow configuration with intermittent functioning of one of the pumps* (P_2) *for preconcentration*

Preconcentration with a gas-diffusion cell entails stopping the acceptor stream, which is used to collect in the microvolume of transferred gas a large volume of the donor solution that is continuously passed through the cell for an accurately preset time. The performance of this step can be enhanced by implementing a derivatizing reaction in the acceptor stream in order to favour continuous removal of the analyte in the vicinity of the membrane, thereby also establishing a favourable concentration gradient. This technical approach can be implemented in two ways. One involves using a six-way valve—two of the ways are cancelled—such as that depicted in Figure 2.4B, which allows its loop to be isolated and hold a static portion of the acceptor liquid while the acceptor stream is driven directly to the detector during the preconcentration step. The other alternative is based on the intermittent pumping mode, which entails using a microprocessor to synchronize the injection or aspiration of the sample with the stop-and-go sequence of the pump (P_2) that establishes the flow of the acceptor stream, while the other pump (P_1) is allowed to work uninterruptedly. In this way, a static microvolume of the acceptor stream collects the analyte from a large sample volume over a programmed stop period.

The configurations depicted in Figure 2.4 are only a schematic simplification of the variety of set-ups used for this purpose.

2.3.5 Detectors

Molecular spectroscopic techniques are the most frequently used for detection purposes in automatic continuous configurations including gas-diffusers. Thus, photometry has been chiefly used to monitor the formation of coloured products and colour changes in acid–base indicators and to measure absorbance decreases. Some chemiluminescence determinations and others based on reflectance measurements with optical fibre, though no fluorimetric determinations, have also been reported.

Atomic spectroscopic detectors in their atomic absorption and atomic emission–ICP varieties are preferred for hydride-generation derivatizing reactions.

Electroanalytical techniques have also been used in this context. Of these, potentiometric techniques based or not on the use of ion-selective electrodes, are the most common. Conductimetry and amperometry have also been occasionally used.

Special mention should be made of the earliest attempts at coupling an automatic continuous system based on FIA principles and furnished with a gas-diffusion cell to a mass spectrometer.[44] This makes a promising hyphenated technique as it combines the advantages of FIA (rapid introduction of samples, continuous removal of interfering species) and the high sensing and discriminating capacity of mass spectrometry. Such a promising association has so far been applied to the determination of arsenic by quadrupole mass spectrometry with chloroform as solvent in a magnetic sector mass spectrometer.

2.3.6 Applications

Although some determinations based on gas-diffusion in automatic continuous systems for analytes such as urea,[32] nitrite,[39] and arsenic[36] were not applied to real samples, most of the work carried out in this context finds practical application. The chief aim is to increase selectivity as a result of the non-permeability of the membrane to other species and the kinetic discrimination between the volatile species. Also, the possibility of implementing preconcentration steps allows the sensitivity to be substantially increased and analytes to be determined at low concentration levels.

The earliest application of these systems in connection with FIA configurations was developed in the field of *clinical chemistry*; it involved the determination of carbon dioxide in plasma by using a straightforward manifold (Figure 2.4A) where the biological fluid was injected into a sulphuric stream and the acceptor stream consisted of an HCO_3^-/CO_3^{2-} buffer containing an acid–base indicator.[16] Configurations such as that depicted in Figure 2.4A have been used for the determination of ammonia in whole blood and plasma with direct potentiometric sensing[23] or photometric detection with the aid of an acid–base indicator,[24] as well as for the determination of ammonia in undiluted urine.[31] Enzymatic systems with integrated separation and detection have been employed to determine such analytes as creatinine in undiluted blood serum.[33]

Environmental analysis has so far been the most frequent field of application of these systems, most of which resemble the configuration shown schematically in Figure 2.4.A. This scheme was used for the determination of chlorine dioxide in water with chemiluminescence detection,[14] inorganic arsenic species in surface seawater by hydride-generation/ICP,[37] sulphide in water with an S^{2-}-selective electrode,[22] and ammonia in water by photometry with an acid–base indicator,[26, 29] UV molecular absorption[28] and chemiluminescence.[27] Special mention should be made in this context of a portable automated spectrophotometric field monitor equipped with a diffusion cell for the determination of ammonia in water.[29] Only two systems of this type included a preconcentration step. One is based on the configuration depicted in Figure 2.4B and was successfully applied to the determination of cyanide in waste water;[21] the other employed the configuration shown in Figure 2.4C and was applied to the determination of residual ozone in water,[13] which included a study of kinetic discrimination between potentially interfering volatile species.

Despite their great potential, these systems have scarcely been applied to *food analysis*. Among the few determinative methods developed in this context three are worth special mention, *viz.* those for oxidized ketone bodies in milk,[38] ethanol in alcoholic beverages such as shakes and whisky,[15] and sulphur dioxide in wines,[17, 19] beer and fruit juice,[18] all of which were developed by using the simplest of all configurations (Figure 2.4A).

These automatic continuous configurations have also been used in *agricultural chemistry* (*e.g.* in the determination of ammonia in Kjeldahl digests from plant tissues, animal feeds, and fertilizers[25]) and in *industrial analyses* (*e.g.* the determination of germanium in GaAs semiconductor material and in acid extracts of a polyethylene terephthalate[35]).

Special mention should also be made here of the systems developed for *simultaneous determinations* of volatile species relying on different principles, though essentially based on chemical selectivity. The manifolds used for this purpose are schematically depicted in Figure 2.5. In fact, Figure 2.5A shows the manifold used for the determination of cyanide and thiocyanate.[45] In this case, the simultaneous determination is facilitated by the selectivity of the indicator reaction taking place in the acceptor

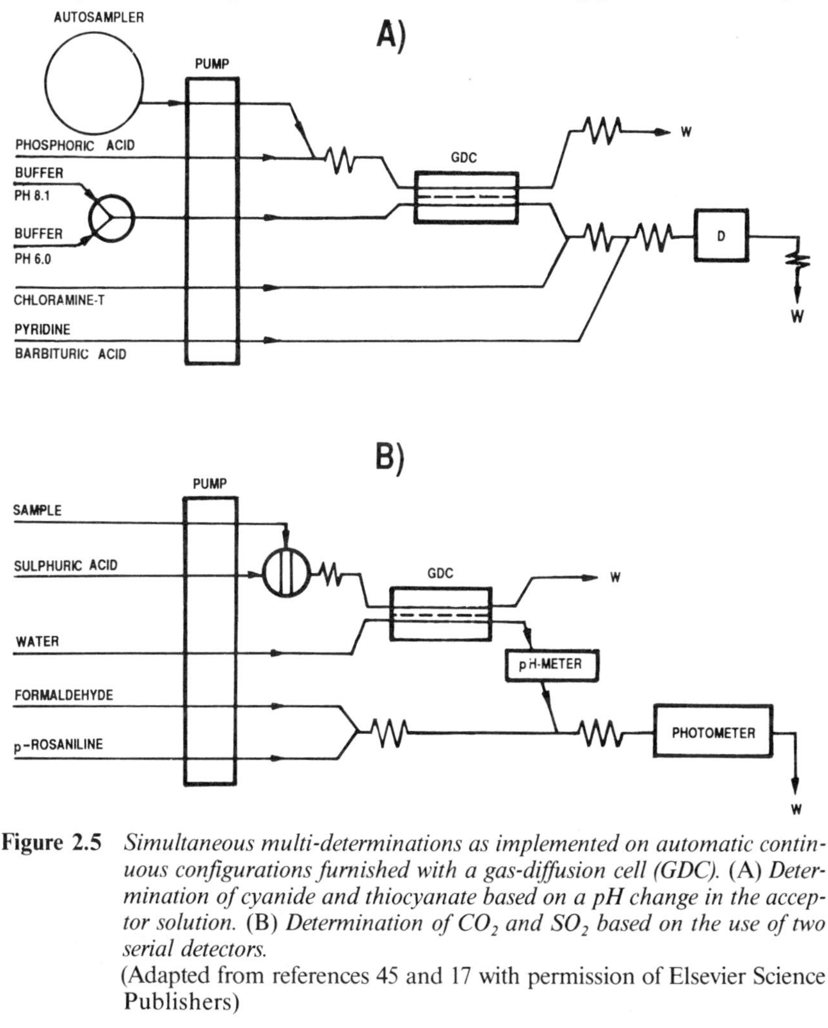

Figure 2.5 *Simultaneous multi-determinations as implemented on automatic continuous configurations furnished with a gas-diffusion cell (GDC).* (A) *Determination of cyanide and thiocyanate based on a pH change in the acceptor solution.* (B) *Determination of CO_2 and SO_2 based on the use of two serial detectors.*
(Adapted from references 45 and 17 with permission of Elsevier Science Publishers)

phase: chloramine reacts very slowly with thiocyanate and very rapidly with cyanide at pH 8.1; on the other hand, both analytes react very rapidly with chloramine at pH 6.0. Thus, by merely changing the buffer used as acceptor solution sequentially (Figure 2.5A includes a selecting valve not present in the original design), one can determine the cyanide concentration (at pH 8.1) and then the overall concentration of the two analytes (pH 6.0). In fact, this is a separation system involving both gas-diffusion (gaseous HCN diffuses across the membrane) and dialysis (the cyanide and thiocyanate ions pass through the pores because of the large pore size—3.5 μm—of the Teflon tube used to construct the tubular cell); in addition, the pH of the two phases plays a major role in the penetration of liquid into the microporous membrane.

Carbon dioxide and sulphur dioxide were determined in urine with the manifold depicted in Figure 2.5B.[17] After the two analytes were simultaneously transferred into the gas-diffusion cell following injection of the sample into a sulphuric carrier and absorption into the acceptor water stream, this was serially passed through two detectors, a potentiometric one used to measure the acidity yielded by the two analytes and a photometric one to monitor the product yielded in the specific reaction of one of the analytes (SO_2) with two derivatizing reagents (*p*-rosaniline and formaldehyde) that were introduced into the system through two channels merging prior to being mixed with the acceptor stream. The former detector provides the overall concentration of the two analytes, while the latter yields the concentration of one.

One atypical application of gas-diffusion involves the use of a non-porous perfluorosulphonate cation-exchange tubular membrane known as 'passive membrane reactor' which is permeable to ammonia. It is used to change the pH of the main stream without diluting it, *i.e.* avoiding the use of an additional channel. One such unit was applied to the enzymatic determination of peroxides.[46, 47]

2.3.7 Integration of Separation and Detection

The incorporation of a spectroscopic measuring system into the compartment of the gas-diffusion cell through which the acceptor solution is circulated allows one to carry out detection simultaneously with the separation process. This results in major advantages such as increased sensitivity (the transferred analyte is not diluted on its way to the detector) and miniaturization, in addition to other, more specific assets. This experimental approach, namely the integration of reaction (and/or separation) and detection, is one of the most recent and promising trends in the use of automatic continuous systems.[48]

Figure 2.6 shows the schemes of two sandwich gas-diffusion cells that were modified with respect to their original designs in order to be used as detection cells as well. In 1985, Ruzicka and Hansen[32] established the principles behind optosensing at active surfaces in FIA integrated

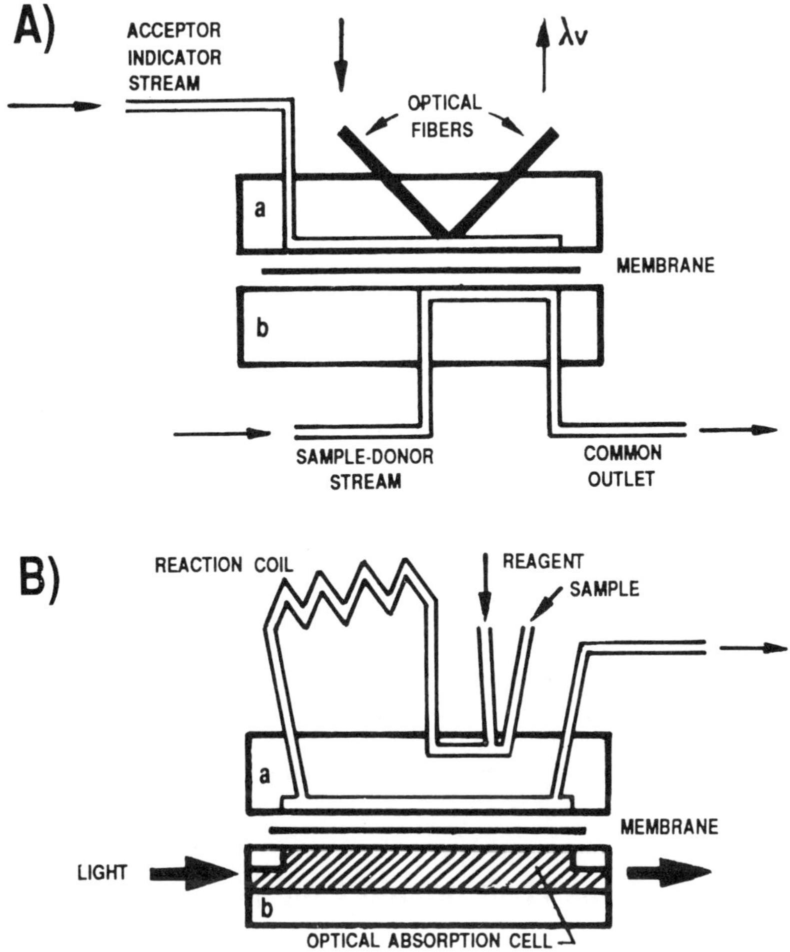

Figure 2.6 *Continuous cells with integrating gas-diffusion and detection.* (A) *Reflectance spectroscopic detection with optical fibres.* (B) *Atomic absorption spectroscopic detection.*
(Reproduced from references 33 and 52 with permission of Elsevier Science Publishers)

microconduits using reflectance spectroscopy and optical fibre. One of the approaches reported by these authors involves the use of a continuous-type optosensor featuring a gas-diffusion membrane. A permeable micromembrane and the two optical fibre branches are located in the zone where the acceptor and donor streams are merged. Later, Hansen *et al.*[33] applied these principles to the determination of endogenous ammonia and creatinine in undiluted blood serum by using ingenious FIA manifolds enabling discrimination between endogenous ammonia and that resulting from the degradation of creatinine, a reaction catalysed by the enzyme creatinine immunohydrolase, which is immobilized and makes the enzyme reactor. The detection/gas-diffusion cell used for this purpose is schematically depicted in Figure 2.6A. It consists of two black PVC blocks (*a* and *b*) with an intervening ammonia-permeable membrane. Part *b* has a central groove connected to the outside through two ports, one which acts as an inlet for the sample-donor stream, while the other functions as an outlet for both the acceptor-indicator and the donor streams. Part *a* has a port allowing passage of the acceptor-indicator stream into the central microzone, as well as two optical fibres focused on the same point of the white gas-permeable membrane, which serves as a reflecting opaque background. Facing the focusing point of the membrane is the inlet for the donor stream. This wall-jet configuration allows pH changes in the indicator solution to be monitored exactly where the maximum transfer of ammonia takes place. The signals yielded by low analyte concentrations can be increased by stopping the donor and acceptor streams at preset times after injection.

In 1983, Andrade *et al.*[49] reported the first diffusion–detection cell for the determination of mercury based on the use of cold mercury vapour generated by sodium borohydride. The original design was subsequently improved by Barnes[50] and Fang.[51, 52] Figure 2.6B shows a scheme of the cell, which is placed in the optical path of an atomic absorption spectrophotometer. Block *a* includes the merging point for the reagent and sample streams and their connection to the central minichamber through an appropriate coil. The mercury vapour generated passes across the membrane and reaches block *b*, the minichamber of which is open to a wider zone that acts as an optical absorption cell as it has two transparent windows on its ends.

2.4 Continuous Hydride and Cold Mercury Vapour Generation

2.4.1 General Considerations

This section deals with continuous separation techniques involving a gas–liquid interface and based on a specific type of reaction intended to yield a volatile species from a short number of analytes among which arsenic, antimony, selenium, bismuth, and mercury are probably the most representative. The reagent most frequently employed for this purpose is sodium or potassium borohydride in an acid medium, where it yields volatile hydrides and cold mercury vapour thanks to the singular physical properties of this metal. A gas-carrier stream is obviously required to transfer the generated hydride (or Hg vapour) and hydrogen to a suitable detector (normally an atomic spectroscopic detector).

Despite its well known advantages, classical batch hydride generation procedures have a number of pitfalls: (*a*) they use much larger sample amounts than other traditional methods for introduction of samples in atomic spectroscopy; (*b*) interferences occasionally pose serious problems despite the attempts at minimizing them; (*c*) their throughputs are relatively modest.

The earliest automatic continuous hydride generation systems connected on-line with atomic emission (ICP) and absorption (AAS) spectroscopic detectors were reported by Thompson *et al.* in 1978[70] and by Astrom in 1982,[53] respectively. Samples were introduced into the system by continuous aspiration in the former case and by FIA injection in the latter. Subsequent work on this topic has shown the significant gains involved in replacing conventional systems by continuous ones, namely: (*a*) reduced sample and reagent consumption; (*b*) substantially increased sampling frequencies; (*c*) markedly increased selectivity; (*d*) increased sensitivity; and (*e*) reduced human participation (automation).

The technical features of this type of continuous assembly (stripping gas, gas–liquid separation modules, manifolds), the essentials of the corresponding automatic analytical methodologies and their applications to real problems are commented on below. Finally, due mention will be made of the earliest attempts at developing other hyphenated methodologies combining continuous separation techniques (ion-exchange and liquid–liquid extraction) and continuous hydride generation systems.

2.4.2 Stripping Gas

The release and transfer to the detector of the generated hydride or cold mercury vapour can be facilitated by using a nitrogen, argon, or hydrogen stream at a flow-rate between 50 and 200 ml min^{-1}. Although this flow-rate is not critical, its should be kept constant as far as possible. The release of large, variable amounts of hydrogen in the reaction zone results in an overpressure of variable magnitude on the inside of narrow-bore tubing. In order to ensure a uniform flow, a throttle in the form of a small copper disk with a tiny aperture (*ca.* 0.2 mm) is fixed to the outlet of the pressure gauge of the gas container. A back-pressure of more than 3 kg cm^{-2} may result from such a low gas flow-rate as 150 ml min^{-1}; however, the flow-rate should not be affected by small back-pressure changes in the flow system.[54]

The gas stream can be introduced into a continuous hydride or cold mercury vapour generation system through different zones, namely (see Figure 2.8):

(a) Between the reaction coil and the gas–liquid separation cell. This is the most widely used alternative.
(b) Through the gas–liquid cell itself. In gas-diffusion cells the gas acts as an acceptor stream, replacing the usual liquid stream of this type of cell, or functions to remove the cold mercury vapour from the optical absorption zone integrated into the diffusion cell itself. In expansion separators, the gas directly introduced facilitates the gas–liquid separation, thereby favouring mixing and sweeping.
(c) Prior to the zone of mixing of sample and reagent, as a means of

increasing sensitivity through the favourable reaction conditions thus set. In this case, the gas is introduced through a tube of the peristaltic pump in order to hinder sample dispersion. This stream by itself does not ensure the complete release and transfer to the detection cell, so it is normally complemented with an additional gas stream circulating between the coil and the separation cell.

2.4.3 Gas–Liquid Separators

This type of continuous unit is intended to collect one or several streams containing a liquid mixed with some gases (volatile analytes, hydrogen, stripping gas) or the stripping gas alone and separate the two phases. The gas phase is transported to the detector, while the liquid phase is driven to waste, either directly or by controlling the emerging flow-rate via a tube of the peristaltic pump.

Separators to be used in continuous separation systems based on the generation of hydrides or cold mercury vapour should ideally meet some basic requirements, *viz*: (*a*) they should work smoothly and regularly in order to avoid such irreproducibilities as those typical of the disaggregation units used in automatic continuous segmented flow systems;[3, 4] (*b*) they should induce minimal dispersion or dilution of the gas analyte, *i.e.* they should have as low dead volumes as possible—this in turn results in increased sensitivity—though not so low as to give rise to incomplete gas-liquid separations; (*c*) they should prevent the liquid from reaching the detection system, where it would give rise to major precision and selectivity problems in addition to deteriorating the detector [*e.g.* a quartz atomizer (AAS) or a plasma torch (AE-ICP)].

There are two general types of gas–liquid separator design that differ in the way in which the isothermal separation is accomplished. Those based on the diffusion of the gas phase across a permeable membrane were described in the preceding section (see Figure 2.3); some authors refer to these cells as 'dual phase gas-diffusion cells' as the membrane separates a liquid carrier stream from a gas stream. Both sandwich and tubular gas-diffusion cells have been used for hydride and cold mercury vapour generation. Occasionally, the acceptor stream outlet is only connected to the detector—it has no inlet—as the excess gas in the carrier phase suffices to carry out the transfer; alternatively, the gas is directly introduced through the acceptor zone, both to transport the gas analyte to the detector and to remove atomic vapour in the above-described integrated diffusion–detection cells.

Gas–liquid expansion units are normally minichambers made of glass. Despite the variety of designs available, they are usually connected to the outside in a similar way. Thus, they normally have one inlet for the liquid–gas mixture emerging from the injection–reaction zone—alternatively, the gas can be directly introduced into this unit—and two outlets: one for the gas and the other for the liquid phase, which are normally located at the top and the bottom of the unit respectively.

Figure 2.7 shows the schemes of various gas–liquid expansion separation units, which have been classified according to the zone through which the stripping gas is introduced. About a quarter of these cells are based on the original U-shaped design of Vijan,[65] which has been modified to improve specific aspects, particularly to reduce dispersion by miniaturization, which is facilitated by the small amount of analyte (sample) and reagents involved. Figure 2.7A depicts gas–liquid separators in which the stripping gas is previously merged with the sample–reagent-gases mixture from the continuous manifold; the first two designs use Pyrex beads that allow for a smooth separation of the liquid and gas. The simple separator illustrated in Figure 2.7A.3 was developed for the separation of cold mercury vapour and consists of two acrylic blocks (*a* and *b*).

Figure 2.7B depicts some gas–liquid expansion separators into which the stripping gas is introduced in a direct manner. In that shown in Figure 2.7B.1, originally designed by Thompson *et al.*[70] and applied for practical purposes at a later stage, the mixed sample plus reagent and argon streams enter the minichamber; the liquid is drained out by a communicating-vessel effect through a U-shaped tube along the bottom of the expansion zone, while the gas is directly driven to the ICP torch. The separator in Figure 2.7B.2 is characterized by its simplicity, while that in Figure 2.7B.3, designed for the generation of cold mercury vapour, is much more complex and somewhat atypical on account of the unusual location of the stripping gas inlet and the presence of a condenser between the expansion minichamber and the gas outlet to the detection cell.

Some other designs are used for specific purposes. Thus, Chan[69] used a mini-impinger partly filled with concentrated sulphuric acid and located between a U-shaped liquid separator and the quartz cell of the detector to eliminate moisture and homogenize the hydride–argon mixture.

2.4.4 Continuous Configurations

The fact that most continuous hydride and cold mercury vapour configurations connected on-line with atomic spectroscopic detectors involve the same type of reaction accounts for their great similarities. Figure 2.8 shows the generic manifold used for this purpose. The sample is introduced into a carrier stream (usually water) by means of an injection valve. An acid (usually hydrochloric) stream is merged with the previous carrier and with one or more reagent streams in order to establish an appropriate pH for the generation of the volatile species and accomplish analyte discrimination (speciation), respectively. Two mixing coils are used to increase the extent of mixing. Finally, a sodium borohydride

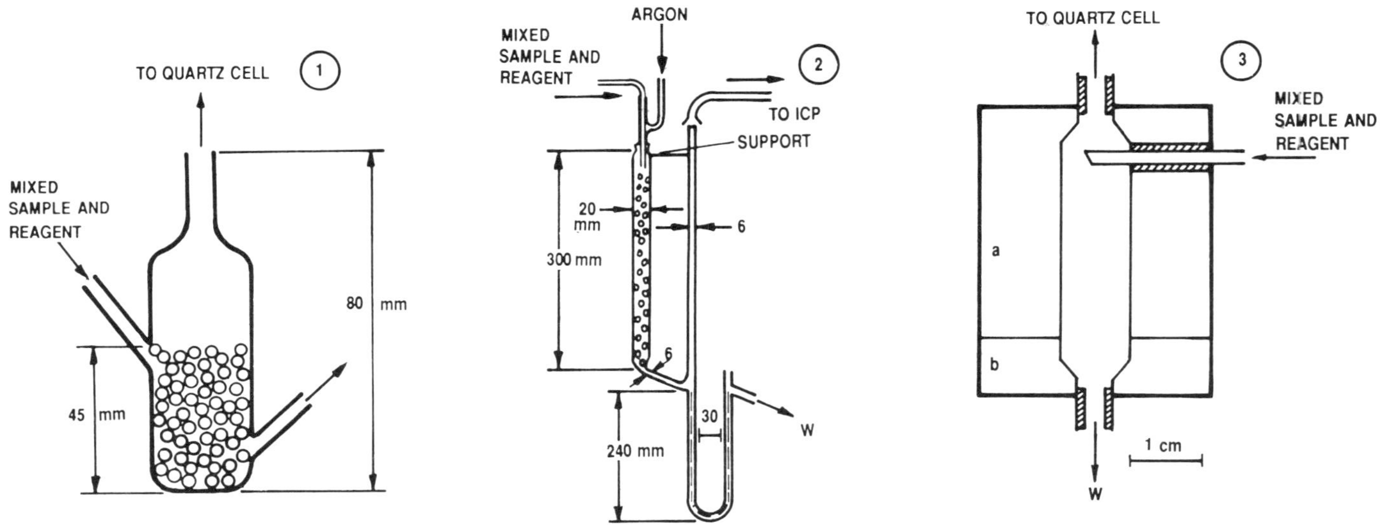

Figure 2.7A *Gas–liquid separators used in continuous hydride and cold mercury vapour generation systems. Assemblies where the stripping gas is previously mixed with the reaction mixture from the manifold.*
(Reproduced from the documentation of the Perkin-Elmer FIAS–200 module (1), and from references 71(2) and 63(3), with permission of the Royal Society of Chemistry and Francis & Taylor Ltd., respectively)

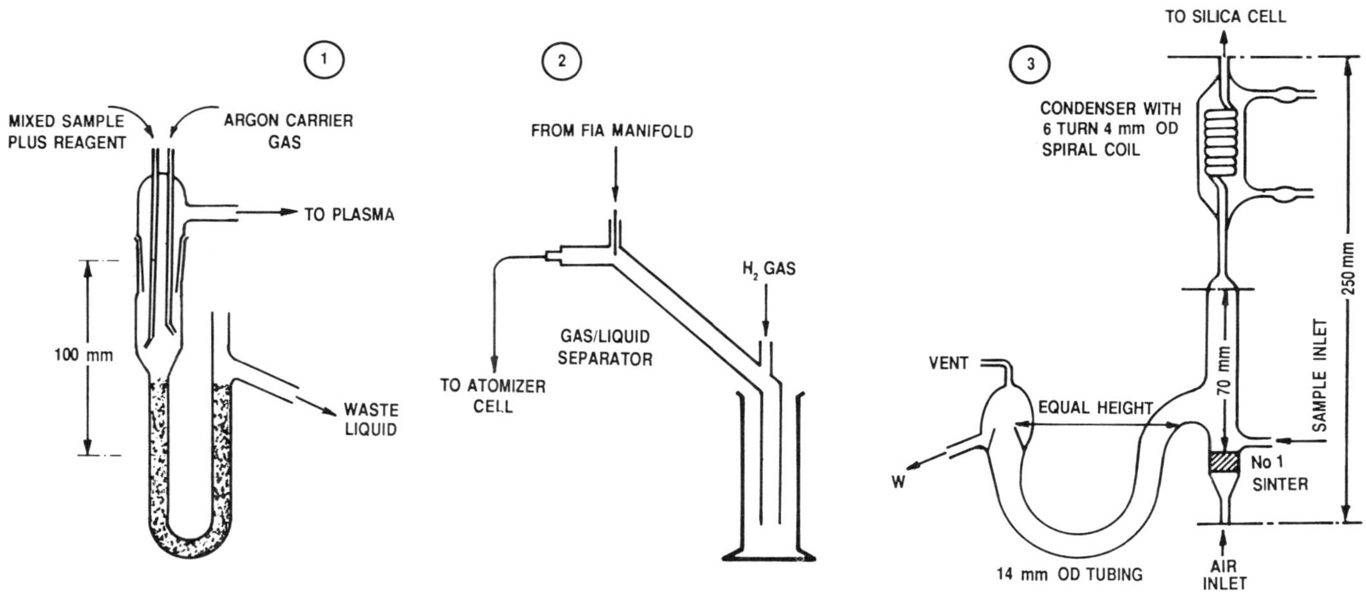

Figure 2.7B *Gas–liquid separators used in continuous hydride and cold mercury vapour generation systems. Assemblies with direct introduction of the stripping gas into the separator.*
(Reproduced from references 70(1), 66(2), and 62(3), with permission of the Royal Society of Chemistry, the American Chemical Society and Francis & Taylor Ltd., respectively)

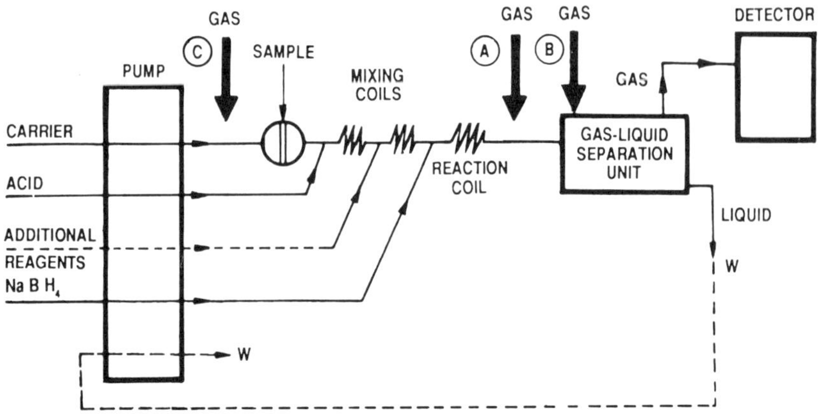

Figure 2.8 *General scheme of automatic continuous systems for hydride (or cold mercury vapour) generation showing the three zones into which the carrier gas stream can be introduced. For details, see text*

stream is also used, and the hydride is generated in the reaction. Cold mercury vapour can also be generated with other reductants (*e.g.* tin chloride[62, 63]). As noted earlier, the stripping gas can be introduced between the reaction coil and the gas–liquid separation unit (A), directly through this unit (B), or prior to the sample injection port (C). The isolated gas phase (hydride or mercury vapour plus hydrogen and the stripping gas) is driven to the detection cell under pressure. The liquid phase is sent to waste, either directly (*e.g.* by overflow in a communicating-vessel minisystem) or by aspiration through the peristaltic pump, which ensures smooth functioning of this unit.

Most of the systems developed to this end respond to this general scheme. However, some peculiar modifications with respect to the generic design are worth special mention. Thus, rotary injection valves have occasionally been replaced with solenoid valves[58, 62] and proportional injectors[56, 63] in order to facilitate automation and to reduce the consumption of sodium borohydride by using dual injections of sample and reagent in a merging-zones configuration, respectively. Sequential merging points have also been substituted by four-way mixing points (three inlets for the sample buffer, $NaBH_4$ and argon, and one outlet for the mixture).[35] A straightforward modification of the manifold connections allows one to alternately introduce the sample and reagent into the system.[59] Samples cannot only be introduced by injection of microvolumes, but also in a continuous fashion.[70, 71] Quite an interesting modification in this respect involves using a minireactor filled with a strong macroporous anion-exchange resin (Amberlyst A-26) to retain the borohydride; this allows one to use a straightforward single-channel manifold. The packed reactor must be regenerated periodically as the borohydride is gradually depleted through continuous use.[66] Due mention should also be made here of attempts at increasing the degree of automation of these configurations. The use of two computer-controlled peristaltic pumps[36] and heating the reaction coil to expedite the kinetics of the chemical and physico-chemical processes involved[62] are two cases in point.

2.4.5 Types of Detector

The physico-chemical features of the analytes to be assayed and the way in which they reach the detector in automatic continuous methodologies based on hydride or cold mercury vapour generation have made atomic spectroscopy the detection technique of choice in most cases.

A heated quartz cell or a similar device placed in the lightpath of an atomic absorption spectrophotometer is connected directly to the gas outlet of the separation cell. This widely used, inexpensive approach has been subjected to various modifications. The continuous introduction of samples into emission spectrometers (ICP) by means of automatic continuous systems offers major advantages over other available alternatives.[72] In this particular case, there is the additional asset of implementing a continuous separation technique with virtually no human participation. Figure 2.9 shows the scheme of a system that uses a proportional injector to perform dual injections of sample and reagent, and a classical U-shaped separator—note the argon is introduced at three different points into the overall integrated system.

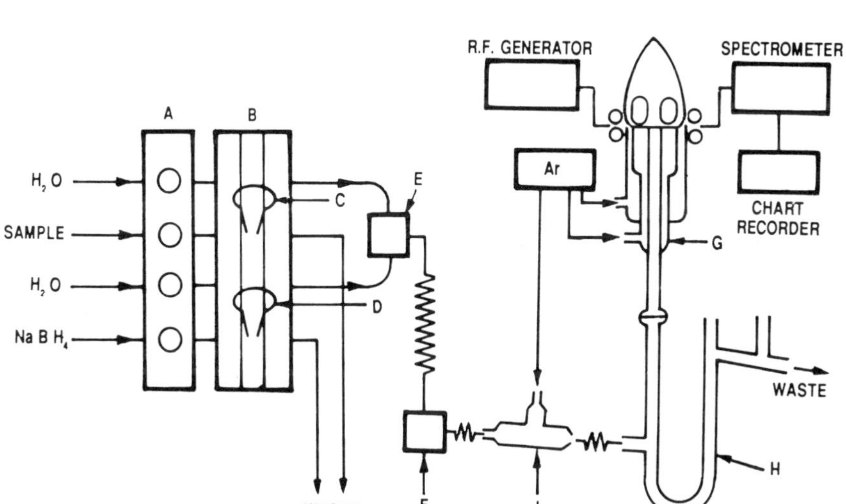

Figure 2.9 *Hydride generation system coupled on-line to an atomic emission spectrometer (ICP).* (A) *Peristaltic pump.* (B) *Proportional injector.* (C) *Sample loop.* (D) *Reagent loop.* (E) *Merging point.* (F) *Mixing point.* (G) *Plasma torch.* (H) *U-tube.* (I) *Nebulizer.*
(Reproduced from reference 56 with permission of Elsevier Sci. Pub.)

Other types of detection such as inductively coupled plasma spectrometry[59] and molecular emission cavity analysis[64] have also been used in this context. The former benefits from the interest aroused by the now well-established ICP–MS association. Figure 2.10 shows the scheme of an instrument of this type coupled to a continuous hydride generation system.

2.4.6 Basic Features of the Analytical Methodology

The figures of merit of the methods reported so far for the determination of metal and non-metal species by hydride or cold mercury vapour generation with on-line atomic spectroscopic detection are rather favourably surprising in terms of quality.

The detection limits achieved (SNR $\simeq$ 3) are extremely low: between 0.01–30 ng ml^{-1}, depending on the element assayed and the detection technique used. Relative standard deviations for the p.p.b. concentration level range between 0.5–5% in most cases. Also, with a few exceptions[62, 69] sample throughputs are always above 100 samples h^{-1} and as high as 200 samples h^{-1} in some cases.[51, 56]

The tolerance to interferents of continuous and batch hydride generation techniques has been analysed in a number of comparative studies. Thus, according to Astrom,[53] the continuous variant is between 100–1000 times more tolerant to cobalt, nickel, and copper than its batch counterpart in the determination of bismuth. Similarly increased tolerated levels have also been reported for the determination of selenium.[61, 69] In addition, the use of various masking agents, whether included in the carrier or introduced through a separate channel, significantly reduces the perturbation of some species.[35, 59, 64, 66] According to Fang,[54] this major gain is a result of the short residence time of the reaction mixture in the continuous

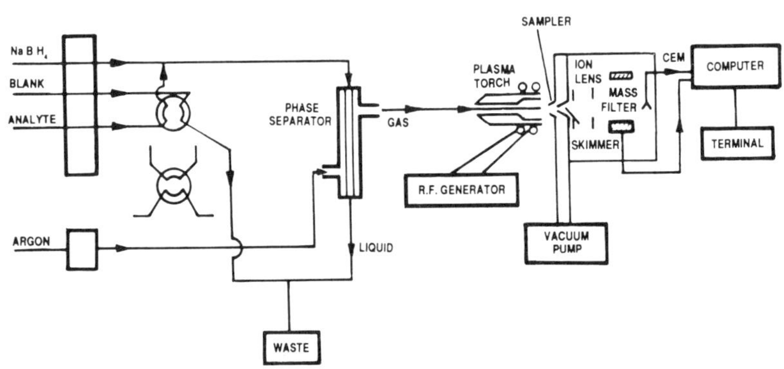

Figure 2.10 *Continuous hydride generation system coupled on-line to an inductively coupled plasma–mass spectrometer system* (FI–ICP–MS). (Reproduced from reference 59 with permission of the Royal Society of Chemistry)

system, which allows full development of the very fast hydride generation reaction while hindering that of slower, potentially interfering reactions, and of the small sample volumes used, which obviously introduce smaller amounts of interferents. Thus, many interferences with hydride generation methods arise in the gas phase from the competition of hydride and active free radicals in the atomization process. Also, the perturbation caused by each interferent is proportional to its absolute amount rather than to its relative concentration with respect to the analyte.[73]

In summation, automatic continuous methodologies based on the generation of hydrides or cold mercury vapour are outstanding in terms of such analytical features as sensitivity, selectivity, precision and rapidity.

2.4.7 Applications

Arsenic (AAS;[34, 37, 55, 56] AE-ICP;[36, 56, 58, 60, 70] MECA[64]), selenium (AAS;[57, 61, 69] AE-ICP[60, 69, 70]) and mercury (AAS[51, 52, 62, 63]) are the elements most frequently determined by automatic continuous hydride and cold mercury vapour generation methods, followed by tellurium (AAS;[55] AE-ICP[70]), antimony (AAS;[55] ICP[60, 70]), bismuth (AAS[53, 55]), germanium (AE-ICP[35]) and lead (ICP-MS[59]).

Many of the above determinations were carried out on standard samples such as steel alloys, coal fly ash, orchard leaves and rice flour.[55–57, 64, 69] In addition, arsenic has been determined in industrial-grade glycerine[58] and surface seawater,[37] germanium in GaAs semiconductors,[35] mercury in waste water[62] and solvay effluents,[63] lead in galena,[59] and selenium in soils, plants and waste water.[61]

Despite its potential interest, speciation has scarcely been addressed by this type of continuous methodology. The few attempts at applying it for this purpose have been limited to As^{III}/As^{V} and Sb^{III}/Sb^{V},[55] and to inorganic/total Hg.[62]

2.4.8 Hyphenated Continuous Separation Techniques

The coupling of a non-chromatographic separation technique to a gas or liquid chromatograph is dealt with in other sections (*e.g.* liquid–liquid extraction with liquid and gas chromatographs, dialysis–HPLC and sorption–HPLC combinations, *etc.*). This section describes the earliest systems based on the combination of two non-chromatographic analytical separation techniques sharing a hydride generation system connected to an atomic spectroscopic detector. The chief aim of this association is to increase the sensitivity (preconcentration) and/or selectivity (interference removal) with respect to the two techniques in isolation.

Figure 2.11 illustrates ion-exchange/hydride generation (A) and liquid–liquid extraction/hydride generation (B) hyphenated methodologies. The former configuration (Figure 2.11A) includes a dual ion-exchange system

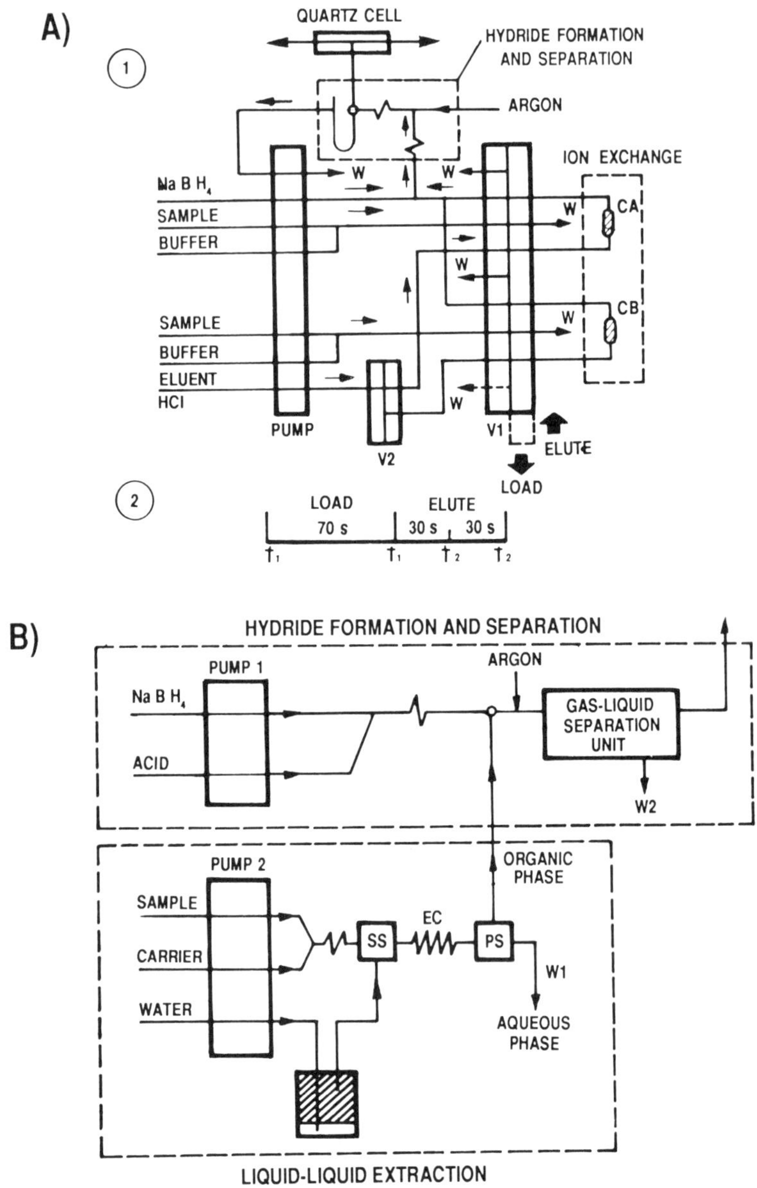

Figure 2.11 *Hyphenated continuous separation techniques.*(A) *Dual-column on-line ion-exchange preconcentration combined with hydride generation–AAS for the determination of selenium.* (1) *FIA manifold* (V_1 *and* V_2 *denote electronically controlled valves, and* C_A *and* C_B *ion-exchange minicolumns).* (2) *Operational sequence.* (B) *Liquid–liquid extraction/hydride generation system for the determination of arsenic, where SS denotes solvent segmenter,* EC *extraction coil and* PS *phase separator. The gas–liquid separation unit is depicted in Figure 2.6.B.*
(Reproduced from referenes 52 and 74 with permission of Elsevier Science Publishers)

intended to increase the sample throughput. It uses a proportional injector rather than an injection valve because of the large number of connectors involved. The figure includes a brief operational scheme. Part 1 of the figure shows the elution position of the exchange column (CA); the eluate is mixed with the borohydride stream and the mixture is then driven to the hydride separator. As the proportional injector (V_2) is switched to a different position, the previous operation is repeated in column CB. In the load position (not shown in the Figure), two samples are directly aspirated, merged with a buffer stream, passed through the two columns and finally sent to waste after being sensed; meanwhile, the borohydride stream is driven to the separator. This system was used to determine selenium at concentrations as low as 2 ng l^{-1} and a sampling frequency of 50 h^{-1} and mercury with a similar detection limit at a sampling rate of 60 samples h^{-1},[52] and bismuth with a detection limit of 1 ng l^{-1}.[75]

A hyphenated on-line liquid–liquid extraction/hydride generation methodology was recently developed by Sanz-Medel *et al.*,[74] following application of its off-line predecessor[71] (Figure 2.11B). In this configuration, arsenic is extracted as AsI_3 into xylene in the continuous extractor, the functioning of which is described in Chapter 3. The outgoing organic phase is merged with the reaction mixture in the hydride generation module, which is similar to those described above. This configuration eliminates interferences with the liquid–liquid extraction process and allows the hydride to be generated from an organic phase, thereby avoiding the customary back-extraction (the borohydride is dissolved in DMF and the 'acid' is glacial acetic acid). The detection limit thus accomplished is 2 μg l^{-1} and the determination is not interfered at all by 25 transition elements at a concentration of 580 mg l^{-1}.

In addition to the type of detector used, these two hyphenated hydride generation methodologies differ in the completely continuous functioning of on-line variant in contrast to the two-step operation of its off-line counterpart. This is reflected in the signals obtained: a plateau with the former and a peak with the latter.

2.5 Distillation

2.5.1 General Considerations

Distillation is based on the compositional difference between a liquid and the vapour formed from it. The efficiency of this separation technique depends on the volatility (vapour pressure) differences between the species present in the liquid. Of the different technical variants of distillation (simple, flash, fractional, rectification[76]), this section deals only with the continuous mode and its modifications for adaptation to different types of analyser.

Distillation systems used in automatic continuous configurations share some basic features, namely: (*a*) they work under continuous feed of liquid mixture and continuous removal of volatile products; (*b*) they are miniature systems normally made of glass—and also occasionally of Teflon; (*c*) the volatility of the analytes is increased by using high temperatures; (*d*) the gas–liquid separation is normally favoured by using a stream of stripping gas that can be introduced into the system at different points—some continuous air-segmented analysers require no additional inert gas stream; (*e*) the gas-analyte is transferred into a liquid stream; and (*f*) the analyte is monitored by means of a continuous detector furnished with a flow-through cell.

Continuous analysers with built-in microdistillation systems usually consist of the following elements: (*a*) the *propulsion module*, which has the two-fold purpose of continuously feeding the microdistiller with the liquid samples, which can optionally be preconditioned by using additional buffer or reagent streams, and of removing the outgoing streams of all the other units, also in a continuous fashion; (*b*) the *microdistillation module* proper, where the gas–liquid separation takes place; (*c*) the *condensation* and/or *absorption module*, which effects the transfer of the gas-analyte to an outgoing liquid stream; and (*d*) the *continuous detection module*.

This type of continuous separation system is formally similar to those of hydride generation in general and of liquid–gas expansion in particular. The only differences lie in the implicit heating and absorption of the gas-analyte into a liquid stream in microdistillation systems, in contrast to room temperature and the direct transfer of the gas to the detector in hydride generation systems.

Below are commented on the most salient features of the above four modules in this context, as are the generic aspects of the configurations used in relation to segmented and unsegmented analysers.

2.5.1.1 Chemical Reactions Involved

The analyte volatility is sometimes enhanced by a prior chemical reaction, which is most often an acid–base reaction (*e.g.* formation of volatile hydrogen cyanide from cyanides,[77, 78] ammonia from ammonium solutions,[79, 80] phenols from phenolates[81, 82]), and only occasionally of a different type (*e.g.* formation of hydrogen sulphide from sulphate[83]).

Whenever the absorbed analyte cannot be sensed directly (*e.g.* colorimetrically in the case of biphenyl[84] and electrochemically in the case of cyanide [77, 78]), its physico-chemical features can be adapted to the continuous detector to be used by means of a variety of derivatizing-indicator reactions (*e.g.* phenol with 4-aminoantipyrine[82] or 3-methylbenzothiazolinone[81] or ethanol with dichromate).[85] These indicator reactions cannot be categorized because of their variety. They are occasionally very complex (*e.g.* the determination of cyanide reported by Caspiere *et al.*,[86] which consists of several steps involving bromination of the analyte, removal of excess bromine with a reductant and coupling of cyanogen bromide with pyridine and *p*-phenylenediamine).

2.5.1.2 Configurations Used

Continuous distillation configurations are usually continuous analysers of great complexity, in that they use a large number of channels, peristaltic pumps and connectors, which hinder smooth operation somewhat. In addition, they feature rather long start-up times.

Another peculiarity of these configurations is the lack of injection valves—and of all types of valve in general: the sample is continuously aspirated into the system. Carry-over is thus normally avoided by frequently passing a blank solution through the system between successively inserted samples.

In contrast to other types of analyser, sample throughput is relatively low as a result of the unusually large sample volumes handled (of the order of a few millilitres).

Figure 2.12 shows the general scheme of an automatic continuous analyser including the microdistillation and condensation/absorption modules that are commented on below. The figure also shows the location of the cooler for the liquid drained from the microdistiller and the expansion trap at the condenser/absorber outlet.

Propulsion units (one or more) are intended to introduce sample, reagent and absorbent streams and aspirate in the opposite direction the solution stream containing the absorbed analyte to drive it to the detector

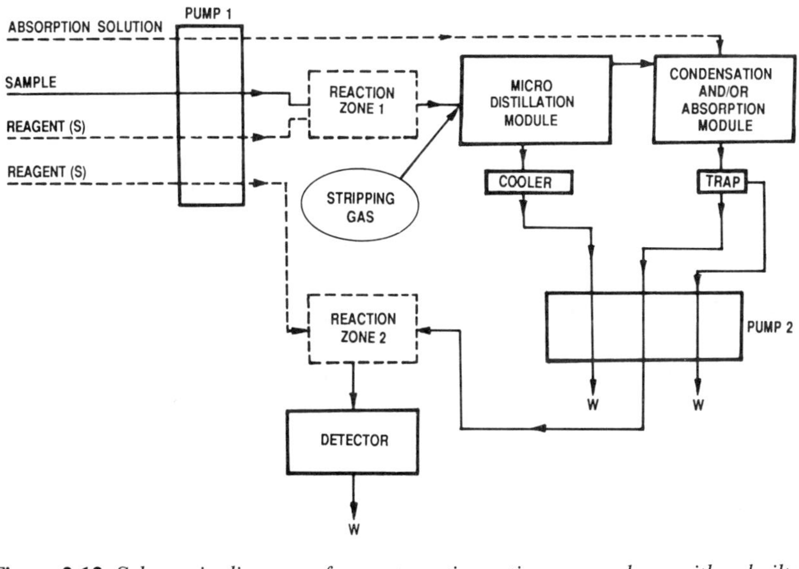

Figure 2.12 *Schematic diagram of an automatic continuous analyser with a built-in microdistillation system. For details, see text*

—after being merged or not with one or more derivatizing-indicator reagents—the waste solution from the microdistillation chamber and the stripping gas from the trap at the condensation/absorption module.

As can be seen, there are two optional reaction zones. In the first, the sample is conditioned to facilitate volatilization of the analyte(s). In the second one or more sequential reactions, intended to facilitate detection of the final product, may take place.

Despite the wide variety of detectors used in this context, most analysers include a built-in colorimeter or UV–visible photometer. Electrochemical detectors are also occasionally used for this purpose.

The examples of microdistillation as implemented in automatic continuous analytical systems exposed below are classified according to the basic difference between such systems, *i.e.* whether or not they are segmented by air bubbles.

2.5.2 Microdistillation Modules

These are small glass systems which collect an incoming segmented or unsegmented liquid stream and split it into two outgoing streams: one with a liquid phase and the other with a gas phase which is passed onto the next module (condensation/absorption). Figures 2.13–2.16 depict some representative modules of the large variety of microdistillers used with automatic continuous analysers.

In some instances no additional stream of stripping gas is necessary as the air in the segmented sample flow can adequately serve this purpose[82] (Figure 2.14). However, most of the analyser–microdistillers reported to date use a stream of gas (air, nitrogen, argon) to facilitate separation of the gas-analyte. The gas can be introduced into the system at different points, *viz.*: (*a*) before the microdistiller and merged with the sample stream (Figures 2.13B and C, and 2.15); (*b*) in the preheating coil (Figure 2.13A); and (*c*) in the expansion chamber, where the phases are separated (Figure 2.16).

The microdistiller can be heated by immersion into a hot water (60–90 °C) or oil bath (> 100 °C) or by using an electrical resistance. It is customary to use a thermal conditioning coil prior to the sample stream inlet, namely in the separation zone proper, which can also be heated (Figures 2.13B, 2.14, and 2.15) or not (Figures 2.13A and C). Alternatively, no coil is used and only the expansion chamber is heated (Figure 2.16).

The chamber where the gas–liquid separation is carried out can occasionally be regarded as a submodule connecting the microdistiller with the condenser/absorber. It is used in a variety of designs which essentially consist of a side inlet for the gas–liquid mixture, a bottom outlet for a liquid stream that is sent to waste and a top outlet for the gas stream (analyte plus stripping gas).

The efficiency of the gas–liquid separation is often increased by filling the microdistiller—usually the precoil (Figure 2.13A) or the expansion

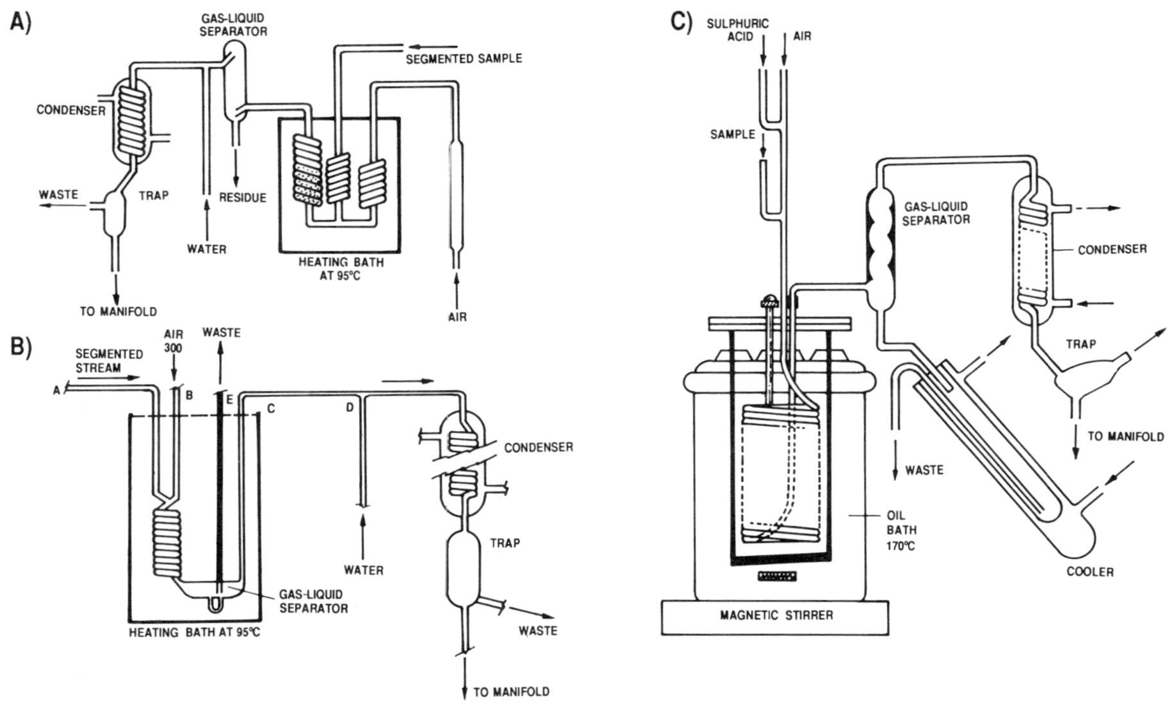

Figure 2.13 *Microdistillation and condensation/absorption modules used in automatic continuous air-segmented analysers.* (A) *Basic configuration of Skalar analysers.* (B) *Modification of the previous system intended to avoid irreproducibility problems.* (C) *Microdistiller for the determination of fluoride.*
(Reproduced by courtesy of Skalar and from references 88 and 87 with permission of United Trade Press Ltd. and by courtesy of Technicon, respectively).

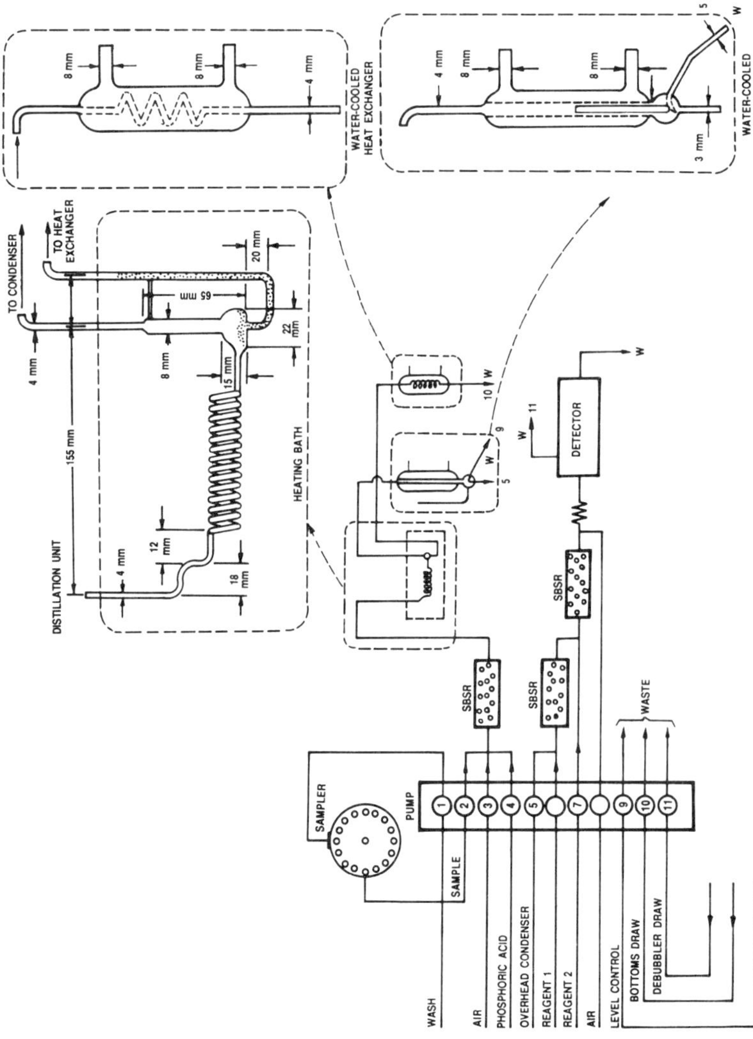

Figure 2.14 *Technicon Autoanalyzer with improved microdistillation/condensation system for the determination of phenol. For details, see text.* (Reproduced from reference 82 with permission of the Elsevier Science Publishers)

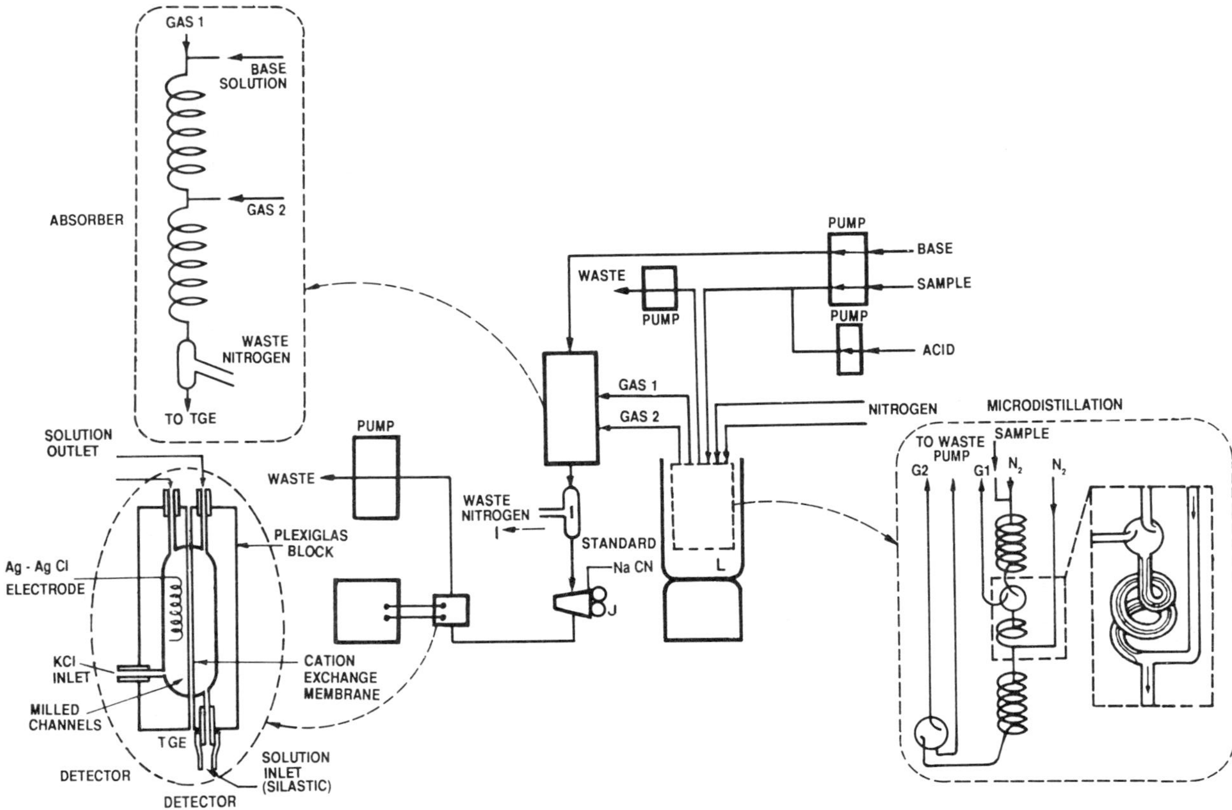

Figure 2.15 *Continuous unsegmented analyser with microdistillation unit for the determination of cyanide. For details, see text.* (Reproduced from reference 77 with permission of the American Chemical Society)

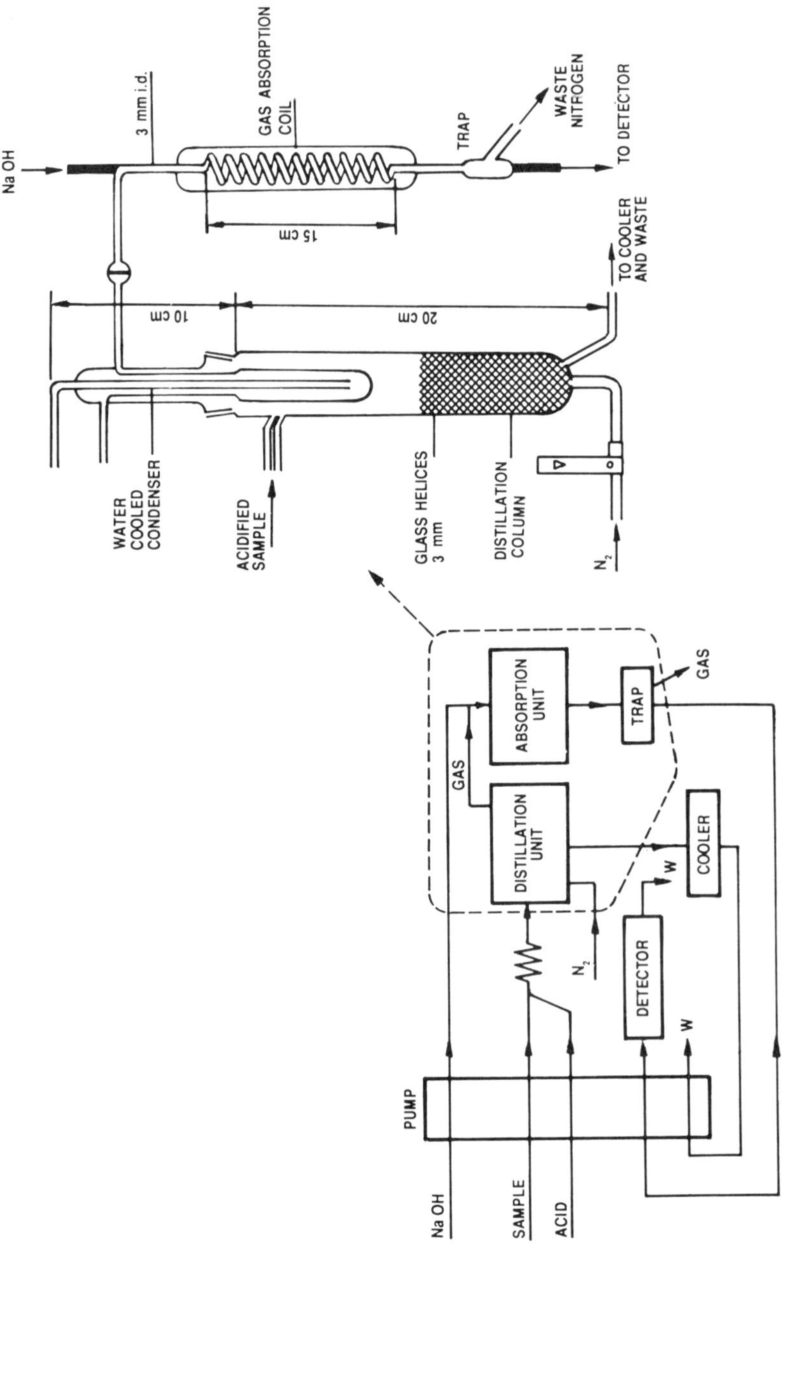

Figure 2.16 *Continuous microdistillation configuration for the determination of cyanide.*
(Reproduced from reference 78 with permission of Elsevier Science Publishers)

chamber (Figure 2.15) with small glass beads or helices that favour mixing and expand the gas–liquid interface.

The stream emerging from the bottom of the microdistiller expansion zone contains the excess reagent and the other components of the sample matrix. This stream is usually drained through a peristaltic pump in order to avoid irreproducibilities arising from passage of some liquid into the gas stream. The problems posed by high temperatures are overcome by using different water cooling devices, two of which are illustrated in Figure 2.13C (cold finger-type) and Figure 2.14 (jacket-type).

This general scheme is occasionally modified for specific purposes. One such modified system is the microdistiller depicted in Figure 2.15, which includes two gas–liquid separators, coils, and outlets for the stripping gas.

2.5.3 Condensation/Absorption Module

The chief aim of this module is to collect the gas-analyte formed and dissolve it in a liquid stream, subsequently driving it to the detector and removing the stripping gas introduced through the previous module in order to facilitate the separation.

There are two general types of condensation/absorption module that differ in whether or not an additional external liquid stream is brought into contact with the gas stream to act as absorbent. Such an absorbent current can be plain water (Figures 2.13A and 2.13B) or contain a reagent (usually an acid or base) to facilitate retention of the gas (Figures 2.15 and 2.16). Alternatively, the condensed solvent emerging from a cooling unit known as condenser is used for this purpose, with the added benefit of manifold simplification (Figures 2.13C and 2.14).

At the end of the condenser/absorber is a trap with two outlets (Figures 2.13A, 2.14–2.16). The bottom outlet collects the liquid containing the interference-free—and occasionally preconcentrated—analyte and drives it to the detector by aspiration through the peristaltic pump at a constant rate. The other outlet, located on one side, drains the stripping gas. The functioning of this unit is sometimes smoothed by aspirating the gas with the peristaltic pump or a moderate vacuum system.

2.5.4 Continuous Air-segmented Analysers

This type of analyser is of great historic significance insofar as it was the earliest to be commercially available—the first model was launched over 30 years ago.[3, 4] However, they are currently on the decline and have been clearly superseded.[104] In any case, many continuous microdistillation systems are associated with this type of analyser, so most of the systems described below were devised a long time ago. This is no special constraint insofar as they have interesting features that can be used as a basis for the development of improved microdistillers for use with modern continuous analytical systems (*e.g.* FIA) in developing new methodologies circumventing well-established shortcomings in this context.[3, 89]

Figure 2.13 shows various alternatives to the coupling of microdistiller/condensers to segmented-flow analysis. The system in Figure 2.13A is a commercial unit marketed by Skalar characterized by its simplicity and by the water stream introduced to dissolve the gas analyte after the gas–liquid separation. Among others, it has been used for the determination of ethanol and volatile acidity in beer and stout samples.[85] The results obtained with this type of unit are poorly reproducible for two basic reasons, (*a*) variations in the composition of the gas–liquid mixture in the coil and the separator resulting from the expansion of the air bubbles on entering the heated zone, which in turn gives rise to surges in the liquid flow lines; and (*b*) the absence of thermostating of the gas–liquid separator, which is customarily placed outside the bath despite the temperature oscillations it is known to undergo. The microdistiller depicted in Figure 2.13B was designed to circumvent these serious shortcomings. It features a shorter coil and uses bead packing to avoid overpressure; in addition, the heating zone of the expansion chamber where the gas–liquid separation takes place is included in the heated zone, and its two outlets are connected to capillaries. This set-up offers smoother operation and hence higher reproducibility in the results.[88]

The microdistiller shown in Figure 2.13C is made of Teflon as it is used for the determination of fluoride in air and plant tissues.[87] The distillation coil is located in a Pyrex glass vessel filled with Primol oil, the temperature of which is kept at 170 °C by means of an immersion heater under magnetic stirring. All the other units are similar to those of the microdistillation and condensation modules described above. No additional absorbent stream is required in this instance. The results obtained in the determination of fluoride on an Autoanalyzer (Technicon) coupled to this system were quite reproducible, and the sample throughput achieved was reasonably good.

In 1983, Goodwin and Marton[82] reported on the use of an improved microdistillation system coupled to an Autoanalyzer for the determination of phenol in water (Figure 2.14). The system includes 14 lines that are identified in the figure. Mixing between streams is favoured by using three single bead string reactors (SBSR), which are customarily employed in FIA for this purpose. Because the expansion chamber is located in a thermostated bath, the temperature of the liquid–vapour interface and the liquid phase from the bottoms draw is more efficiently controlled. The surface area of the expansion chamber is further increased by partially filling it with 1 mm glass beads that reduce bubbling at higher temperatures. The overhead condensate is successively mixed with a stream of reagent 1 (4-aminoantipyrine) and reagent 2 [hexacyanoferrate(III) in a phosphate buffer], and is subsequently air-segmented and passed through a colorimetric detector to monitor the formation of the reaction product. It is interesting to note that the system includes aspiration through channel no. 11 in order to ensure smooth functioning of the bubbler built into the detection cell typically used in these systems. The results provided by this automatic methodology were consistent with those obtained by applying its conventional counterpart.

The above examples are but a taste of the large variety of automatic methodologies implemented on continuous air-segmented analysers coupled to microdistillers. One such methodology allows carbon dioxide to be determined directly in serum,[90] antibiotics to be quantified indirectly by measuring the CO_2 released by bacteria,[91,92] and L-lysine[93] and serum oxalic acid[94] to be assayed after enzymatic decarboxylation. By distilling ammonia, nitrogen can be determined in metal uranium[79] and in faeces and soil samples.[89] Other determinations in this context include those of alkaloids in tobacco and smoke,[95] simple and complex cyanides,[96] sulphur in leaf sample digests, faeces and solid extracts,[83] and volatile aldehydes and ketones.[97]

2.5.5 Continuous Unsegmented Analysers

These analysers introduce samples by continuous aspiration rather than by means of injection valves. Strictly speaking, they are not FIA set-ups. As noted earlier, they have been scarcely applied in this context. Notably, the two examples described below deal with the same analyte.

The system reported by Easty *et al.*[77] can be regarded as one of the earliest examples of continuous unsegmented configurations, which its proponents applied to the determination of cyanide released from plant tissues. Plant homogenizates are continuously acidified and hydrogen cyanide is volatilized in two stages by high-velocity nitrogen streams. Following absorption by a stream of dissolved base, the cyanide is measured amperometrically with a tubular gold electrode. The set-up is shown schematically in Figure 2.15. It has some peculiar features such as the use of four peristaltic pumps, one with two ways intended to aspirate the liquid sample and the absorbing solution, respectively, and three with a single way which are used to introduce the stream for sample acidification, to aspirate the liquid drained from the distillation unit, and to control the detector outgoing stream, respectively. The microdistiller is immersed in a thermostated ethylene glycol bath and has a peculiar design: in fact, it is rather a serial combination of two microdistillers which accommodate the heating coil and expansion minichamber, respectively, these two elements being connected through the liquid drain of the former—that of the latter is sent to waste. Two nitrogen streams are introduced through the top of each subunit. The two outgoing gas streams (N_2 + HCN) lead to the absorption unit, which is continuously fed with a base stream. A trap allows the nitrogen to be separated and the channel leading to the electrochemical detector to be continuously fed. A plain stopcock enables calibration with standards of NaCN. The detector is also depicted in detail in Figure 2.15.

In 1980, Pihlar and Kosta[78] designed a microdistillation system incorporated into an automatic continuous unsegmented analyser characterized by its great simplicity compared to other similar analysers. In the system (Figure 2.16) the sample is continuously aspirated into the analyser, where it is mixed with an acid stream and then led to the microdistillation unit, which is made of borosilicate glass, packed to half-height with glass helices and wrapped with a heating wire not shown in the figure. Condensation and absorption are effected separately: a cold-finger located in the centre of the microdistillation unit condenses the steam formed in the distillation process. Thus, the absorber only receives a gas stream (N_2 + HCN), from which the analyte is transferred to the liquid phase to be driven to the detector through an intervening trap intended to remove the nitrogen. After cyanide has been isolated, the reflux from the condenser and sample solution is cooled and pumped to waste. Discrimination between total and tightly bound metal cyanide complexes is accomplished by UV photodecomposition of the complexes. Cyanide is thereby recovered quantitatively except when its concentration is beyond the operation range of the microdistillation and/or absorption unit. The standard-addition method is used for calibration. By alternating samples and a washing solution every 30 s, up to 60 waste water samples can be analysed for cyanide in an hour.

2.6 Evaporation

Removing the solvent from a liquid phase containing the sample, whether partially or completely (to dryness), can be of great interest to solving some analytical problems in relation to (*a*) the removal of volatile interferents, (*b*) preconcentration, and (*c*) solvent changeover. The first two are aimed at increasing sensitivity and selectivity, while the third is intended to facilitate or improve application of the determinative analytical technique. Unlike in the other gas–liquid separation systems described in this chapter, the analyte is not volatilized, but should rather be very scarcely volatile in order to avoid losses.

The automation of this analytical gas–liquid separation technique poses major technical problems.[104] The only choice for automatic continuous methodologies in this respect is the use of laboratory mini-robots.[97] Thus, a Zymark robot station uses a chloroform evaporation system simultaneously for five samples in preparing oral contraceptive tablets.[99]

Evaporation has also scarcely been used in continuous analysers. In this context, the digestors included in Technicon Autoanalyzers have been employed for the evaporation of organic extracts of chloroform,[99] ether,[101] and acetonitrile,[102] among others.

A few years ago, Technicon launched a device called the 'evaporation-to-dryness module' (EDM), which is schematically depicted in Figure

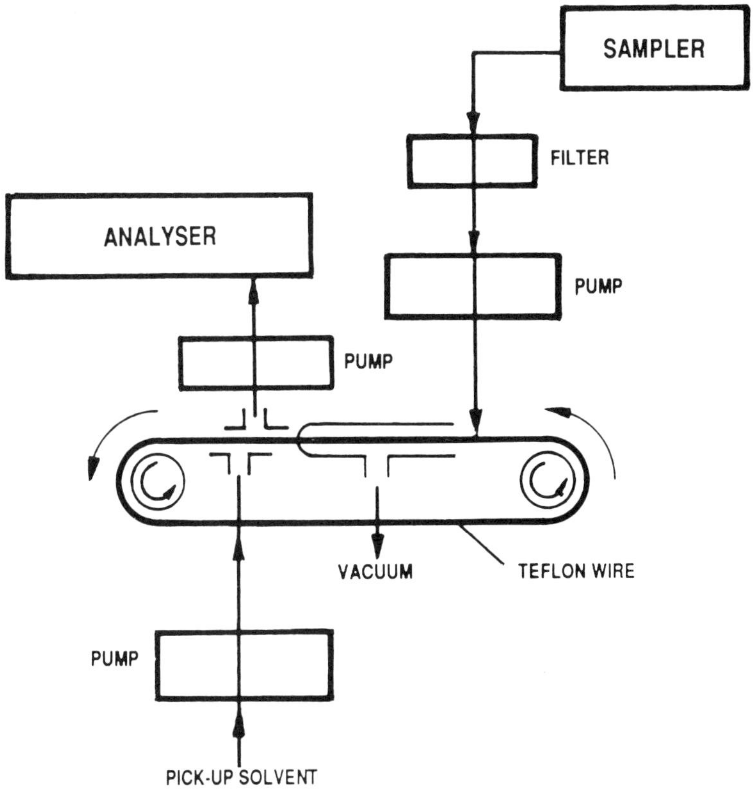

Figure 2.17 *Module for continuous evaporation to dryness and dissolution of the resulting residue marketed by Technicon under the trade name 'Evaporation-to-Dryness Module'.*
(Reproduced by courtesy of Technicon)

2.17. The liquid sample is aspirated through a peristaltic pump and slowly poured over a Teflon conveyor belt. A zone of the module is evacuated thoroughly in order to evaporate the original solvent—the zone is occasionally irradiated with an IR lamp to favour the separation by increasing the local temperature. The dry sample depositing in the conveyor belt crosses a zone where a new solvent is circulated upstream to dissolve the sample and transfer it to the analyser proper, in which a derivatizing reaction can be implemented prior to sensing.

The module is very useful for solvent changeover in HPLC (*e.g.* to suit the sample for analysis by reversed-phase chromatography) and for introduction of samples into instruments with serious constraints on some solvents (*e.g.* aqueous samples into mass spectrometers).

Dolan *et al.*[103] developed a hybrid continuous liquid extraction/evaporation-to-dryness system connected on-line with a liquid chromatograph for the determination of theophylline and four anticonvulsants and their metabolites in serum (Figure 2.18). The serum sample is aspirated into

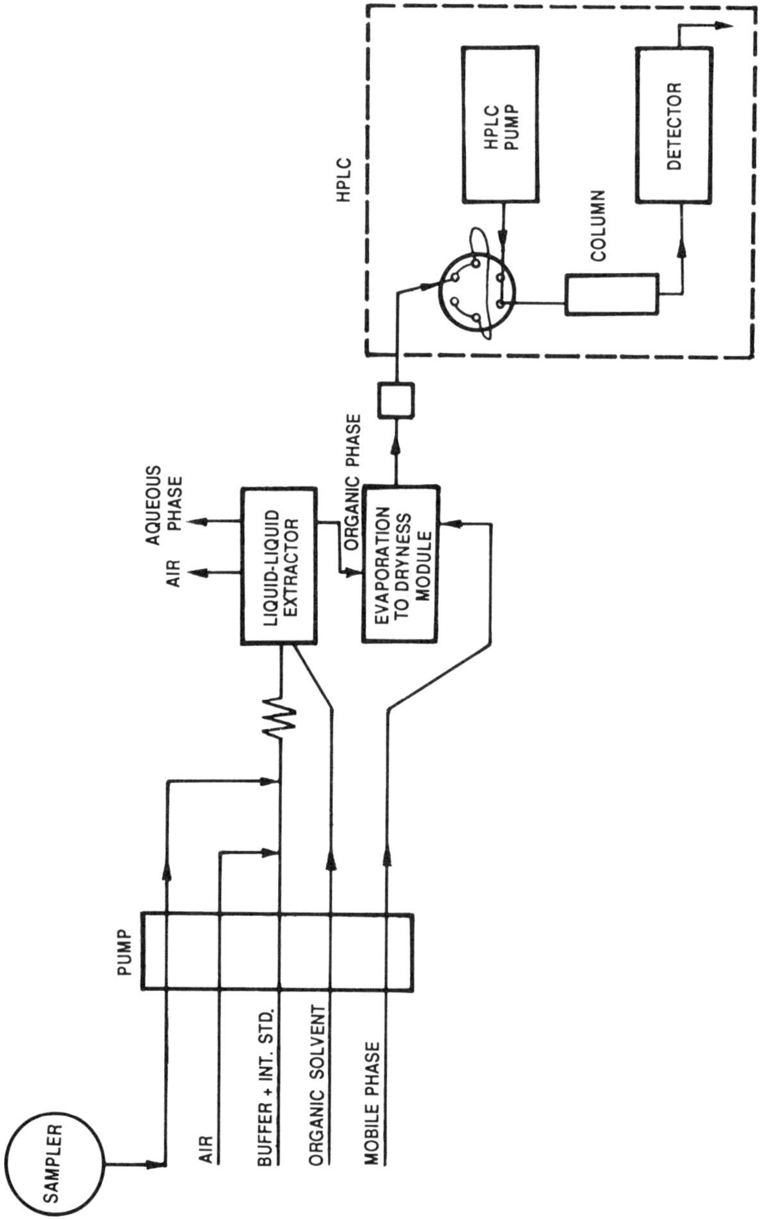

Figure 2.18 *Hyphenated continuous separation technique (liquid–liquid extraction–evaporation/dissolution) coupled on-line with liquid chromatography*

the system and mixed with a buffer containing an internal standard in a continuous air-segmented analyser. The resulting aqueous stream is introduced into a continuous liquid–liquid separator which also collects a stream of the organic extractant phase (chloroform/isopropyl alcohol). The drugs and their metabolites are extracted in the continuous extractor, which is described in detail in the next chapter. The stream of the aqueous phase is sent to waste, while that of the organic phase is driven to the EDM as it is too non-polar to be directly introduced into a liquid chromatograph with a reversed-phase column. A suitable mobile phase stream (water/methanol/isopropyl alcohol) is introduced into the EDM to dissolve the deposited analytes. This stream is used to fill the loop of the conventional injection valve of the chromatograph.

A pervaporation module, which combines continuous evaporation and gas diffusion, was recently developed to overcome problems arising from the use of biosensors in on-line fermentation monitoring.[105] Pervaporation is a process whereby volatile substances such as ethanol and diacetyl contained in a heated donor stream evaporate through a porous hydrophobic membrane (PTFE, 0.45 μm pore size). The vapour condenses on the surface of a cool acceptor stream on the other side of the membrane. Surface tension forces withhold the liquids from the pores and prevent direct contact between them. The temperature difference, which results in a vapour pressure difference across the membrane, provides the driving force for the separation process. Evaporation occurs at the solution interface if the vapour pressure is higher than that at the condensate surface.

References

1. M. Valcárcel and M.D. Luque de Castro, 'Continuous Separation Techniques in Flow Injection Analysis', Chapter 7 in 'Selective Sample Handling and Detection in HPLC' (Part. B), ed. K. Zech and R.W. Frei, Elsevier, Amsterdam, 1989.
2. M. Valcárcel and M.D. Luque de Castro, *J. Chromatogr.*, 1987, **393**, 3.
3. W.B. Furman 'Continuous Flow Analysis. Theory and Practice', M. Dekker, New York, 1976.
4. W.A. Coakley, 'Handbook of Automated Analysis: Continuous Flow Techniques', M. Dekker, New York, 1982.
5. M. Valcárcel and M.D. Luque de Castro, 'Flow Injection Analysis. Principles and Applications', Ellis Horwood, Chichester, 1987.
6. J. Ruzicka and E.H. Hansen, 'Flow Injection Analysis', 2nd Edn., Wiley, New York, 1988.
7. B. Karlberg and G. Pacey, 'Flow Injection Analysis. A Practical Guide', Elsevier, Amsterdam, 1989.
8. S.M. Ramasamy and H.A. Mottola, *Anal. Chem.*, 1982, **54**, 283.
9. A. Ríos, M.D. Luque de Castro, M. Valcárcel, and H.A. Mottola, *Anal. Chem.*, 1987, **59**, 666.
10. F.W. Nyasulu and H.A. Mottola, *J. Autom. Chem.*, 1987, **9**, 46.

11. F. Cañete, A. Ríos, M.D. Luque de Castro, and M. Valcárcel, *Anal. Chim. Acta*, 1989, **224**, 127.
12. F. Cañete, A. Ríos, M.D. Luque de Castro, and M. Valcárcel, *Anal. Chem.*, 1988, **60**, 1540.
13. M.R. Straka, G. Gordon, and G.E. Pacey, *Anal. Chem.*, 1985, **57**, 1799.
14. D.A. Hollowell, J.R. Gord, G. Gordon, and G.E. Pacey, *Anal. Chem.*, 1986, **58**, 1524.
15. T. Imato, H. Ohura, Y. Asano, S. Yamasaki, and N. Ishibashi, *Bunseki Kagaku*, 1988, **37**, T105.
16. H. Baadenhuijsen and H.E.H. Seuren-Jacobs, *Clin. Chem.*, 1979, **25**, 443.
17. P. Linares, M.D. Luque de Castro, and M. Valcárcel, *Anal. Chim. Acta*, 1989, **225**, 443.
18. J. Moller and B. Winter, *Fresenius' Z. Anal. Chem.*, 1985, **320**, 451.
19. M. Granados, S. Maspoch, and M. Blanco, *Anal. Chim. Acta*, 1986, **179**, 445.
20. C. Okumoto, M. Nagashima, S. Mizoiri, M. Kazama, and K. Akiyama, *Eisei Kagaku*, 1984, **30**, 7.
21. Z. Zhu and Z. Fang, *Anal. Chim. Acta*, 1987, **198**, 25.
22. K. Brunt, *Anal. Chim. Acta*, 1984, **163**, 293.
23. W.E. Meyerhoff and Y.M. Fraticelli, *Anal. Lett.*, 1981, **14(B6)**, 415.
24. G. Svensson and T. Anfalt, *Clin. Chim. Acta*, 1982, **119**, 7.
25. C. Pasquini and L.C. de Faria, *Anal. Chim. Acta*, 1987, **193**, 19.
26. T. Kuwaki, M. Akiba, M. Oshima, and S. Motomizu, *Bunseki Kagaku*, 1987, **36**, T81.
27. P.B. Kraus and S.R. Crouch, *Anal. Lett.*, 1987, **20(2)**, 183.
28. T. Aoki, *J. Flow Injec. Anal.*, 1988, **5**, 95.
29. J.R. Clinch, P.J. Worsfold, and F.W. Sweeting, *Anal. Chim. Acta*, 1988, **214**, 401.
30. G. Schulze, C.Y. Liu, M. Brodowski, O. Elsholz, W. Frenzel, and J. Moller, *Anal. Chim. Acta*, 1988, **214**, 121.
31. R. Nakata, T. Kawamura, H. Sakashita, and A. Nitta, *Anal. Chim. Acta*, 1988, **208**, 81.
32. J. Ruzicka and E.H. Hansen, *Anal. Chim. Acta*, 1985, **173**, 3.
33. M.T. Jeppesen and E.H. Hansen, *Anal. Chim. Acta*, 1988, **214**, 147.
34. G.E. Pacey, M.R. Straka, and J.R. Gord, *Anal. Chem.*, 1986, **58**, 502.
35. F. Nakata, H. Sunahara, H. Fujimoto, M. Yamamoto, and T. Kumamaru, *J. Anal. At. Spectrom.*, 1988, **3**, 579.
36. X. Wang and R.M. Barnes, *J. Anal. At. Spectrom.*, 1988, **3**, 1091.
37. M. Yamamoto, K. Takada, T. Kumamaru, M. Yasuda, S. Yokoyama, and Y. Yamamoto, *Anal. Chem.*, 1987, **59**, 2446.
38. P. Marstorp, T. Anfalt, and L. Andersson, *Anal. Chim. Acta*, 1983, **149**, 281.
39. G.B. Martin and M.E. Meyerhoff, *Anal. Chim. Acta*, 1986, **186**, 71.
40. F. Lázaro and M.D. Luque de Castro, *Analusis*, 1988, **16**, 216.
41. W.E. van der Linden, *Anal. Chim. Acta*, 1983, **151**, 359.
42. W.E. van der Linden, *Anal. Chim. Acta*, 1983, **155**, 273.
43. J.S. Canham, G. Gordon, and G.E. Pacey, *Anal. Chim. Acta*, 1988, **209**, 157.
44. J.S. Canham and G.E. Pacey, *Anal. Chim. Acta*, 1988, **214**, 385.
45. A. Tanaka, K. Mashiba, and T. Deguchi, *Anal. Chim. Acta*, 1988, **214**, 259.
46. H. Hwang and P.K. Dasgupta, *Mikrochim. Acta*, 1985, **III**, 77.
47. H. Hwang and P.K. Dasgupta, *Anal. Chem.*, 1986, **58**, 1521.
48. M. Valcárcel and M.D. Luque de Castro, *Analyst (London)*, 1990, **115**, 699.

49. J.C. Andrade, C. Pasquini, N. Baccan, and J.C. van Loon, *Spectrochim. Acta*, 1983, **38B**, 1329.
50. R. M. Barnes, *Biol. Trace. Elem. Res.*, 1984, **6**, 93.
51. S. Zhang, Z. Fang, and J. Sun, *Guangpuxue Yu Guangpu Fenxi*, 1987, **7**, 57.
52. Z. Fang, Z. Zhu, S. Zhang, S. Xu, L. Guo, and L. Sun, *Anal. Chim. Acta*, 1988, **214**, 41.
53. O. Astrom, *Anal. Chem.*, 1982, **54**, 190.
54. Z. Fang in Chapter 4 of 'Flow Injection Atomic Spectroscopy' ed. J.L. Burguera, M. Dekker, New York, 1989.
55. M. Yamamoto, M. Yasuda, and Y. Yamamoto, *Anal. Chem.*, 1985, **57**, 1382.
56. R.R. Liversage, J.C. van Loon, and J.C. de Andrade, *Anal. Chim. Acta*, 1984, **161**, 275.
57. X. Wang and Z. Fang, *Fenxi Huaxue*, 1986, **14**, 738.
58. N.H. Tioh, Y. Israel, and R.M. Barnes, *Anal. Chim. Acta*, 1986, **184**, 205.
59. X. Wang, M. Viczian, A. Lasztity, and R.M. Barnes, *J. Anal. At. Spectrom.*, 1988, **3**, 821.
60. G.S. Pyen and R.F. Browner, *Appl. Spectrosc.*, 1988, **42**, 508.
61. X. Wang and Z. Fang, *Kexue Tomgbao*, 1986, **31**, 791.
62. S.E. Birnie, *J. Autom. Chem.*, 1988, **10**, 140.
63. C. Pasquini, W.F. Jardim, and L.C. de Faria, *J. Autom. Chem.*, 1988, **10**, 188.
64. M. Burguera and J.L. Burguera, *Analyst (London)*, 1986, **111**, 171.
65. P.N. Vijan and G.R. Wood, *At. Absorpt. Newslett.*, 1974, **13**, 33.
66. S. Tesfalidet and K. Irgum, *Anal. Chem.*, 1989, **61**, 2079.
67. G.S. Pyen, S. Long, and R.F. Browner, *Appl. Spectrosc.*, 1986, **40**, 246.
68. G.S. Pyen and R.F. Browner, *Appl. Spectrosc.*, 1988, **42**, 262.
69. C.C.Y. Chan, *Anal. Chem.*, 1985, **57**, 1482.
70. M. Thompson, B. Pahlavanpour, S.J. Walton, and G.F. Kirkbright, *Analyst*, 1978, **103**, 568.
71. A. Menéndez-García, J.E. Sánchez-Uría, and A. Sanz Medel, *J. Anal. At. Spectrom.*, 1989, **4**, 581.
72. M. Valcárcel, 'Flow Injection Analysis' Chapter 2 of 'Sample Introduction in Atomic Spectroscopy', ed. J. Sneddon, M. Dekker, New York, 1990.
73. A. Meyer, C. Hofer, G. Tolg, S. Raptis, and G. Knapp, *Fresenius' Z. Anal. Chem.*, 1979, **296**, 337.
74. A. Menéndez-García, J.E. Sánchez-Uría and A. Sántez Medel, *Anal. Chim. Acta*, 1990, **234**, 133.
75. S. Zhang, S. Xu, and Z. Fang, *Quím. Anal.*, 1989, **8**, 191.
76. J.R. Fair, 'Distillation', Chapter 5 of 'Handbook of Separation Process Technology', ed. R.W. Rouseau, Wiley, New York, 1987.
77. D.B. Easty, W.J. Blaedel, and L. Anderson, *Anal. Chem.*, 1971, **43**, 509.
78. B. Pihlar and L. Kosta, *Anal. Chim. Acta*, 1980, **114**, 225.
79. J.R. Prall in 'Advances in Automated Analysis, Technicon International Congress, 1970', Vol. II, Thurman Associates Miami, 1971, pp. 395.
80. J. Keay and P.M.A. Menage, *Analyst (London)*, 1969, **94**, 895.
81. H.O. Friestad, D.E. Ott, and F.A. Gunther, *Anal. Chem.*, 1969, **41**, 1750.
82. A.E. Goodwin and J.L. Marton, *Anal. Chim. Acta*, 1983, **152**, 295.
83. J. Keay, P.M.A. Menage, and G.A. Dean, *Analyst (London)*, 1972, **97**, 897.
84. F.A. Gunther and D.E. Ott, *Analyst (London)*, 1966, **91**, 475.
85. R. Sanyer and E.J. Dixon, *Analyst (London)*, 1968, **93**, 680.
86. P. Casapieri, R. Scott, and E.A. Simpson, *Anal. Chim. Acta*, 1970, **49**, 188.

87. R.H. Mandl, L.K. Weinstein, J.S. Jacobson, D.C. McCune, and A.E. Hitch, Proc. Technicon Symposium, New York, 1965, pp. 270.
88. R.G. Lidzey, R. Sawyer, and P.B. Stockwell, *Lab. Practice*, 1971, **20**, 213.
89. J.K. Foreman and P.B. Stockwell 'Automatic Methods of Analysis', Ellis Horwood, Chichester, 1975, pp. 284.
90. L.T. Skeggs, *Am. J. Clin. Pathol.*, 1960, **33**, 181.
91. J.R. Gerke, T.A. Haney, J.F. Pagano, and A. Ferrari, *Ann. N.Y. Acad. Sci.*, 1960, **87**, 782.
92. A. Jones and G. Palmer, *Analyst (London)*, 1970, **95**, 463.
93. G.D. Schaiberger and A. Ferrari, *Ann. N.Y. Acad. Sci.*, 1960, **87**, 890.
94. C.F. Knowles and A. Hodgkinson, *Analyst (London)*, 1972, **97**, 474.
95. C.W. Ayres in 'Automation in Analytical Chemistry, Technicon Symposia, 1966' Vol. II, Mediad. Inc., White Plains, New York, 1967, pp. 107.
96. P.D. Goulden, B.K. Afghan, and P. Brooksbank, *Anal. Chem.*, 1972, **44**, 1845.
97. R.E. Duncombe and W.H.C. Shaw in 'Automation in Analytical Chemistry. Technicon Symposia, 1966', Vol. II, Mediad. Inc., White Plains, New York, 1967, pp. 15.
98. M.J. Hurstand and J.W. Mortimer, 'Laboratory Robotics', VCH Pu., New York, 1987.
99. E. Siebert in 'Advances in Laboratory Automation Robotics, 1984', ed. by G.L. Hank and J. Strimaitis, Zymark Co. Inc., Hopkinton, MA, USA, 1985, pp. 257.
100. S. Ahuja, C. Spitzer, and F.R. Brofaci, *J. Pharm. Sci.*, 1970, **59**, 1833.
101. C. McNeil, C.R. Berret, N. Rich, S. Gaufin, M.A. Warr, and H. Hutchinson, in 'Advances in Automated Analysis. Technicon Instrumentation Congress 1970', Vol. I, Thurman Associates, Miami, 1971, pp. 547.
102. W.D. Horman, G. Formica, K. Ramsteiner, and D.D. Eberle, *J. Assoc. Off. Anal. Chem.*, 1972, **55**, 1031.
103. J.W. Dolan, S.J. van der Waal, S.J. Bannister, and L.R. Snyder, *Clin. Chem.*, 1980, **26**, 871.
104. M. Valcárcel and M.D. Luque de Castro, 'Automatic Methods of Analysis', Elsevier, Amsterdam, 1988.
105. U. Prinzing, I. Ogbomo, C. Lehn, and H.L. Schmidt, *Sensors Actuators*, 1990, **B1**, 542.

CHAPTER 3

Liquid-Liquid Systems

3.1 Introduction

This chapter deals with automatic continuous methodologies involving mass transfer between two liquid phases. The only two main variants available in this context (dialysis and liquid–liquid extraction) are of clear significance in the field of non-chromatographic continuous separations on account of both the large number of systems of this type reported to date and the enormous variety and complexity of analytical problems they solve. This has fostered the development of a much wider range of commercially available devices than currently available for any other methodology dealt with in this book.

The main difference between the two basic ways in which mass transfer between two liquid phases can be effected lies in one of their essential features, namely miscibility, which in turn conditions the way in which they come into contact. In dialysis processes, the two phases—usually aqueous—are miscible, so they call for a barrier to isolate them and allow the transfer of matter (generally the analytes) across. In continuous liquid–liquid extraction, the phases (aqueous and organic) are immiscible and are brought into contact by formation and subsequent separation of a stream of regular and alternate segments of the two where mass transfer takes place.

The interface in dialysis processes is thus rather well defined: a semipermeable membrane in contact with the two phases acts as a well delimited separation surface. This is not the case with continuous liquid–liquid extraction, in which the interface (the zone of contact between the segments of the two phases plus the area of the film formed by one of the phases to cover the walls of the tubing through which the segments are circulated) is variable.

The continuous separation modules typically used to implement the two alternatives are rather different; however, they also share a common feature: the two phases involved enter and leave the modules, within which mass transfer takes place.

A dialyser is a fairly simple continuous operation unit that bears formal resemblance to the gas diffusers described in Chapter 2. The donor and acceptor streams merge within and are separated by the membrane. Conversely, a continuous liquid–liquid extractor consists of three elements, namely a segmenter, an extraction coil and a phase separator; all three work in a sequential fashion, which may hinder constant, regular functioning. It is interesting to note that sandwich membrane separators, by far the most commonly used for phase separation, are rather similar to dialysers; however, they do not effect the analyte transfer, but merely separate the previously mixed phases. In addition to the typical designs of these units, there are other, more specific alternatives that are commented on in their respective sections. Occasionally, dialysis and liquid–liquid extraction are integrated (*e.g.* in liquid supported membranes in a sandwich separator).

Dialysis is always enacted in a continuous fashion for chemical analytical purposes, while liquid–liquid extraction is commonly used batchwise. In addition, both include more dynamic and efficient separation alternatives, namely partition and size-exclusion chromatography, respectively.

Dialysis has a narrower scope of application than liquid–liquid extraction on account of the fact that separations are based on marked differences in ionic or molecular mobility across a membrane and is widely applied to clinical and biological samples to isolate small molecules or ions from macromolecules. Liquid–liquid extraction, in conjunction with derivatization reactions, has a much broader scope of application because of the more universal nature of the property involved, *i.e.* relative solubility. Nevertheless, the basic aims of the two techniques are the same: interference removal and analyte preconcentration.

This chapter discusses these two continuous separation modes separately in relation to their foundation, the components and automatic configurations they use and their most outstanding applications. Due mention is also made on some atypical modes and latest trends.

3.2 Dialysis

Dialysis is one of the techniques based on barrier separation processes. The separation of the different solutes relies on the difference between the rates at which they are transferred across a semi-permeable membrane located between the two liquid phases. Such phases, donor and acceptor, need not be immiscible—unlike in liquid–liquid extraction processes. Dialysis is clearly different from other barrier separation processes such as osmosis or ultrafiltration. In osmotic processes it is the solvent rather than the solute which crosses the membrane—it is still unclear, though, whether this type of process occurs concomitantly with dialysis. In ultrafiltration, occasionally referred to as reverse osmosis, a solution is forced under pressure across a membrane with concomitant separation of its components. The driving force in this case is the energy resulting from the pressure difference and it is the solvent rather than the solute which crosses the membrane against the concentration gradient. Unwanted solutes and impermeable colloidal matter are rejected. On the other hand, dialysis is always kinetically controlled and separations are based on the occurrence of a concentration gradient between the two liquid phases isolated by a selective membrane that discriminates solutes on the basis of pore size.

Dialysis processes can be classified according to various criteria.

Depending on the dynamic state in which one or the two phases involved occur, the process may or may not be allowed to develop until mass transfer equilibrium is reached. If the two phases are quiescent, then the process must develop to completion; however, if one or both phases are in motion—or even if either is stopped over a given interval to increase the efficiency of the process—equilibrium is never reached and scrupulous timing must be employed. In this latter instance, the process can be carried out with efficient agitation of a constant volume of the two phases or by using a stream of the two phases which may flow concurrently. Analytically, the concurrent flow alternative is also the most commonly used.

The nature of the barrier (membrane) used determines the features of the transfer process that takes place across it, which allows dialysis processes to be classified as passive or active (Donnan). The former type of process involves diffusion of particles within a given range of molecular weights across a neutral membrane. Active or Donnan dialysis involves the transfer of ions with a given type of charge across an ion-exchange membrane. Liquid (hydrophobic) membranes are a special type of barrier which are conditioned with a suitable organic solvent to allow passage of one or several specific analytes in a combined liquid–liquid extraction/dialysis process.

There is a clear difference between the above dialysis alternatives and electrodialysis, in which the solutes (electrolytes) are transferred across a membrane on application of electrical energy. Although seemingly just another type of ordinary dialysis on account of the involvement of membranes, its peculiar equipment, objectives, and membrane requirements make it a class of its own. As it uses external energy, electrolytes can be transferred from a dilute to a more concentrated solution; this is the root of its most significant advantage over ordinary dialysis insofar as it allows electrolyte traces to be exhaustively separated. Ultracentrifugation and electrodialysis are formally similar as regards the presence of an external source of energy (pressure and electric field, respectively), even though the species to be isolated in each case are potentially different.

The chief aim of dialysis is separation (normally of the analyte, which is transferred to the acceptor stream free from all interferences). Unlike other techniques in which the separation process may entail analyte preconcentration, dialysis involves analyte dilution as it is partitioned between the donor and acceptor phase and only rarely does the system involve a preconcentration process simultaneous to the separation, with the exception of electrodialysis, active dialysis and modes involving liquid membranes, where the process can be completely or nearly exhaustive.

Although continuous dialysis processes have been developed for a long time,[1,2] many of the methods reported in connection with them use segmented flow configurations, the most serious pitfalls of which arise from the presence of air bubbles that decrease the efficiency of the process and are the source of major irreproducibilities. The use of no air bubbles in unsegmented flow systems has enabled development of methodologies clearly improving on their earlier segmented counterparts as regards efficiency, reproducibility, and procedural and instrumental simplicity—and hence reduced instrumental costs.

3.2.1 Theoretical Foundation

The efficiency of a dialysis process depends on four factors, namely (*a*) the features of the solute (*e.g.* size, charge, and molecular or ionic diffusion coefficient), (*b*) its relative concentration in the two phases (*i.e.* its concentration gradient), (*c*) the ratio between the flow-rates of the two streams—or agitation—and, (*d*) the rate constant of solute transfer across the membrane, which in turn depends on the type of membrane used, its area, the geometry of its pore structure, and the flow-rate of the liquid that its circulated across it.

Figure 3.1 shows the scheme of a dialyser through which a donor and an acceptor stream are circulated at a flow-rate q_d and q_a, respectively, and a concentration C_d^1 and C_a^1, and from which another two streams of concentrations C_d^2 and C_a^2 at the original flow-rates emerge. The overall rate of solute transfer across the membrane of a dynamic dialyser (v_o) can be expressed in terms of the flow-rate and the difference between the incoming and outgoing concentrations (C_d^1 and C_d^2) of the donor or the reverse difference for the acceptor:

$$v_o = q_d(C_d^1 - C_d^2) = q_a(C_a^2 - C_a^1)$$

or, as a rate equation:

$$v_o = kA\Delta C$$

where k is the overall rate constant of transfer, which is a function of the properties of the membrane and the stationary liquid layers on its two sides, A is the effective surface area of the membrane, and ΔC is the mean of the concentration differences (gradients) between the two fluids:

$$\Delta C = (C_d^1 - C_a^1) + (C_d^2 - C_a^2)$$

or

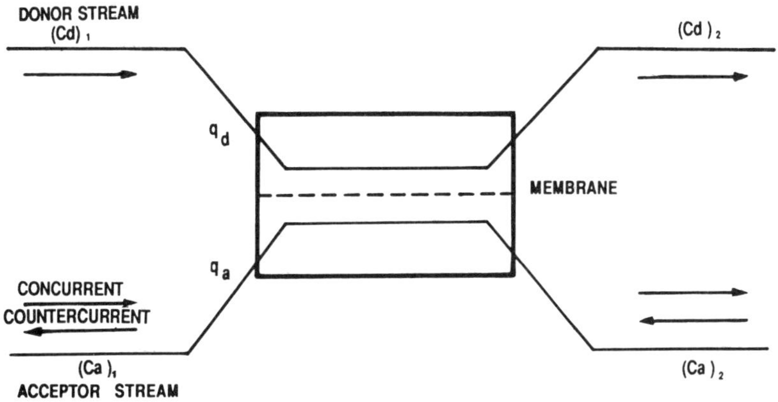

Figure 3.1 *Scheme of a dialyser working in concurrent and countercurrent modes.* C_d *and* C_a *denote the donor carrier and acceptor carrier, respectively,* q *the flow-rate, and* 1 *and* 2 *the dialyser incoming and outgoing streams, respectively*

$$\Delta C = \log (C_d^1 - C_d^2) = \log (C_a^1 - C_a^2)$$

The operating efficiency of a dialyser can be expressed in terms of term *D*, *viz*. the ratio between the rate and the concentration gradient:

$$D = v_o/(C_d^1 - C_a^1)$$

However, as the acceptor stream normally contains no dissolved solute, then:

$$D = v_o/C_d^1$$

Thus, *D* has the same dimensions as a flow-rate.

Alternatively, the efficiency of a dialysis process can be expressed in terms of the non-dimensional parameter D/q_d:

$$D/q_d = v_o(C_d^1 - C_a^1)/v_o(C_d^1 - C_d^2) = (C_d^1 - C_a^1)/(C_d^1 - C_d^2)$$

If $C_a^1 = 0$, then:

$$D/q_d = (C_d^1/C_d^2)/C_d^1$$

which relates the efficiency (in absolute value) to the flow-rate of the donor phase.

A combination of the above expressions taking account of the logarithmic relationship between C and the incoming and outgoing concentrations yields the following equations:

$$D/q_d = 1/(1 + z)\{1 - \exp[- R(1 + z)]\}$$

$$D/q_d = \{1 - \exp[R(1 + z)]\}\{z - \exp[R(1 - z)]\}^{-1}$$

for concurrent and countercurrent flows, respectively, and where z denotes the flow-rate ratio ($z = q_d/q_a$) and $R = kA/q_d$. Obviously, the greater z is (*i.e.* the higher the flow-rate of the donor stream in relation to the acceptor stream), the lower the efficiency of the process will be.

The temperature and the relative direction of the donor and acceptor streams also influence the dialysis efficiency. In fact, the effect of the temperature on the dialysis efficiency is significant enough to warrant control of this variable if reproducible results are to be obtained. Studies in this context have shown a temperature change of 0.1 °C to result in an efficiency loss of 0.26%.

The possibility of using the relative direction of the donor and acceptor streams as a means of increasing the dialysis efficiency has been the subject of much debate; while countercurrent flow results in a higher efficiency, the rise time (*i.e.* the time phase required for equilibration) is shorter with concurrent flow, which also results in less marked cross-contamination thanks to more efficient washing.

The composition of the acceptor stream is a key to the efficiency of the dialysis process; however, no generalizations can be made in this respect as the effect depends on the particular features of the species to be separated.

Theoretical studies on solute transfer in on-line analytical flow analysers allowed equations describing and predicting the behaviour of these systems to be formulated. Thus, Cooley *et al.*[3,4] derived equations describing mass transfer across membranes in semi-infinite parallel-plate dialysers, while Bernhardson *et al.*[5] obtained such equations for infinite-plate dialysers. In any case, a detailed description of either is beyond the scope of this section.

3.2.2 Equipment

The basic units making up on-line dialysis described here have been divided according to whether they are characteristic of the separation unit (dialyser and membrane) or are incorporated into the continuous configuration for this specific purpose. The reader is referred to Chapter 2 for a detailed description of the typical units of continuous systems (pumps, valves, *etc.*).

3.2.2.1 Dialysers

Dialysis units are available in a variety of designs according to their origin and application; hence we shall classify them into conventional dialysers (commercially available or laboratory-made) and special dialysers (*i.e.* those developed for a specific purpose such as incorporation into biotechnological sampling systems).

Conventional dialysers are generally of the sandwich type and made of Perspex or another chemically and mechanically resistant material. They are made up of two halves which are mirror images, each of which has a semi-circular or occasionally triangular or square groove on one side.

Figure 3.2 depicts schematically various laboratory-made dialysers. The most general and hence the easiest to construct (Figure 3.2A_A), consists of a straight channel running across each half. Winding or squarely arranged channels (Figures 3.2A_B and 3.2A_C) provide higher efficiency insofar as the exchange surface is increased by a factor of four or greater with respect to the previous design, though at the expense of increased dispersion or dilution of the solute-analyte. The efficiency of the rather complex design in Figure 3.2A_D is even much higher.

Between the two sides of the dialyser, which are firmly screwed to each other, is placed a semi-permeable membrane to facilitate the process.

The channels are normally between 0.2 and 1.0 mm deep and 0.5–1.5 mm wide, their length varying depending on the particular design between 5 and 15 cm (Figure 3.2A_A) and 2.7 m (Figure 3.2A_D). The inner volume of the dialyser and hence its dispersion can be reduced by packing the channels with beads of inert material (usually glass).

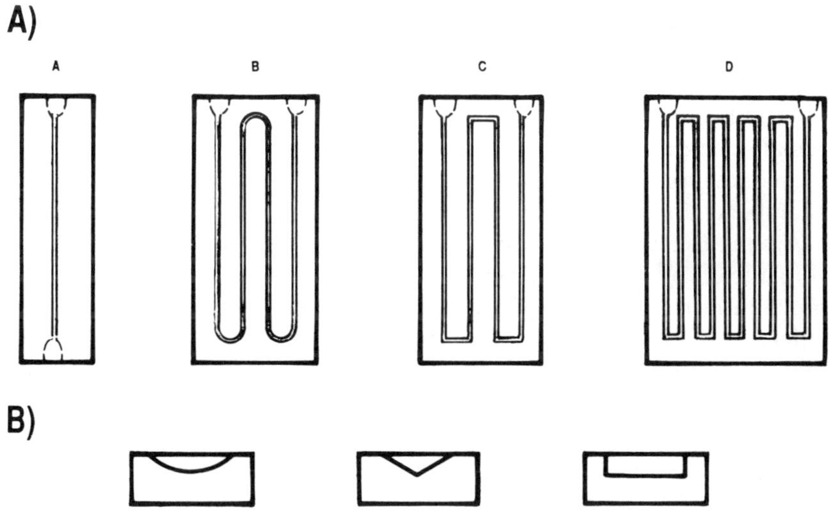

Figure 3.2 (A) *Different designs of laboratory-built dialysers.* (B) *Cross-sectional view of the most common types of channels used. (See text for further details)*

Figure 3.3 shows two commercially available dialysers. The scheme in Figure 3.3A corresponds to the dialyser incorporated into the ASTED™ (Automated Sequential Trace Enrichment of Dialysates) system marketed by Gilson. It consists of two polymethylmethacrylate blocks between which a cellulose (Cuprophan) membrane is sealed. The membrane only allows passage of low-molecular weight molecules (molecular cut-off, 15 kD). The dialyser is manufactured in two sizes according to the capacity of the donor phase channel (100 or 370 μl) and makes up the first separation unit of the ASTED™; the second of such units is a micro-precolumn, namely a TEC (Trace Enrichment Cartridge) which can be readily replaced as required and is located in the loop of the injection valve of a liquid chromatograph.

The Tecator 5106 Dialysis Module, which is schematically depicted in Figure 3.3B consists of the following parts: (*a*) two nitrile rubber plates with engraved channels between which the membrane is placed. The channel is 50 cm long, 1.5 mm wide, and 0.5 mm deep, and features an inner volume of 161 μl. (*b*) A socket into which the plates are placed and squeezed together. (*c*) A Plexiglass unit for reagent merging and mixing (Chemifold III). (*d*) A device for separating the rubber plates for membrane changeover. This module is mounted on a tray that can be placed in the Analyzer 5020 just like an ordinary Chemifold.

With a few exceptions, conventional dialysers can work in the concurrent and countercurrent mode.

Tyler Research Instruments markets microdialysis and flow microdialysis manifolds machined from high-density inert acrylics and polycarbonates with stainless steel fittings. They typically consist of 1–36 two-stage chambers separated by a semi-permeable membrane. In use, the sample chambers are open only to the membrane surface. The dialysate chamber can be configured for continuous buffer flow in either a serial array (all the dialysate chambers are connected in series) or a parallel array (the chambers have independent inlets and outlets). Parallel configured arrays can be readily adapted to serial use and, in either case, the sample chambers remain completely isolated. The systems are used to clear bubbles from the chambers in a semi-automatic fashion. Large enough individual chambers can occasionally accommodate magnetic stirring bars.

Among the above-mentioned special dialysers, the one designed and constructed by Craig and Stewart[6] deserves special mention as it combines thin-film dialysis and the continuous countercurrent mode. The chief aim of this dialysis system is to provide an exchange fast enough so that the flow (and the overall volume) of the diffusate stream do not have to be more than two or three times that of the retentate stream. The module is schematically depicted in Figure 3.4 and consists of the following elements: (*a*) An inside spacer tube (106 cm long, 15 mm OD) with a standard ground glass male adapter joint at the top and sealed to a 1 mm capillary tube running the length of the inside of the tube. (*b*) An 8 cm glass collar with a retentate exit tube and an ID of 18 mm except at the

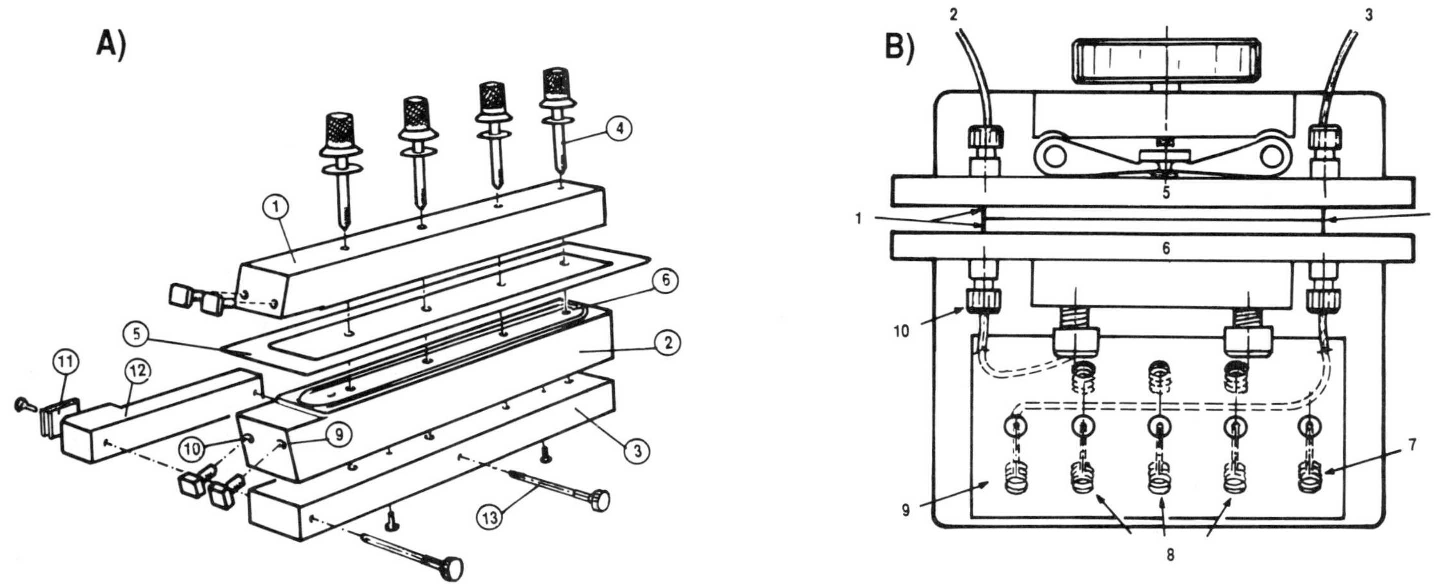

Figure 3.3 (A) *Dialyser marketed by Gilson as part of the ASTED system:* (1) *donor half-block,* (2) *recipient half-block,* (3) *base plate,* (4) *clamping screws,* (5) *dialysis membrane,* (6) *holes for locating studs,* (7) *donor channel inlet,* (8) *donor channel outlet,* (9) *recipient channel inlet,* (10) *recipient channel outlet,* (11) *mounting clamp and screw,* (12) *dialyser side holder,* (13) *securing screws.* (B) *Tecator 5106 dialysis module:* (1) *rubber plates,* (2) *donor channel outlet,* (3) *donor channel inlet,* (4) *dialysis membrane,* (5) *donor half-block,* (6) *recipient half-block,* (7) *channel to detector,* (8) *connection points for reagents and mixing coils,* (9) *recipient channel inlet,* (10) *recipient channel outlet*
(Courtesy of Gilson and Tecator)

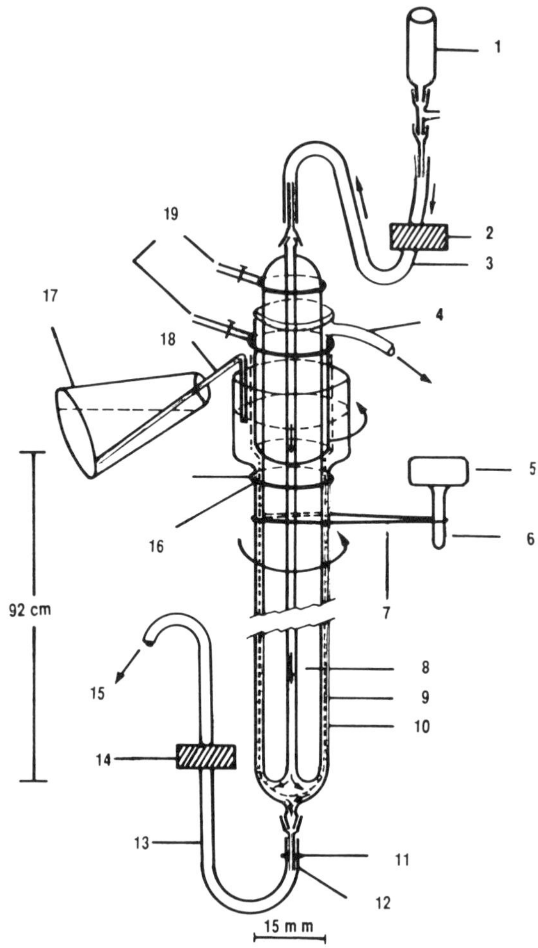

Figure 3.4 *Scheme of a thin-film dialyser.* (1) *Solution reservoir,* (2) *solution pump,* (3) *Teflon tube,* (4) *retentate exit,* (5) *motor,* (6) *shaft,* (7) *belt,* (8) *inside spacer tube,* (9) *membrane,* (10) *outside tube (rotating),* (11) *clamp,* (12) *bearing,* (13) *Teflon tube,* (14) *diffusate pump,* (15) *diffusate exit,* (16) *bearing,* (17) *solvent reservoir,* (18) *syphon,* (19) *support clamps*

top 3 cm, where it is 24 mm (not shown in the figure). This collar accommodates the membrane, the area of which is *ca.* 435 mm^2. (*c*) A 97 cm outside rotating tube with an ID of 15.6 mm except at the top 5 cm (ID, 23 mm). The bearing at the bottom of the tube consists of a stainless steel needle of about 3 cm inserted into the capillary Teflon tubing. The distance between the inside spacer tube and the outside tube is critical. (*d*) A bearing and support for the outside tube in the form of a small Teflon-covered wire ring. (*e*) A drive belt. It uses O-rings despite their short lifetime, as rubber bands are completely useless for this purpose. (*f*) A constant-speed motor to spin the outside tube at 150 r.p.m. (*g*) Retentate and diffusate pumps. One is used to propel both the retentate and the diffusate solution. When the dialysers are operated at temperatures other than ambient, the dialyser is wrapped in a water jacket with its bottom bearing passing through a rubber stopper. The excellent performance of this module was exhaustively checked by its designers, who prompted application of the principle behind thin-film countercurrent flow to the development of artificial kidneys.

Another special dialyser is depicted in Figure 3.5. It consists of two concentric tubes of which the inner one is of a microporous material (PTFE or TFE). As the conditioned sample is inserted into the external tube, the species with a suitable size in relation to the pores of the inner tube diffuse into it, through which an acceptor stream is circulated. The inner diameter of the tubes ranges between 0.3–1.0 mm (inner tube) and 0.5–2.0 mm (outer tube), while their lengths range between 3–15 cm. Hence, the volume of the dialysate may vary between 117–210 ml.

Continuous on-line determination of the major components of a fermentation broth is a prerequisite for efficient data monitoring and computer control of bioreactors. At present, there is a lack of reliable sterilizable and specific probes for measurement of organic and inorganic components of interest directly in the fermenter. Until relatively recently, continuous sampling could not be implemented routinely for a number of reasons, namely: sampling devices were only suitable for specific processes; many sampling devices had poor long-term stability owing to membrane fouling; signal features were not too brilliant because of time delays, difficulties arising from calibration of the analysing system together with a sampling device; fragile or technically complicated sampling devices were prone to contamination or were commercially unavailable; *etc.*

Attempts at overcoming the above pitfalls and improving the performance of both laboratory and industrial fermenters have been reported in the literature. Two of the approaches to this problem are illustrated by way of example in Figure 3.6.

Figure 3.6A shows a cross-sectional view of the probe designed by Mandenius *et al.*[7] All parts are made of autoclavable materials: acid-

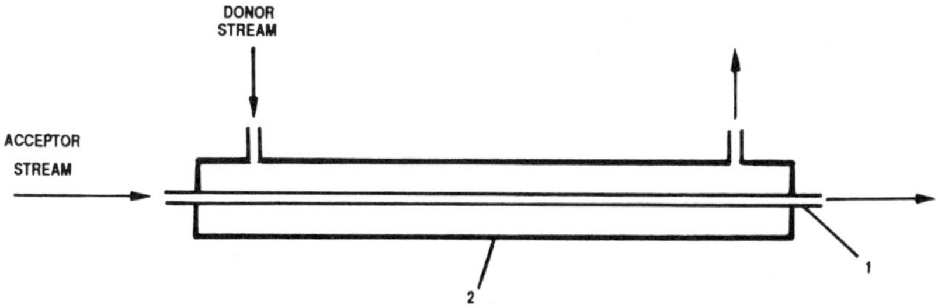

Figure 3.5 *Cylindrical dialyser consisting of two concentric tubes of microporous* (1) *and non-porous material* (2)

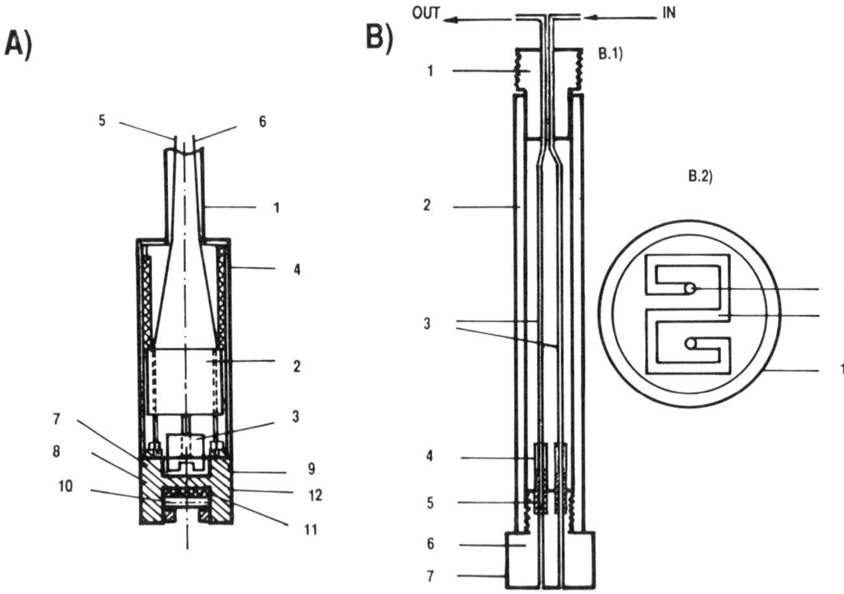

Figure 3.6 *Schemes of laboratory-built dialysis probes for fermenter sampling.* (A) *Cross-sectional view of a design developed by Mandenius* et al.: (1) *stainless steel shaft,* (2) *autoclavable d.c. motor,* (3) *permanent magnet,* (4) *stainless steel housing,* (5, 6) *incoming and outgoing dialysis solution,* (7) *polycarbonate membrane holder,* (8) *membrane,* (9) *membrane support with spiral groove,* (10) *magnetic stirring bar,* (11) *thin Teflon washer,* (12) *channels for air escape and improved circulation.* (B) *Side* (B.1) *and end view* (B.2) *of Gibson and Woodward probe:* (1) *threaded end-piece to fit into the Applicon fermenter,* (2) *0.75 in Perspex shaft,* (3) *1.00 mm ID nylon tube,* (4) *silicone rubber connectors,* (5) *nylon screw connectors (1.0 mm bore),* (6) *dialysis probe head,* (7) *silicone rubber O-ring,* (8) *inlet and outlet holes,* (9) *1.5 mm wide, 1.0 mm deep groove,* (10) *silicone rubber O-ring*
(Reproduced from references 7 and 8 with permission of Elsevier Science Publishers)

proof steel, polycarbonate and PTFE. The probe is furnished with an autoclavable d.c. motor that is powered from an external speed control variable from 0 to 100 r.p.m. The system is intended to minimize the effect of fouling by creating a tangential flow over the membrane surface either by means of a built-in impeller or by directing the existing turbulent flow within the fermenter over the membrane surface. By choosing a membrane with an appropriate surface area and the flow-rate of the fluid to be assayed, variable dilution can be accomplished in addition to the separation. The probe remains operative for a week with no need for membrane changeover. In addition, all parts are made of materials such that it can be steam-sterilized and mounted inside a fermenter.

A more recent design constructed from Perspex and nylon was reported by Gibson and Woodward.[8] The probe can be screwed into the head plate of a fermenter vessel and its end is covered with a dialysis membrane held in place by an O-ring and has a channel cut into it intended to allow passage of a sterile fluid to remove solutes that diffuse across the membrane. The prototype probe could not be sterilized within the fermenter, so it was attached after sterilization of the vessel in a laminar-flow cabinet. A stainless steel probe that can be sterilized *in situ* was subsequently constructed. The dialysis probe was operated by drawing sterile recipient fluid through it. The results obtained by using the probe in a semi-automated system compared favourably with those yielded by samples withdrawn directly from the fermenter. The incorporation of automated valving enabled washing cycles to be alternated with the sampling cycles in order to flush the analytical system.

B. Braun Melsungen AG manufactures an aseptically and continuously operating sampling device called BIOPEM™ which includes a

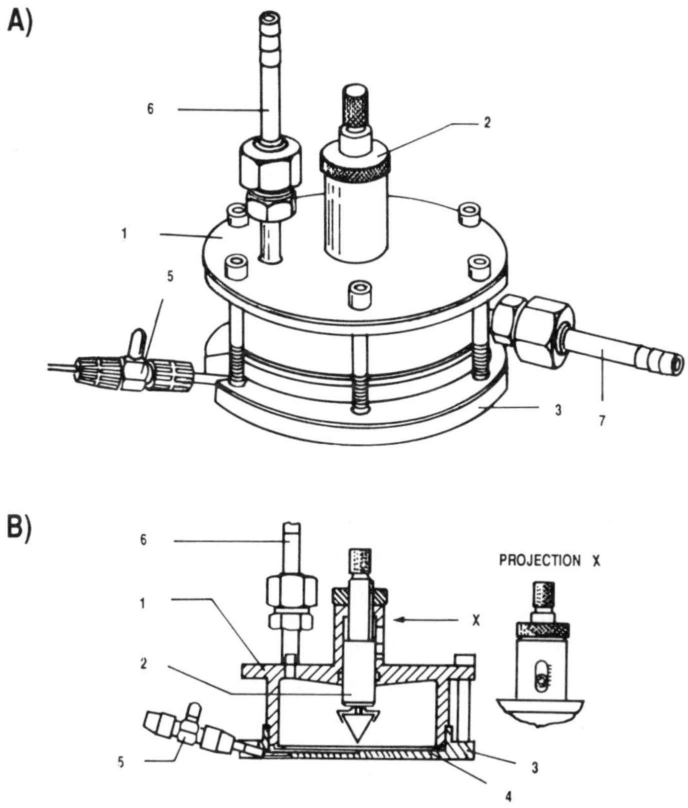

Figure 3.7 *BIOPEM™ sampling module for the continuous aseptic sampling of bioprocess fluids.* (A) *Perspective,* (B) *cross-section, and* (C) *projection.* (1) *Top plate,* (2) *adjustable magnetic stirrer assembly,* (3) *bottom plate,* (4) *channels,* (5) *shut-off valve,* (6) *outlet line,* (7) *inlet line*
(Courtesy of B. Braun Melsungen AG)

dynamic filtration/dialysis system suited to the features of the membrane used. The system can be readily sterilized or autoclaved *in situ* and consists of three main parts as shown in Figure 3.7A, namely (1) a top plate, (2) an adjustable stirring assembly, and (3) a bottom plate. All metal parts are manufactured from stainless steel. The bottom plate (Figure 3.7.B) accommodates a channel array (4) for the controlled removal of the filtrate via the shut-off valve (5) and is covered by the separation membrane, which can be any type of sterilizable microfiltration, ultrafiltration, or dialysis membrane with a diameter of 90 mm. The inlet and outlet for the process fluid and the adjustable support for the magnetic stirrer are mounted on the top plate in order to accomplish perfect mixing. The position of the stirrer above the membrane can be adjusted throughout the process and is indicated on an external scale (x). Cut-off and the membrane material (adsorptive features) can be selected as required by the process conditions. The membrane is fixed and sealed by means of an O-ring in conjunction with a support plate. The cell capacity is 200 μl and the membrane surface area is 45 cm^2 (ø = 90 mm). Application of the dynamic filtration/dialysis principle in combination with an optimized design of the filtration/dialysis unit resulted in the development of this dependable apparatus for the aseptic interfacing of a fermenter to an analytical system. It is also of help in monitoring other operations of bioprocesses such as cell disintegration and product recovery.

Pressurized membrane reactors, where a microporous tube acts as the membrane proper, represent a rather different approach from dialysis insofar as their aim is not separation but rather the introduction of a reagent into a continuous analyser while avoiding the dilution involved in adding a reagent channel to be merged with that of the sample. The most marked differences from a typical separation unit are as follows: (*a*) the acceptor solution (sample) is in a dynamic regime, while the donor stream is quiescent; and (*b*) the device is pressurized in order to ensure constant reagent delivery to the analytical system—thus, the overall process is a sort of dialysis–filtration hybrid. Figure 3.8 depicts the two most commonly used units of this type, the complexity of which varies according to whether the reagent is fed continuously (A) or intermittently (B). In the former case, the microporous system through which the main stream containing the sample is circulated is connected to the pressurized reactor containing the reagent and consists of an air-tight reservoir that is pneumatically pressurized through one of the three holes of its stopper—the other two are the inlet and outlet for the microporous tube. The pressure thus applied forces the reagent to pass continuously across the microporous material into the main stream.

The modules typically used for the intermittent feeding of reagent operate in much the same way as intermittent pumping. This can be accomplished by connecting the third opening of the module depicted in Figure 3.8 to a common part of an electrically actuated three-way valve, so that the normally open port of the valve is connected to control the nitrogen

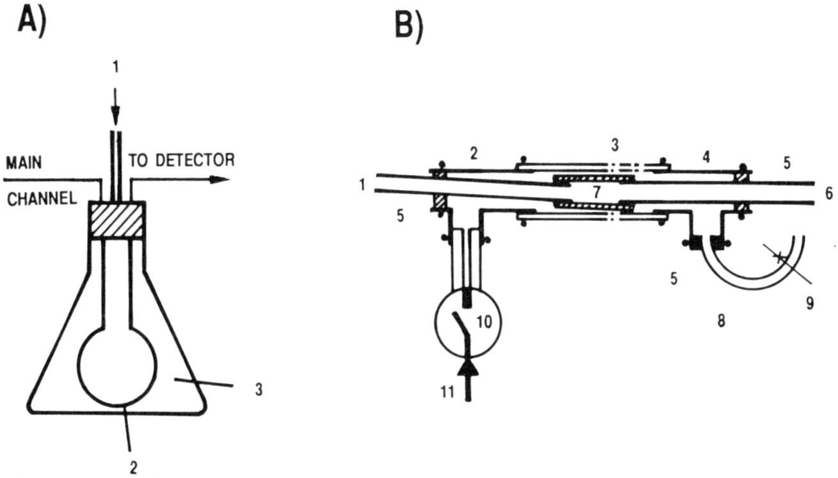

Figure 3.8 *Schemes of pressurized membrane reactors for introduction of reagents into a continuous analyser.* (A) *Continuous introduction of reagent:* (1) *air pressure,* (2) *pressurized membrane reactor,* (3) *reagent.* (B) *Pulsed membrane reagent introduction system:* (1, 6) *in, out tubes,* (2, 4) *polypropylene tees,* (3) *PTFE jacket tube,* (5) *PTFE tube segment,* (7) *membrane tube,* (8) *flexible tubing,* (9) *on/off clamp,* (10) *fast-acting on/off valve,* (11) *pneumatically pressurized reagent*
(Adapted from references 54 and 55 with permission of the American Chemical Society)

pressure. As the valve is briefly energized, the reservoir is momentarily pressurized and a reagent pulse enters the main stream across the membrane. This same purpose can be served by the device shown in Figure 3.8B where the reagent reservoir is separated from the main stream. The membrane reactor it uses is separated from the reagent reservoir as shown in the Figure. The annular tubular assembly is constructed from two miniature polypropylene tees (2 and 4 in the Figure) and PTFE jacket tube (3). The main flow stream enters through tube (1) and leaves through tube (6)—their entries into the tees are sealed with PTFE tubing segments (5) and crimped with nichrome wire represented by dots in the Figure. The inner flow channel contains the porous PTFE membrane segment (7), with an active length of about 1 cm. The entry and exit tube are push-fitted into the membrane tube. The short arm of one of the tees is connected to a flexible poly(ethylvinyl acetate) tube (8) furnished with a pinch clamp (9), while the short arm of the other tee is connected to the common port of an internally PTFE-coated three-way valve (10). The normally closed port of the valve is connected to the liquid outlet of a pneumatically pressurized reagent delivery bottle. The normally open port of the valve is sealed off, so it functions as an on/off valve. Initial operation is started by pumping the main flow stream through the inner channel, opening the clamp (9), and energizing the valve to allow the reagent to flow into the annular channel and fill it up with the reagent liquid. Then both the clamp (9) and the valve port (10) are shut off. As the latter is briefly energized, a pulse of reagent enters the inner stream. Operation of the valves is controlled by a microprocessor.

The location of a dialysis unit in a continuous analytical system depends on its specific purpose (Figure 3.9). By placing it prior to the point where the sample is inserted into or merged with the main stream the analyte is diluted to a greater extent than normal, which is useful for some types of sample—this is the typical location of dialysis probes. By accommodating the dialyser within the sample insertion system, the separation process is rendered more versatile as shown below in Section 3.2.4. However, the dialysis unit is most commonly located within the transport (and reaction) system, so it can act both as a separation unit and as a way in for reagent insertion.

On the other hand, dialysis membranes are rarely located within the detector and virtually exclusively for kinetic determinations, a peculiar case of integrated reaction-detection.

3.2.2.2 Membranes

The membrane is no doubt the key part to every dialysis unit. In order to perform its intended function efficiently, a membrane should be semi-permeable, *i.e.* it should allow passage of some chemical species while preventing or substantially delaying that of others. In addition, the rate of transfer of the permeable species should be high enough to ensure reasonably fast separation. The membrane should also be mechanically and chemically resistant. Dialysis membranes are usually one of the following three types: microporous, homogeneous, and exchange.

A microporous membrane resembles a conventional filter very closely, both operationally and structurally. It should be rigid and highly voided with randomly distributed interconnected pores of a much smaller size than those of conventional filters (10–100 Å *versus* 10 000 Å). All particles larger than the largest pores are rejected by the membrane, while those smaller than its smallest pores find their way across it and those of intermediate sizes are partly rejected and partly allowed across it to a varying extent depending on the pore size distribution. A microporous membrane thus only allows particles of rather different sizes to be separated.

Homogeneous membranes are homogeneous films or interfaces across which species are transferred by molecular diffusion. The efficiency with which the components of a mixture are separated is proportional to their transfer rates across the interface, which in turn is determined by the diffusivities and concentrations of the individual components in the film. Thus, homogeneous diffusion membranes allow particles of the same size to be separated provided their solubilities in the membrane are sufficiently different.

Exchange membranes, also of microporous nature, have positively or negatively charged ions fixed to the pore walls and hence function as anion or cation exchangers, respectively. Their separation efficiency

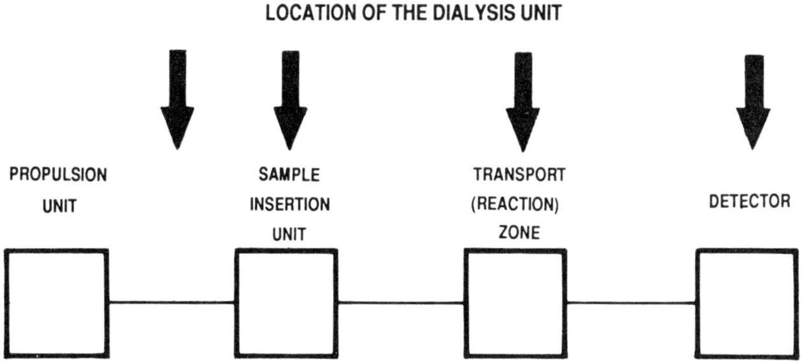

Figure 3.9 *Different possible locations for placement of a dialysis module coupled on-line to a continuous analyser*

depends not only on pore size, but also on the ease with which the membrane co-ions can be excluded from it. This type of membrane is usually made of regenerated cellulose or Cellophane in the form of hydrogels containing up to 50% water after equilibration with water, their permeability increasing with the extent of hydration. In order to maintain the expanded solvation state in the dry membrane and hence its permeability during storage, this type of membrane usually contains 5–40% glycol by weight. Once wet, drying the membrane in the absence of a plasticizer causes its structure to break down irreversibly and its permeability to be impaired, which is a serious constraint to its use for some applications.

Membranes made of glassy polymers, methacrylates, polysulphones, polycarbonates, and polyacrylonitriles are not hydrogels, so they are not irreversibly altered by drying. However, insofar as they are usually hydrophobic, their permeability can be reduced by repeated drying as a result of air being trapped within the pores; the trapped air can be removed by wetting the membrane with a low-surface tension solvent (*e.g.* alcohol) prior to exposure to the aqueous solutions.

Ion-exchange membranes can be of two chief types: homogeneous and heterogeneous. Homogeneous exchange membranes are fabricated by evaporation of solutions of film-forming polymers containing the exchange particles or by dispensing the ion-exchange material in partially polymerized film-forming polymers and allowing polymerization to complete. On the other hand, homogeneous ion-exchange membranes are made by graft polymerization of anionic or cationic halves into preformed films, casting of the film from a solution of a linear film-forming polymer and a linear polyelectrolyte, and allowing the solvent to evaporate, or, alternatively, by polymerization of mixtures of reactant that can be polymerized by addition or condensation.

The choice of a membrane for a given application is dictated both by the type of separation to be carried out and by the working environment.

Other factors such as the temperature and pH should also be taken into account in making the final choice. Glassy polymer membranes can only be used below 80°C, while cellulose membranes remain operative at higher temperatures. As far as the pH is concerned, most glassy polymer membranes can be used over a wide range; those of polycarbonate are rapidly deteriorated above pH 7.5, while those of cellulose are slowly hydrolysed below pH 2 and swell markedly in very alkaline solutions (over 40% NaOH).

The above types of membrane are normally used as flat sheets, tubes or hollow fibres. While flat sheet membranes are simpler to design and construct, tubular membranes provide larger dialysis areas in smaller spaces, particularly when coiled. Hollow-fibre dialysis membranes were introduced more recently and provide compact devices with very large surface areas; in addition, their peculiar structure allows them to be used without a support.

3.2.2.3 Other Units

In addition to the indispensable units of a continuous analyser (*e.g.* propulsion system, reactors, injection unit if required, *etc.*), configurations using on-line dialysers may include other units intended to facilitate application of the analytical methodology concerned. Such units are also occasionally of separative nature (*e.g.* a dialysis–liquid chromatograph assembly, in which the dialysis module acts as a pre-column system, an ion exchanger intended to increase selectivity or a gas-diffusion module for the simultaneous or sequential analysis of two analytes in the same sample or the introduction of gas reagents in addition to liquids into the main flowing stream). Improving the dialysis process by stopping the acceptor stream calls for a device allowing the standstill period to be precisely timed and synchronized with the circulation of the sample as the donor stream.

The sensing systems most frequently used in this context are optical detectors in general and photometers in particular. Detection is normally preceded by a derivatizing reaction between the analyte and a reagent dissolved in the acceptor stream; this increases the efficiency of the separation process as the dialysis equilibrium is gradually shifted as the free analyte is removed from the acceptor stream. Other optical techniques such as fluorimetry and atomic absorption spectroscopy have also occasionally been used for detection purposes.

The efficiency of ion-selective electrodes is often increased to specificity levels by the use of dialysers. However, amperometry remains the electroanalytical technique most frequently used in conjunction with continuous dialysis. When a dialyser is coupled on-line to a chromatograph, the detector of the latter is used in the overall automatic system.

3.2.3 Configurations

The basic types of manifold coupled on-line to a dialysis module are depicted schematically in Figure 3.10. Figure 3.10A is illustrative of an air-segmented system, the configurational and operational complexity of which arises from the presence of the air bubbles.

Those configurations into which samples are introduced by aspiration (completely continuous systems, Figure 3.10B) are no doubt the simplest. On the other hand, they require much larger amounts of sample and yield drastically reduced sample throughputs as a result of the need to flush the system with an appropriate solution or the sample itself prior to each determination.

Sample injection systems (Figure 3.10C) additionally require a four- or six-way valve for injection. In contrast to this slight complication, they require smaller sample volumes (of the order of a few microlitres) and feature extremely high sampling frequencies and enormous flexibility as shown below in the following section.

3.2.4 Applications

By way of example, this section deals with a series of methodologies that exploit continuous dialysis. The discussion is started with methods developed in segmented-flow configurations as they were the first on which this continuous separation technique was applied with or without determinative purposes. Unsegmented configurations are more versatile, so in addition to determinative methods and theoretical studies, their discussion includes the use of other separation techniques such as aqueous–aqueous extraction and dialysis probes as sampling/separation systems. Finally, we shall review the use of dialysis for non-separative purposes (*e.g.* the introduction of reactants into continuous analysers).

The determination of glucose in serum or plasma based on a hexokinase/glucose-6-phosphate method reported by Neeley[9] is implemented on a configuration similar to that depicted in Figure 3.10A. The method requires only 50 µl of sample and an incubation period of 7 min prior to the photometric measurement of the reduced form of the co-enzyme NADPH which is accomplished by using long coils and a low flow-rate. The dialyser used removes interferents such as bilirubin, hemoglobin, lipemia, and common drugs as checked on applying the method to uraemic patients (pre-dialysis) from the excellent correlation between parallel assays by the glucose oxidase and *o*-tolidine methods.[10] The dialyser membrane must be replaced weekly. A glass pressure chamber is placed in the flow-cell pull-through line to smooth sample peaks.

Simultaneous determinations with segmented analysers call for the use of as many configurations as there are analytes to be determined. Such is the case with the simultaneous determination of carbon dioxide and chloride in serum by using the Autoanalyzer I furnished with the dialysis

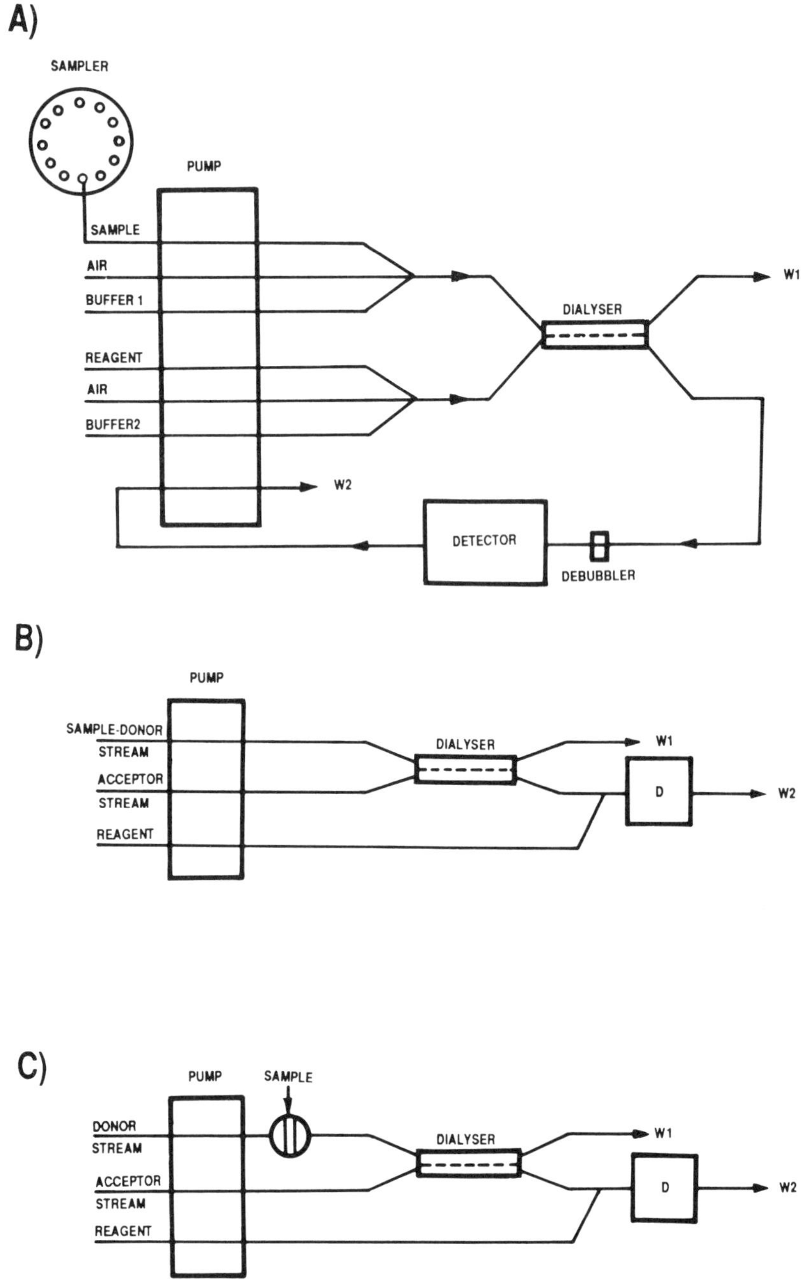

Figure 3.10 *General continuous configurations including an on-line dialysis module.* (A) *Segmented-flow configuration;* (B) *with sample aspiration;* (C) *with sample injection*

and gas-diffusion modules of the Autoanalyzer II. The sample is aspirated into the system from the sampler and split into two parts that reach two sub-systems similar to that in Figure 3.10A where the separation modules are furnished with silicone rubber membranes for CO_2 and type C membranes for chloride (the module pathlengths are 15 and 30 cm, respectively). Two filter colorimeters connected to an ordinary recorder detect the products from the derivatizing reactions (Cresol Red for CO_2 and Hg^{II}/SCN for chloride). The performance of this method was checked by using it in an emergency laboratory for 5 months.[11]

Equilibrium dialysis[12,13] is a provenly useful technique for the obtainment of quantitative data on the reversible interaction between a dialysable species and a macromolecule. By appropriate treatment of the binding data, the equilibrium formation constant of the complex can be determined and the number of sites on the macromolecule to which the small species is bound be estimated.[14–16] A serious pitfall of the experimental procedure is the long interval required for equilibrium to be reached. In fact, periods of 24 hours or longer are involved, and if a labile material is being dealt with, decomposition products may give rise to spurious results. These shortcomings can be overcome by applying a method based on the difference in the dynamic, short-term dialysis behaviour of the species of interest in the presence and absence of non-dialysable molecules. The dialysis can be implemented at the chosen conditions, so the amount of material diffusing across the membrane will be directly proportional to the free concentration in the original solution stream. In conjunction with an automatic analytical system like the one depicted in Figure 3.10A, it allows the elucidation of the binding profile concerned within a few hours.[17] The rapidity, reproducibility and general applicability of the method make it a useful tool for this type of study; its practical performance was checked by reproducing the work of Klotz *et al.*[14] on the binding of Methyl Orange to bovine serum albumin.

Configurations using no air bubbles are much simpler to construct and operate, as shown in Figures 3.10A and 3.11, both of which are used for the determination of glucose. The manifold in Figure 3.11 was conceived for the continuous monitoring of this analyte in blood by amperometric monitoring of dissolved oxygen after an enzymatic reaction catalysed by glucose oxidase. Blood is continuously sampled by means of a double-lumen catheter inserted in a vein of the patient's arm that is connected to the flow system. The operational sequence is as follows: the proportioning pump uses a four-tube manifold, each channel of which functions as a separate pump. One 6 V motor is coupled to four Teflon wheels rolling on a cast aluminium surface. The fastest pump head is coupled directly to the output shaft of the motor. Other desired flow-rates are obtained by means of fine gears. The pumping action is precise: rates as low as 0.1 ml min^{-1} can be readily achieved. Reservoir A holds a dilute heparin isotonic saline solution which is pumped into the inner lumen of the dual catheter at a rate of 0.3 ml min^{-1}. The herapin solution does not enter the

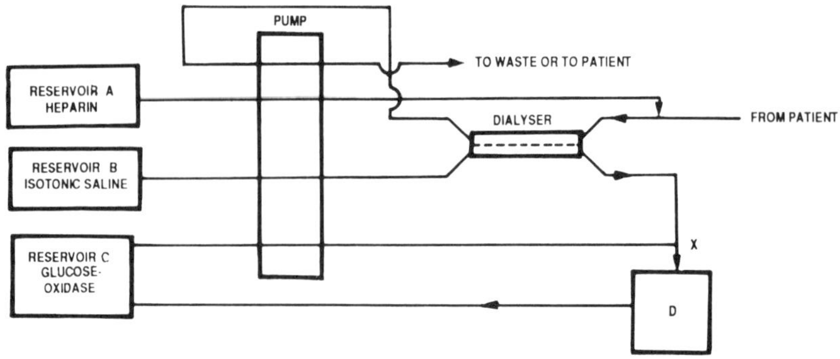

Figure 3.11 *Completely continuous configuration for the enzymatic determination of glucose with catalyst recirculation*

patient's vein at any time, but mixes with the blood at the tip of the dual catheter. The blood–heparin mixture (donor stream) circulates at a rate of 0.4ml min^{-1} through the dialysing chamber. On the other hand, the patient's blood is circulated at a rate of 0.1 ml min^{-1}. However, such blood can be returned to the patient's vein when lengthy studies are considered. Reservoir B holds an isotonic saline solution that is pumped through the other side of the dialyser in the opposite direction to that of the blood flow in order to obtain a countercurrent effect at a rate of 0.4 ml min^{-1}. The flow-rates and osmotic pressures on both sides of the dialysis membrane are the same. Reservoir C is vented and contains oxygen-saturated glucose oxidase that is pumped at a rate of 3.9 ml min^{-1} to merge with the dialysate at point X, after which the reagent–dialysate mixture is passed through a 1-min delay coil within which the reaction is allowed to take place. The mixture, now depleted of oxygen in proportion to the amount of glucose present, is passed over a polarographic oxygen sensor. Since the enzymes present are not directly involved in the reaction, the glucose oxidase solution is returned to the reagent pool, where it is aerated for recycling with fresh sample. In the event that blood is returned to the patient, contamination with the reagent is prevented by using a separate circulating system for this. The system was employed for the continuous monitoring of blood glucose in twelve patients previously studied by another automatic method of analysis and provided the same curves. The response time of the system was four minutes, *i.e.* about one half that obtained with the other automatic system (Technicon Autoanalyzer); this time can be reduced even further by using smaller dialyser volumes. The two main reasons leading the system proponents to choose countercurrent dialysis were a higher process yield and a shorter delay time resulting from the fact that the dialysate sample introduced into the sensor was more recently obtained from the patient.[18]

Continuous unsegmented systems with sample aspiration are extremely useful for theoretical studies.[35] Martins *et al.* used one such system consisting of two channels (donor and sample plus acceptor) carrying a buffer and connected to a flow-through detector furnished with a hanging mercury drop electrode[19] to study the effect of the ionic strength, pH, and chelating ligands on the dialysis of metal ions, particularly Zn^{II}, across acetate cellulose membranes. Some of the conclusions drawn by the authors were as follows: (*a*) the dialysis factor, which is dependent on the mass transfer and the membrane permeability, is independent of the ionic strength over the range 0.05–0.3 M and increases only slightly with the pH between 4.6 and 7.0; (*b*) the addition of chelating agents normally decreases the rate of metal–ion transfer; (*c*) the coefficients of partition between the polymer membrane and the buffer were evaluated and compared with the dialysis factors and no significant differences with respect to the variation range, effect of the concentration, or pH were found; (*d*) the time scale of uptake or release of metal ions by the polymer is much longer than that of a steady-state transfer in dialysis. After all, partition and dialysis might not be directly related to each other.[20]

Despite the fact that configurations with sample injection are more complex than their unsegmented counterparts with sample aspiration, this apparent disadvantage is offset by the inherent advantages of FIA configurations commented on above. The straightforward FIA configuration shown in Figure 3.10C was used to develop a series of methods where dialysis was exploited for a number of purposes.

Thus, earlier methods for the determination of serum calcium on segmented analysers involved a preliminary sample digestion step, which decreased the sampling rate to only 20 samples h^{-1}.[21] The FIA method developed for the same purpose by Basson and van Staden, which relies on the use of cresolphthalein complexone in a 2-amino-2-methylpropan-1-ol basic medium[22] employs an on-line dialyser located immediately after the injection port which facilitates dilution and deproteination and avoids manual digestion and the dilution step, thereby raising the sampling frequency to 180 h^{-1}. The method was compared with its segmented counterpart (standard atomic absorption procedure) and was found to yield perfectly consistent results.[23]

The photometric determination of such inorganic ions as sulphate[24] and phosphate and chloride[25] in biological fluids (urine and serum) involves not only the isolation of potentially interfering macromolecules, but also sample dilution. Both processes can be readily implemented on a dialysis module within an FIA system.

The use of ion-selective electrodes for the analysis of biological samples entails a separation step intended to isolate all those species that might interfere with the determination through their response at the electrode or through fouling of the active sensor surface. Dialysis is clearly the separation procedure of choice in this case provided the analyte is concentrated enough, as shown by the method for the determination of lithium in serum reported by Xie and Christian.[26] A volume of 200 μl of undiluted serum is injected into a stream of de-ionized water flowing at a rate of 0.4 ml min^{-1}, while the acceptor stream (borax buffer) is circulated at a flow-rate of 0.3 ml min^{-1}. The ratio of Li/Na transfer across the membrane (Technicon type C, cellulose acetate) was 0.57, which required the use of an electrode with a high Li/Na selectivity coefficient.

The use of an enzymatic reactor coupled on-line with the dialyser and located after it within an FIA system increases selectivity without increasing the complexity of the experimental set-up or its analytical costs. Gorton and Ogren reported an enzymatic photometric method for the determination of glucose using both types of module. The method relies on the use of two enzymes, glucose oxidase and glucose peroxidase, both of which are immobilized on CPG and make up two reactors located serially along the transport system. The reaction product, hydrogen peroxide, forms a coloured compound with 4-aminophenazone and *N,N*-dimethylaniline that is monitored at 555 nm. The proponents used two different dialysis membrane (Cuprophan) thicknesses, *viz*. 11.5 and 17.0 μm, which provided a process yield of 16 and 13.4%, respectively, at a flow-rate of 1.0 ml min^{-1} in both channels.[27] A similar method for the determination of glucose in blood but based on amperometric detection of the hydrogen peroxide formed uses two separation systems to remove interferences from the sample matrix, namely a dialyser and a column containing Cu^{II}-diethyldithiocarbamate immobilized on CPG—the separation module is located prior to the enzymatic detector in both cases. The two systems provide adequate separation of interferences; however, the dialysis system calls for low flow-rates (1.0 ml min^{-1} in this case) in order to achieve adequate yields (13% under the working conditions used), which results in a throughput of 70 samples h^{-1}. This constraint does not affect the separating column, which allows up to 300 samples to be determined.[28] The results obtained were consistent with those yielded by commercial immobilized tubes in a Beckman II analyser.

The use of dialysers as sample loops is quite an interesting novelty in this field. Figure 3.12A depicts an FIA configuration based on this innovation. It was used by Chang and Meyerhoff to determine salicylate in serum by means of an electrode sensitive to this ion. A fresh buffer plug is used for each measurement. The upper channel of the dialyser serves as the injection loop of a conventional flow-injection set-up, while the bottom channel acts as a second loop into which the test sample is loaded manually as a static plug. After a preset trapping time, the solution in the upper channel is flushed to the electrode and the salicylate levels are measured potentiometrically. Diffusion of the salicylic acid molecules across the silicone rubber membrane is facilitated by the fresh acceptor solution in the upper channel, which traps salicylic acid as salicylate ions, thus maintaining a large chemical gradient of salicylic acid across the membrane. The efficiency of transfer of salicylic acid from the sample to the acceptor solution ranged between 4 and 7% for a trapping time of two minutes. Such an efficiency was increased by keeping the dialyser at a high temperature, which, however, resulted in outgassing of the solutions and bubble generation. Thus, at 25°C, the Nernstian range of the

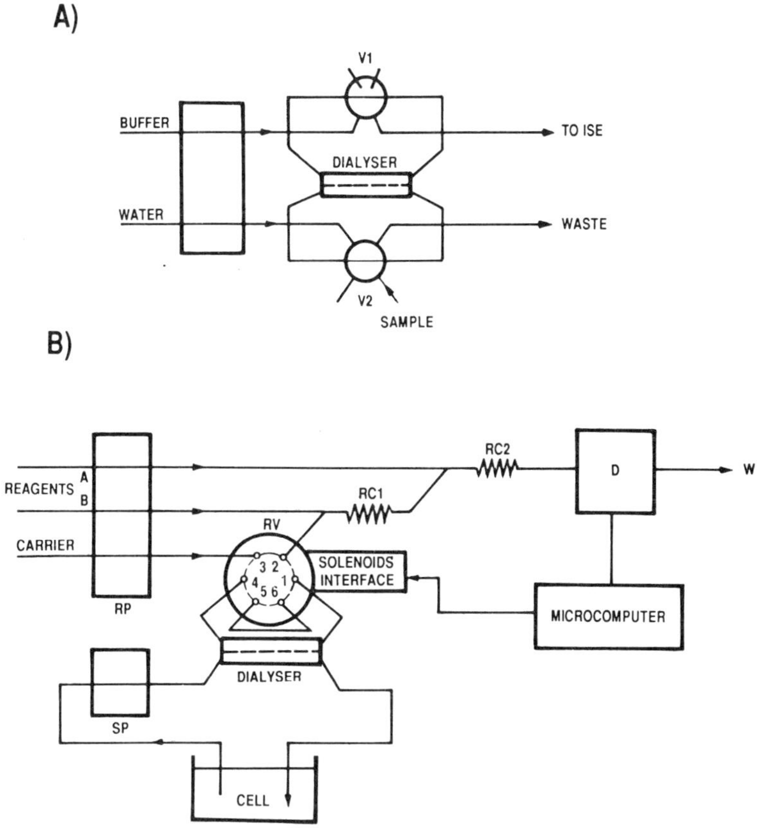

Figure 3.12 (A) *Scheme of a dialyser/flow set-up for the determination of salicylic acid. V_1 and V_2 denote flow injection valves.* (B) *Scheme of a flow-injection serial dynamic dialysis system: RP, reagent pump; SP, sampling pump; RC_1 and RC_2, reaction coils; RV, rotary valve; W, waste*
(Reproduced from reference 29 with permission of Elsevier Science Publishers)

electrode response was between 10^{-4} and 10^{-2} M salicylate for a trapping time of 2 min; such a range was from 2×10^{-6} to 10^{-2} M in the absence of the membrane. The selectivity of the membrane-dialyser/flow-injection assembly was excellent: it was not interfered with by most of the organic and inorganic ions to which the membrane is not permeable, as shown by the assays carried out in its absence. Even chloride concentrations above 0.1 M gave rise to a virtually negligible electrode response. Perchlorate ion, one of the most serious interferents with the type of electrode used (porphyrin membrane) seemingly does not cross the silicone rubber membrane—at least not during the two minute trapping period. The high selectivity provided by the dialysis unit allows the determination of the analyte in serum.[29]

A microporous tubular PTFE membrane module with an outer casing similar to that depicted in Figure 3.5 was used for the separation of cyanide and thiocyanate and subsequent simultaneous determination by the pyridine/barbituric acid method on the basis of the different rate at which these two anions react with chloramine T. The membrane module was laboratory-made from microporous PTFE membrane tubing (Gore-Tex TB001), T-joints and PTFE tubing, with a nominal sizing of the tubular membrane of 1.8 mm OD, 1.0 mm ID, maximum pore size 35 μm, and porosity 70%. The tubing was inserted into larger PTFE tubing (3 mm OD, 2 mm ID). The distance over which diffusion occurred was 10.5 cm. The cyanide–thiocyanate solution was introduced into the donor (outer) side of the membrane module and transferred to the acceptor (inner) side. A 30 m long tube was connected to the donor side of the membrane module to provide back-pressure. The transfer mechanism on which the separation process is based, a combination of gas diffusion and dialysis, was accounted for by the authors on the basis of a concept previously established by Durst,[30] according to which the pores of the microporous membrane are filled with the solution, thereby enabling the transfer. In this case, the buffer solution flowed through the membrane pores from the acceptor to the donor side, depending on the pressure difference between the outer and inner solutions. This was checked from the flow-rates of the two waste solutions, which were 1.5 and 1.2 ml min^{-1} for the donor and acceptor side, respectively. By contrast, cyanide and thiocyanate passed from the donor to the acceptor side. There must be a solution/membrane interface along the pore lips to the inside of the pore; such an interface must consist of an aqueous layer fixed to the membrane frame, forming a path that connects the two sides of the membrane. This neighbouring layer allows the entry of solute and the ease of entry differs between solutes. Cyanide and thiocyanate can readily enter this layer because they are less hydrophilic. Cyanide crosses the membrane only partly as cyanide ion and partly as dissolved hydrogen cyanide because the logarithmic acidity constant of hydrocyanic acid is 9.2.[31] By contrast, the logarithmic acidity constant of thiocyanic acid is −1.84,[32] so thiocyanate is transferred across the membrane in its ionic form. This is the reason why thiocyanate penetrates the membrane, whereas it is not distilled by the standard distillation method. The percent transfer across the membrane [T = (amount of analyte transferred/overall injected amount) × 100] was calculated as a function of the pH, the module length, and the direction of transfer across the pores of the tubular membrane. Cyanide was transferred to the acceptor side more readily when the buffer solution was of a lower pH because the buffer solution penetrated into the donor side (pore size 3.5 μm) and the pressure on the acceptor side was higher than on the donor side. The pH on the donor side was further increased by penetration of the buffer with the higher pH. As the pH on the donor side affected the transfer process, the percent transfer into the phosphate buffer was higher than that into the carbonate buffer. The percent transfer of thiocyanate could not be measured at pH

8.1, where thiocyanate could not be detected. The percent transfer of cyanide was found to depend on the direction of the solute flow through the wall of the tubular membrane, *i.e.* on whether the donor solution was pumped inside the tubular membrane or in the channel outside the membrane. When the donor solution was inside the tubular membrane, the percent transfer of cyanide to the outer acceptor solution was 32%, but when the donor solution was passed through the outer channel, the transfer to the inner acceptor solution rose to 68%. The increase in the percent transfer presumably arose from the increased number of inlet pores available for cyanide and thiocyanate; the nominal porosities and mean pore sizes were the same on the inner and outer surfaces of the tubular membrane, but the inner and outer surface areas were 3.3 and 5.9 cm^2, respectively. Accordingly, the direction of solute diffusion chosen for further work was from the outside to the inside of the tubular membrane, *i.e.* the sample solution was pumped through the outer channel. The percent transfer of cyanide in a 76.4 cm membrane module was almost twice that in a 10.5 cm module; about 32% of the cyanide was transferred in the first 20.5 cm of the overall length, and *ca.* 35% in the remaining 65.0 cm. The transfer process was thus rather slow.[33]

The FIA/dialysis association has also been used for non-determinative purposes. Most interesting in this respect is the work on binding by macromolecules (drug–protein binding interactions). As with the previous method, the separation module is located in the loop of the injection valve of the FIA/dialysis configuration used; however, this system is automated and a little simpler and perfected (Figure 3.12B). The overall set-up was assembled by interfacing a customized flow-injection photometric analyser[34] to a conventional four-slot dialysis unit (Technicon). The analyser consists of two peristaltic pumps (sample and reagent), a rotating valve operated by a pneumatic two-position actuator, an appropriate manifold of Teflon tubing, a conventional photometer, and a microcomputer to control the whole system. The dialysis unit, constructed from Plexiglass, has a dialysis surface of 670 mm^2 and two identical chambers of 1.0 ml each which are separated by a semi-permeable Cuprophan membrane with a pore size of 20 μm. The upper chamber, connected to ports one and four of the rotary valve (the 'sample loop' of the analyser), serves as the acceptor chamber. The lower chamber is connected to the sample pump, which propels the protein–drug solution from the 50 ml cell, thermostated at 25 ± 0.1 °C. Periodically, the sample zone is injected into the manifold and fresh solution is loaded into the sample loop. The drug concentration in the protein compartment is changed by successive manual additions of small volumes of standard drug solution. This methodology, called 'flow-injection serial dynamic dialysis', provided automation of dialysis, sampling, and spectrophotometric determination. The binding of some sulphonamides to bovine serum albumin was studied with the aid of this automatic system and yielded results similar to those obtained by other procedures. The time required for a complete run is typically 100 min, including calibration. The system can also be used to calculate dialysis rate constants.[35]

The determination of octanol–water partition coefficients by batch extraction methods is currently the only acceptable procedure for application to structure–activity relationships. Work on the development of flow-injection systems for the determination of the aforementioned partition coefficients was reported by Gluck,[36] who used a pressure-balanced separation apparatus and two commercially available and five laboratory-made dialysis-based extractors that were evaluated for their efficiency of mass transport of a Cresol Red solution. A 'system efficiency' was calculated for each dialyser which included not only the efficiency of mass transport across the membrane, but also the effect of dilution arising from the sample dispersion in the system. To determine the system efficiency, a 1.0 mM solution of Cresol Red was injected and passed through the dialyser, and the peak-height response for Cresol Red was measured on the opposite side of the membrane. For comparison, 10 μM Cresol Red was pumped directly through the detector; this gave an absorbance of 0.402, which corresponds to 1.0% efficiency of the dialyser. The percent efficiency was calculated by dividing the peak-height response by 0.402. As far as partition studies are concerned, even though these devices provided successful extraction, the determination of octanol–water partition coefficients was inaccurate because of the partitioning process between some components of the membrane.

The association of two continuous separation techniques has usually been exploited to increase selectivity and, less often, also for other purposes. One such combination was reported by Olsson *et al.*, who used it for the determination of D-galactose based on its oxidation in a reactor containing immobilized galactose oxidase which yielded hydrogen peroxide that was determined photometrically after an oxidative colour-forming reaction in a peroxidase reactor. The separation step involved in the determination of the analyte in serum was effected in three serially arranged modules (Figure 3.13). The sample, which was injected into a water stream, first reached the dialyser, which was furnished with a 0.025 mm thick cellulose acetate membrane whose channels were filled with glass beads 150–250 μm in diameter intended to decrease dispersion. At the exit of the acceptor channel was located a Bond-Elut-NH_2-Cu column intended to remove the ascorbic acid contained in the dialysate by quantitative oxidation to dehydroascorbic acid plus hydrogen peroxide. This last reaction product, together with that already present in the dialysate (*e.g.* as a result of aerobic oxidation) were removed at the enzymatic reactor with the aid of immobilized catalase. A column packed with such an absorbent material as porous arylamino glass placed countercurrent in relation to the reagents prior to their merge with the main channel removed potential interferents. In this way, the analyte was determined in serum at concentrations between 10 μM and 14 mM with excellent results. The sampler, pump, and valves of the flow manifold were con-

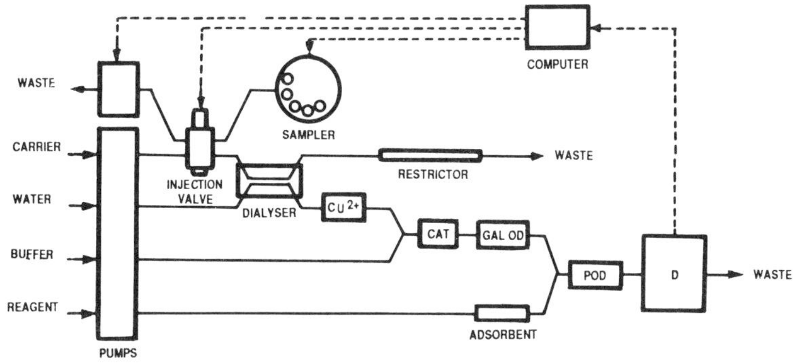

Figure 3.13 *Scheme of a flow-injection system for the spectrophotometric determination of galactose. Reactors: Cu^{2+}, Bond-Elut-NH_2-Cu; Cat, catalase; Galod, galactose oxidase; POD, peroxidase immobilized on adsorbent arylamino porous glass; D, detector*
(Reproduced from reference 37 with permission of Elsevier Science Publishers)

trolled via a personal computer which also performed data collection and processing.[37] The same authors later reported a simpler system for this determination based on the use of an amperometric detector.[38]

Dialysis is most commonly combined on-line with liquid chromatography. In fact, a large number of methods rely on dialysers as pre-column separation devices.

Every method for the analysis of amino-acids in serum by liquid column chromatography requires prior deproteination of the sample. This is typically accomplished by using sulphosalicylic or perchloric acid for ion-exchange chromatography,[39] or acetonitrile when using reversed-phase chromatography.[40] These techniques are difficult to automate and call for sophisticated robots. Deproteination by dialysis overcomes these shortcomings but the dialysate obtained can contain under-analysable amino-acid concentrations. This problem can be circumvented by using a suitable reagent on the acceptor channel as a pre-column derivatizer (*e.g. o*-phthalaldehyde/2-mercaptoethanol) and a high-pressure reversed-phase liquid chromatograph[40,42]; in addition, the sensitivity is increased and the dialysis process is favoured by shifting the equilibrium through formation of the fluorescent product as shown by Turnell and Cooper[43] by using a laboratory-made automated system.

An original, also laboratory-made, design was reported by Nordmeyer and Hansen.[44] The dialysis chamber consists of a length of 0.022 ID TFE tubing with a single strand of hollow fibre dialysis membrane inside. The acceptor solution, which is transferred to the sample loop of an auxiliary injection valve after equilibration, is located inside the hollow dialysis fibre. The sample solution is drawn into the dialysis chamber through the lower 0.015 ID TFE tubing until it fills all or part of the annular space inside the 0.022 ID TFE tubing. The dialysing injection system also uses slider valves (A, B, and C in Figure 3.14) with pneumatic actuators led by

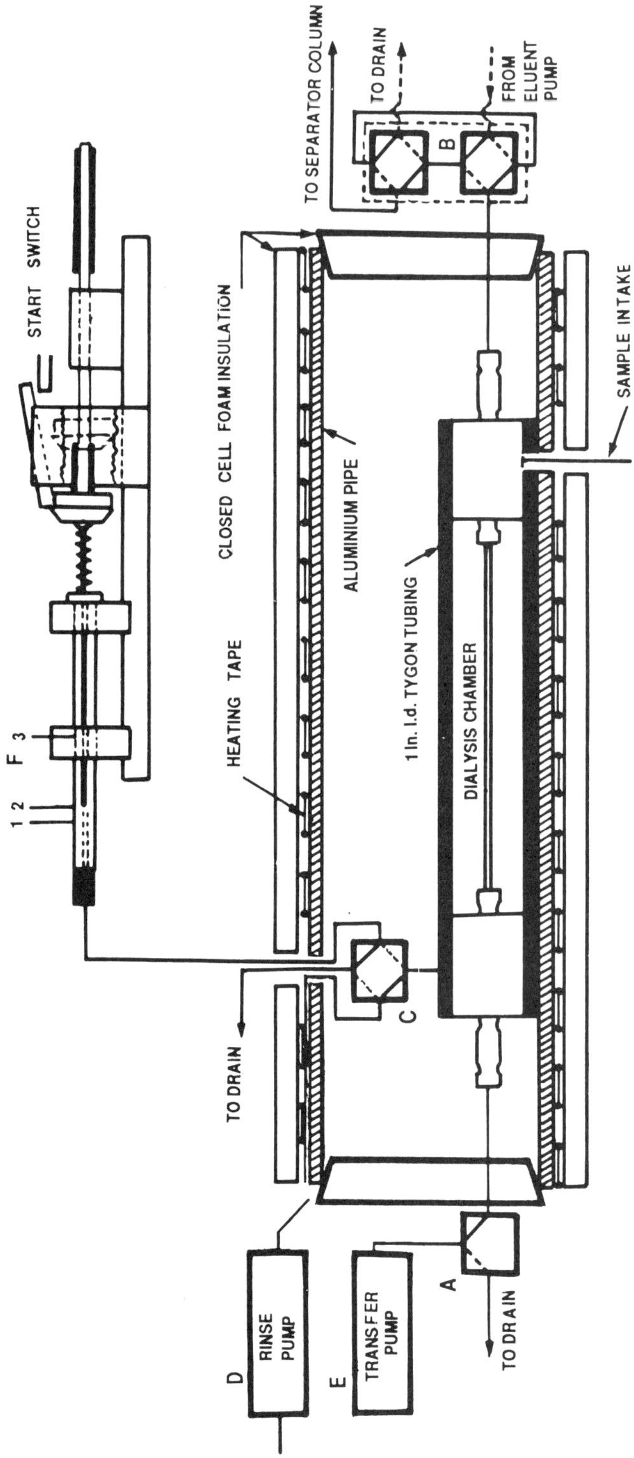

Figure 3.14 *Dialysing injection system in which valves and pumps are shown diagramatically. (For details, see text)* (Reproduced from reference 44 with permission of the American Chemical Society)

solenoids that in turn are operated by a digital controller. A syringe clamped in a device with mechanical stops measures reproducible sample volumes. Depressing the plunger of the water-filled 100 μl gas-tight syringe to position 1 expels the contents of the annular space. A 10 μl air bubble is inserted by removing the liquid drop from the tip of the sample intake tubing and allowing the spring-driven plunger to retract to the trigger stop (position 2). The syringe is shown in this position. A preset volume of sample is introduced into the dialysis chamber by positioning the sample at the intake tubing and triggering the syringe. Once the dialysis time has elapsed, this pump transfers the acceptor solution through the 7 cm of tubing to the 50 μl sample loop of valve B. Injection of the acceptor solution is accomplished by means of valve B, which is in the eluent stream just ahead of the normal sample injection valve of the ion chromatograph. The rinse pump (D) is a peristaltic pump set to deliver 2.5 ml min^{-1} that flushes the old sample solution and rinses the annular space with water or a solution. Water from pump E is again pumped through the inside of the hollow fibre as the annular space is rinsed. After the rinse cycle is finished, dialysable solutes present in the rinse solution equilibrate with the water inside the fibre to provide the acceptor solution for the next sample. These dialysable solutes can be omitted from the rinse solution if it is desired to have pure water as the acceptor medium. The dialysis–injection cycle is programmed via a digital controller specially constructed for this purpose.

The ASTED system, described earlier in the equipment section, has been used on-line with HPLC instruments for sample preparation. A case in point, employed in the development of a number of methodologies, is illustrated in Figure 3.15A.

The potential of this dialysis-trace enrichment association was put to the test by applying it to the determination of total anticonvulsants and theophylline in serum. The difference between the chromatograms of different anticonvulsants recorded in the absence (*a*) and presence (*b*) of the pre-column system (Figure 3.15B) shows the usefulness of the incorporation of a prior dialysis/preconcentration unit. In addition, orthogonal regression revealed the ASTED method to provide results consistent with those obtained for the reference method applied to the same drugs.[45] Similar equipment was used for the determination of cortisol and associated glucocorticoids in serum with a brilliant performance.[46]

Robotic samplers increase the degree of automation of pre-column systems, as shown by Cooper *et al.*, by coupling the ASTED to a liquid chromatograph and a Gilson Cartesian robotic sampler for the complete automation of the preparation and HPLC analysis of biological samples.[47] The system was applied to the determination of free and total analytes quite successfully. Figure 3.15C illustrates the sensitivity enhancement accomplished by using the TEC (Trace Enrichment Cartridge) in conjunction with a dialysis module.

Aqueous/aqueous extraction by use of a liquid membrane is a singular variant of dialysis that enables analyte preconcentration in addition to the separation. The procedure is based on the use of non-porous membranes whose mechanism of action is a combination of solubilization and molecular diffusion that endows this type of membrane with special selectivity. Despite their great use for the treatment of samples that cannot be directly introduced into a chromatograph or their direct determination after the extraction process, they have scarcely been applied to date. Melcher[48] studied the critical parameters determining the obtainment of sharp, well defined FIA peaks that can be readily measured and returns to its baseline rapidly enough to allow a fresh injection to be made in quite a short time. For this purpose, he divided the overall process into three stages: the first involves bringing the aqueous sample containing the analyte into contact with the membrane, whereby part of the analyte is extracted at the membrane surface. Secondly, the analyte dissolved in the membrane diffuses across its walls to the membrane/extractant interface. Finally, part of the analyte is isolated by the extractant. Under static conditions, this system is roughly equivalent to an equilibrium in which the ratio between the analyte in the sample layer, membrane wall, and extractant layer is determined by the coefficients of partition between the two interfaces of the membrane. Conversely, in a dynamic system where the sample and extractant continuously emerge from opposite sides of the membrane, a given amount of the analyte permeates the membrane and is carried away by the flowing extractant. At steady state, the permeation rate (*e.g.* in $\mu g\ min^{-1}$) is constant and the concentration in the sample can be regarded as the enrichment or concentration factor. In the flow-injection system, where the sample is in contact with the membrane for only a short period, the extraction parameters required to obtain a sharp, well defined peak rapidly returning to the baseline are critical. Such parameters are related to one another; one of them is the coefficient of partition between the sample and the membrane, which controls the amount of analyte extracted into the membrane; another critical factor is the diffusion rate of the analyte from the sample/membrane interface to the membrane/extractant interface; a third critical factor is the coefficient of partition between the extractant and

Figure 3.15 (A) *Scheme of the ASTED used as a pre-column system.* (B) *Chromatograms obtained by using (A) direct injections of 20 µl, and (B) the ASTED. The arrows indicate the times of injection. The HPLC conditions were identical, and the serum samples were taken from the same serum pool. Peak identification:* (1) *carbamazepine epoxide,* (2) *phenobarbitone,* (3) *carbamazepine,* (4) *hexobarbitone, and* (5) *phenytoin.* (C) *Recovery of analyte from dialysate, with and without trace enrichment. Peak area (recovery) of phenobarbitate from a 20 mg l^{-1} solution exhaustively dialysed for different times and then loaded onto the HPLC instruments using a 50-µl loop (■) and TEC (▲). The dialysate flow-rate was 1.1 ml min^{-1}*
(Reproduced from reference 46 with permission of Elsevier Science Publishers)

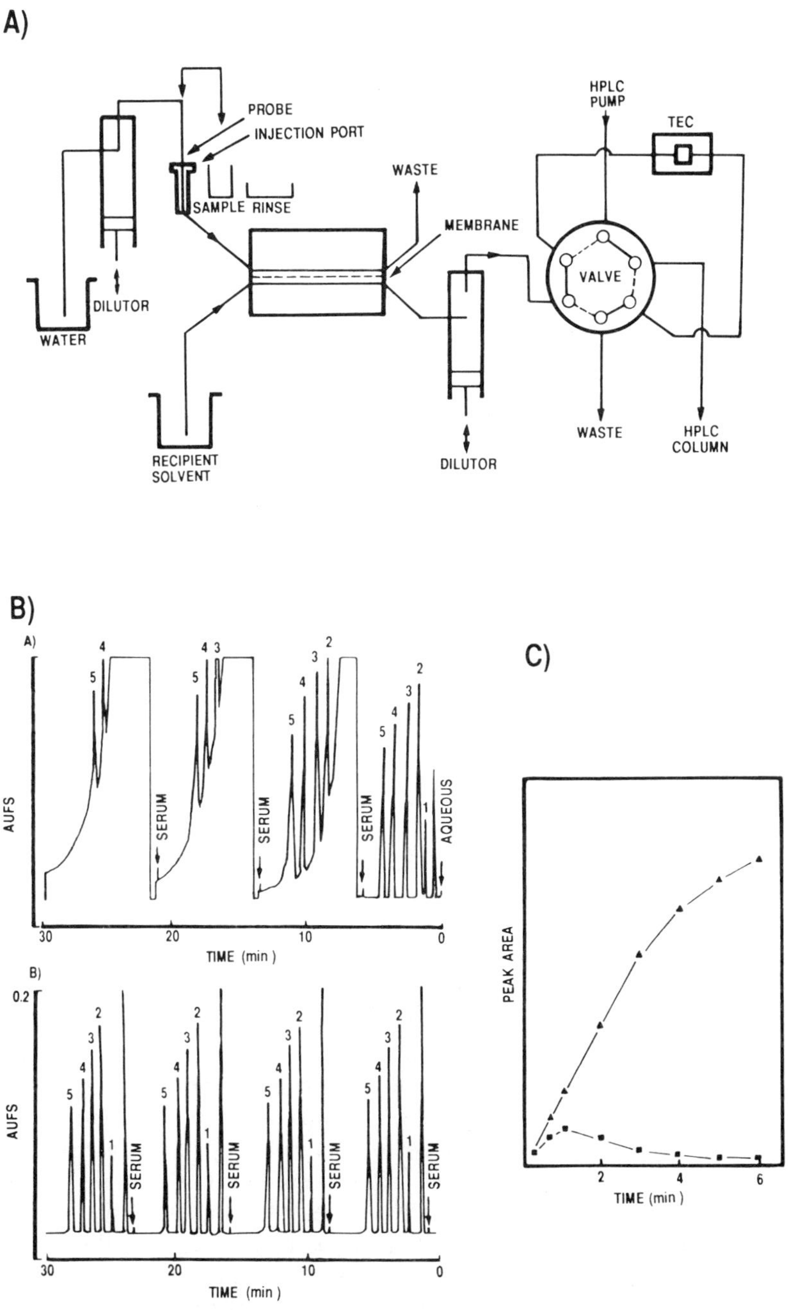
A)
PROBE
INJECTION PORT
SAMPLE
RINSE
WASTE
MEMBRANE
DILUTOR
WATER
RECIPIENT SOLVENT
DILUTOR
HPLC PUMP
TEC
VALVE
WASTE
HPLC COLUMN
B)
A)
AUFS
SERUM
SERUM
SERUM
AQUEOUS
TIME (min)
B)
0.2
AUFS
SERUM
SERUM
SERUM
SERUM
TIME (min)
30
20
10
0
C)
PEAK AREA
TIME (min)
2
4
6

the membrane, and a fourth factor is the shape and contact time of the injected sample plug with the membrane. Figure 3.16 depicts the membrane/flow injection system and a detail of the membrane/cell assembly used in the above study and proposed for on-line applications requiring adequate sensitivity and selectivity. By selecting the appropriate extraction conditions and detector, single compounds or classes of compound (*e.g.* total aromatics) can be determined quite rapidly. The main advantages of this system for on-line applications are its simplicity and the ability of the membrane to exclude any interferents such as inorganic compounds and particulates.

The principle of aqueous/aqueous extraction with a liquid membrane was used by Audunsson for sample clean-up and preconcentration of amines in a flow system[49] where the sample is passed through a conventional tube similar to that depicted in Figure 3.1 and the analyte is separated and trapped within a stagnant acceptor phase on the other side. A comprehensive study of the behaviour of the different variables influencing the process (*e.g.* sample volume, support matrix for the liquid membrane, type of immobilized solvent in the liquid membrane, flow-rate and pH of the donor and acceptor streams, and coefficient of partition between the donor phase and the membrane phase) showed them to affect the enrichment and separation factors markedly and allowed the author to compare the experimental results with theoretical predictions.

Probe-type sampling systems involving dialysis are rather commonly used in biotechnology to accomplish samples free from the macromolecules and suspended particles usually present in bioreactors and interfering with the determination following the sampling. The risk of the sampling device being clogged by microbial cells or solids from the broth and the need to maintain aseptic conditions at the sampling point and obtain responses in near-real time make configurations used for this special type of on-line monitoring additionally complex. The growing development of biotechnology calls for concomitant developments in the devices required for appropriate monitoring and control of bioreactors. As far as sampling is concerned, the probes typically used for this purpose have been the subject of much interest from researchers and commercial manufacturers. The dialysis probes developed in the last few years are good proof of the said interest.

The probe reported by Gibson and Woodward,[8] which is depicted in Figure 3.6B, has been exhaustively tested on the determination of ethanol and glucose in a model fermentation system using enzymatic spectrophotometric and direct peroxide detectors. Enzymatic spectrophotometric monitoring was accomplished by using a semi-automatic configuration including an Mk-II AutoAnalyzer from Technicon. Discrete samples from the fermenter were collected into sample cups and sampling was done from the same cup within a few minutes. The sample line was washed for one minute with distilled water between successive samples. When peroxide probes were used, the peroxide produced by the immobilized enzyme was measured directly with a probe which was placed

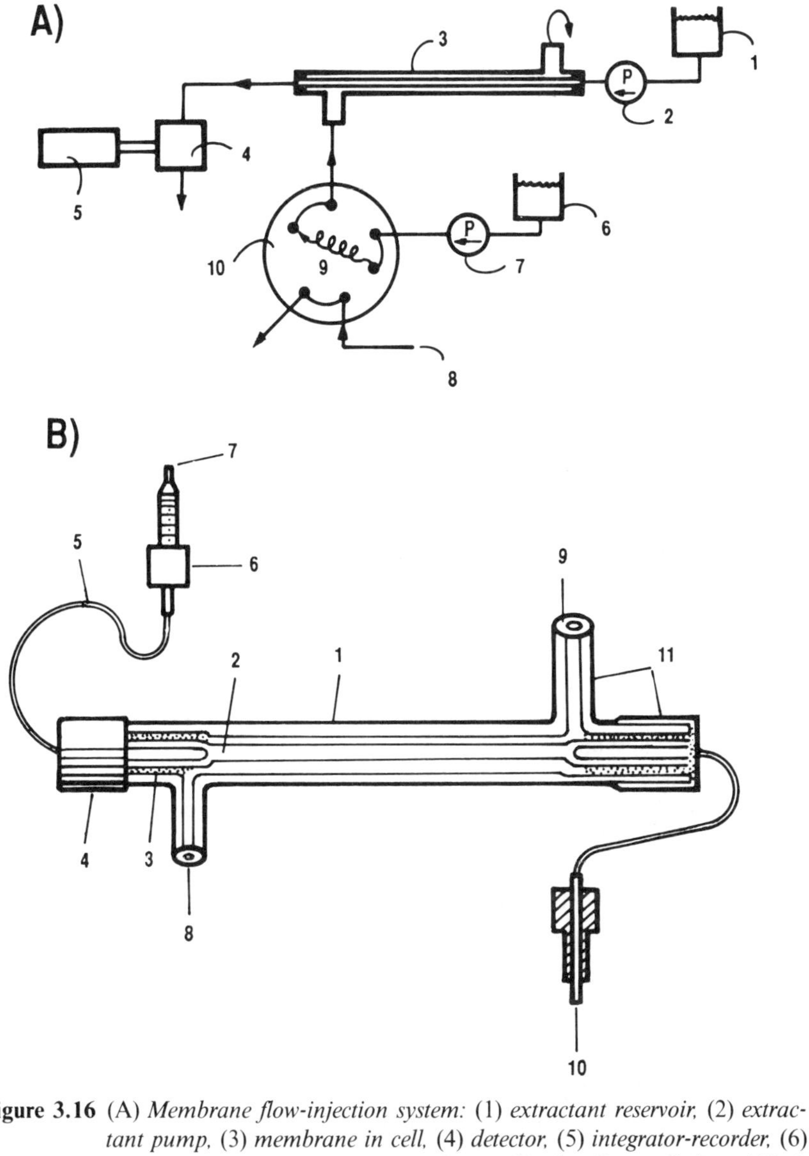

Figure 3.16 (A) *Membrane flow-injection system:* (1) *extractant reservoir,* (2) *extractant pump,* (3) *membrane in cell,* (4) *detector,* (5) *integrator-recorder,* (6) *carrier reservoir,* (7) *carrier pump,* (8) *sample inlet,* (9) *sample loop,* (10) *6-port injection valve.* (B) *Membrane cell assembly:* (1) *cell body,* (2) *membrane connected to Teflon tubing,* (3) *silicone sealant,* (4) *plastic column end-cap,* (5) *1/32-in Teflon tubing,* (6) *1/16-in liquid chromatographic fitting,* (7) *outlet to detector,* (8) *carrier/sample outlet,* (9) *carrier/sample inlet,* (10) *extractant inlet,* (11) *connection side-arms (2 mm ID)*
(Reproduced from reference 48 with permission of Elsevier Science Publishers)

downstream of the coils in the AutoAnalyzer system. The results obtained showed the dialysis probe to reduce the risk of contamination of the fermentation broth through the sample line and to largely eliminate the need to dilute samples for analysis. Some sensitivity was lost at the lower solute concentrations; in fact, only ethanol concentrations higher than 1.0 g l^{-1} could be detected with the probe. However, up to 50 g l^{-1} ethanol could be measured with no sample dilution. The results obtained by using the dialysis probe in the semi-automatic system were compared with those yielded by samples withdrawn directly from the fermenter.

Kroner and Kula[50] used two types of device for on-line measurement of extra-cellular enzymes during fermentation by using membrane separations as the main separation unit, both to achieve continuous sampling from the bioreactor prior to analysis and to measure hydrolytic activities by separating the cleavage products during the reaction of enzymes with high-molecular weight substrates by means of an ultrafiltration/dialysis membrane (Figure 3.17). The continuous sampling device used, which included dynamic filtration/dialysis, consisted of a commercially available cell that was connected to the bioreactor via a by-pass (Figure 3.17A). The broth was continuously propelled through a filter by means of a peristaltic pump at a moderate feeding rate (0.5 ml min^{-1}) and through a microporous or ultrafiltration membrane that produced a continuous filtrate/dialysate stream (0.2–1.0 ml min^{-1}) that could be connected directly to the flow injection system of a Technicon AutoAnalyzer. Membranes of different porosity as required by the barrier membrane process involved were used. A second method applied was a kinetic assay based on the reaction of the enzymes with their natural high-molecular weight substrates. An ultrafiltration/dialysis membrane was used to split the reaction volume continuously into two streams that differed in the molecular size of their solutes, which in turn depended on the membrane cut-off. The changes in the solute concentrations caused by the enzyme activity were monitored (*e.g.* with a spectrophotometer) in the dialysate stream. The general scheme of this procedure is illustrated in Figure 3.17B. In this assembly, the pump (1) is connected to a reaction vessel (5) and the ultrafiltration/dialysis cell (2); (6), (7), and (8) denote devices used to monitor or regulate the flow, pressure, and temperature, respectively. The cell (2) is connected to the detector (3) and the latter is connected in turn to a computer (4) or recorder (5). The process can be implemented either batchwise (in cycles for automated operation) or in a continuous fashion, *i.e.* either the substrate or the enzyme is pumped continuously through the cell and the enzyme or substrate, respectively, is added intermittently to compare measurements with a reference cell, using an enzyme or substrate standard. Alternatively, detection can be done in the retentate stream. Both types of configuration were tested by coupling them to a 70 or 30 l Chemap bioreactor during production of alkaline protease from *Bacillus licheniformis*. With the continuous sampling device, the time course of the protein production was followed by taking samples from the

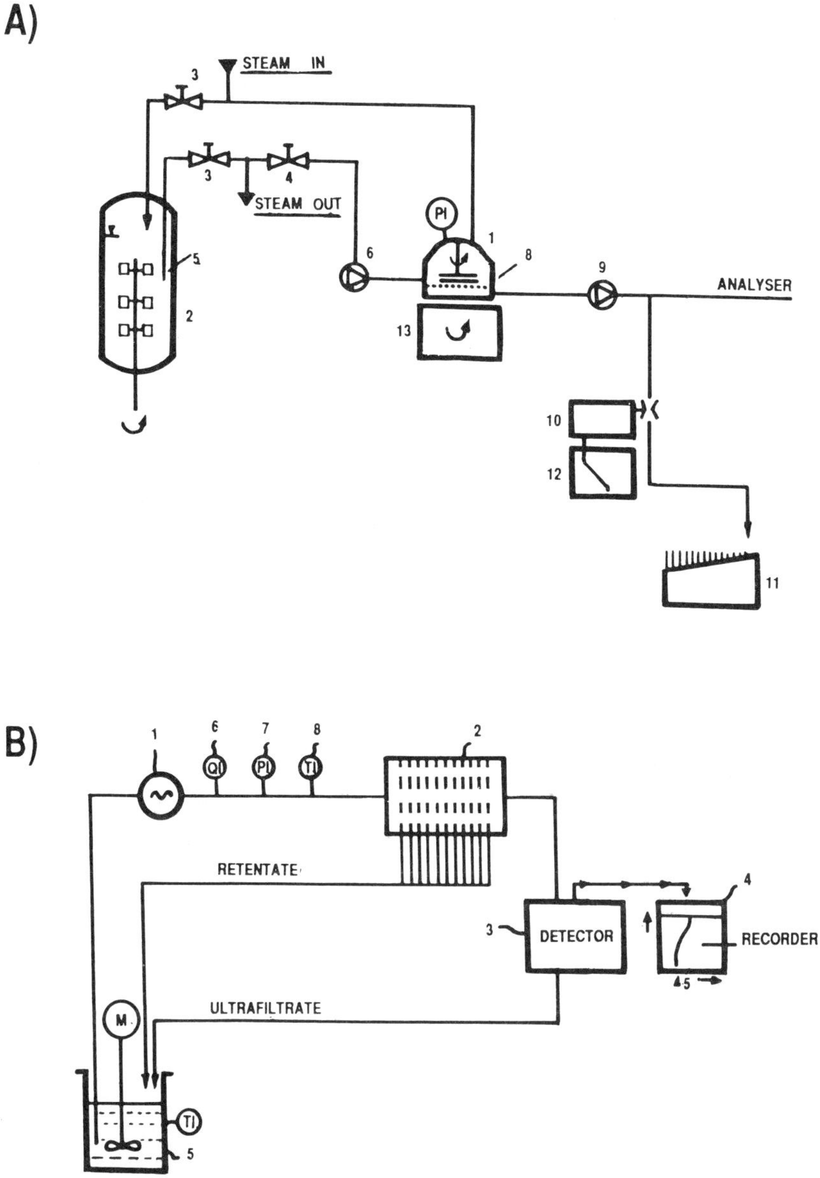

Figure 3.17 (A) *Scheme of continuous sampling system:* (1) *stirred cell,* (2) *fermenter,* (3, 4) *membrane valves,* (5) *sampling tube,* (6) *recirculating pump,* (7) *pressure gauge,* (8) *membrane,* (9) *filtrate pump,* (10) *detector (optional),* (11) *fraction collector,* (12) *recorder (optional).* (B) *General scheme of kinetic membrane analysis. (For further details, see text)*
(Reproduced from reference 50 with permission of Elsevier Science Publishers)

bioreactor and the fraction collector connected to the sampling device. The results obtained were consistent with those provided by the standard method and no infection of the bioreactor was detected over the entire cultivation period. The two systems were found to work smoothly and be subjected to no interferences for periods of at least 48 h. Membrane analysis operating in the flow-injection mode is easier to achieve and more economic than the standard AutoAnalyzer technique.

The commercially available dialysis cell depicted in Figure 3.7A was coupled on-line to a simple FIA configuration for monitoring enzyme inducer L-phenylalanine. The flow diagrams of the fermentation process and FIA system are shown in Figure 3.18A; also, a detailed view of the FIA manifold is shown in Figure 3.18B. This last is a two-channel system into which 2 μl volumes are injected at a rate of 90–100 samples h^{-1}. The linear response thus obtained ranges between 0.1–1.12 mM, and the r.s.d. is less than 1%. The combination of the FIA manifold with the BIOPEM™ sampling device decreases the L-phenylalanine concentration that can be monitored on-line over the required period (87 h). It is worth noting that the flow-rates at points X and Y are 50 l h^{-1} and 0.4 ml min^{-1}, respectively.[51]

A segmented-flow/dialysis system using no dialysis probe was used by Ahlmann *et al.*[52] for the continuous on-line monitoring of intracellular enzyme activity required for optimal cultivation of recombinant *E. coli*

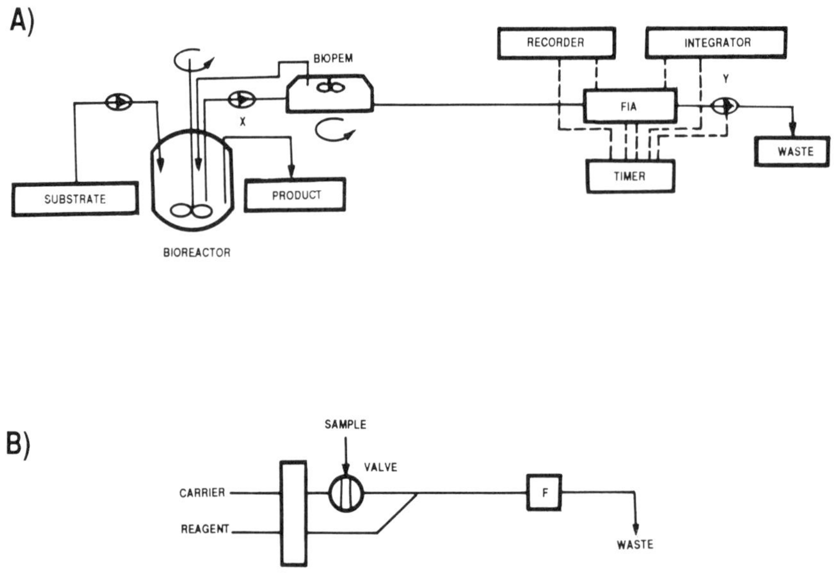

Figure 3.18 (A) *Flow diagram of fermentation and analytical system. Flow-rate: 50 l h^{-1} at point X and 0.4 ml min^{-1} at point Y.* (B) *Scheme of the flow injection system. F: fluorimetric detector*

cells. A continuous sample stream containing whole cells and fermentation media is drawn off from the reactor by a multi-channel peristaltic pump. This sample flow is passed through a glass trap to prevent growth of the bacteria back into the fermenter. A magnetic valve allows a small amount of the sample stream to be taken from the forward-run flow for the analyser during the measurement periods. The reaction coils are followed by a dialysis unit with a Serva membrane. The sample flow travels on a flat spiral path through the lower part of the dialysis cell, while distilled water flows through the upper part. The coloured products of the enzymatic reaction pass into a flow-cell where the absorbance is measured at 410 nm. Usually, air is added to the sample stream at a branching point in the flow lines. This avoids back-mixing and gives visual confirmation of proper sample flow. Clogging can be detected earlier and bursting of tube combinations can thus be avoided. The process includes flushing to prevent cells from settling in the tubes and dialysis unit. Every 90 minutes, 0.1 M NaOH is passed through the system for 30 minutes. Zero shifts can be measured and corrected for in the final data evaluation.

The introduction of reagents into continuous systems is a representative example of non-separative uses of dialysis membranes; this has been accomplished by using two types of configurations, namely: (*a*) those using annular membranes developed by Dasgupta *et al.* to insert the re-agent into the main channel, through which the analyte is circulated and where the derivatizing reagent takes place prior to reaching the detection system; (*b*) those using flat membranes within the detection cell (microporous membrane flow-cell, developed by Nieman *et al.*).

(*a*) In a preliminary study preceding the development of determinative methods,[53] Dasgupta dealt with the behaviour of annular membranes, both linearly and helically arranged. Although Dasgupta's work was focused on the efficiency of mass transfer of different species in terms of the annular gap and helix diameter in helices of nylon monofilament filled Nafion perfluorosulphonate membrane tubing, the results obtained apply to other systems. Thus, in designing dialysers with porous membranes, the density of pore distribution is analogous to the exchange site density of anionomeric membranes, while the selectivity coefficient will be controlled by the pore size distribution and the size of the molecule to be dialysed.

A practical example of application of Dasgupta's work lies with the fluorimetric methods reported for the determination of peroxides[54] and sulphite, sulphide and methanethiol,[55] which use two types of membranes serially arranged within an FIA system that give rise to dialysis and gas diffusion processes (PTFE membrane or TFE filament fillet porous polysulphone membrane and Nafion 811x perfluorinated membrane, respectively). The FIA configurations used were quite similar and are depicted schematically in Figure 3.19. Peroxides (H_2O_2 and CH_3HO_2) were determined on the basis of the *p*-hydroxyphenylacetate/peroxide/

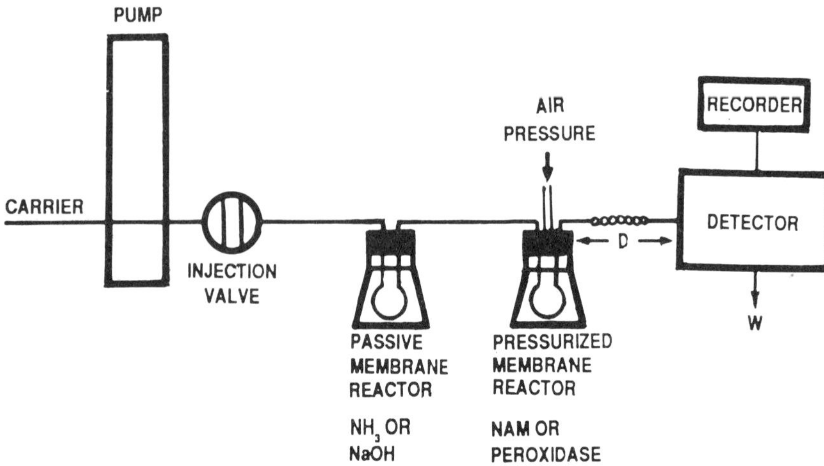

Figure 3.19 *Membrane-based FIA system for the determination of peroxides or sulphur derivatives*
(Reproduced from references 54 and 55 with permission of the American Chemical Society)

peroxidase reaction at a rate of 50 samples h^{-1}. The pressurized PTFE membrane reactor introduced the enzyme and the pH of the flowing stream was altered by introducing ammonia through the non-porous cation exchange membrane reactor. A separation unit consisting of a reactor packed with MnO_2 grains and located prior to the injection system functioned to separate residual peroxide from the reagent water.[56] The reproducibility and sensitivity thus achieved in the overall determination of the two peroxides with the configuration depicted in Figure 3.19 were excellent; on the other hand, the differential determination of the two analytes called for the use of a nested loop system.[56,57] An overall and differential determination of sulphite, sulphide, and methanethiol based on the use of the same FIA configuration relies on the reaction between the analytes and *N*-acridinylmaleimide to yield a fluorescent product. In both cases, the reagents are continuously introduced across the pressurized membrane. More recently, the same principle was used to introduce the reagent(s) by pulses into the continuous system.[58] The experimental set-up used for this purpose is shown schematically in Figure 3.8B.

As a rule, the introduction of reagents across this type of membrane results in more complicated flow injection designs and in major irreproducibilities which are possibly not offset by the smaller dilution undergone by the sample. With catalysts (*e.g.* enzymes) which can be immobilized on solid supports, dilution can be avoided by using an enzyme reactor, with the additional advantage that analytical costs are reduced as the catalyst is not consumed in the process.

(*b*) More is gained by placing the microporous membrane within the cell of a chemiluminescence detector as a way of integrating reaction and detection, which is of great relevance to this type of system. Work in this field conducted by Nieman *et al.* since 1979[59,60] shows the potential interest of this approach, an immediate application of which is the determination of glucose in serum by a reaction catalysed by glucose oxidase that yields hydrogen peroxide this is then measured upon reaction with luminol in the presence of $[Cu(1,10\text{-phen})_3]^{2+}$ as a catalyst for the derivatizing reaction. The optimal working conditions for the analytical and derivatizing reaction are rather different (pH 5 and 11, respectively), so the membrane also allows passage of the protons required for the enzymatic reaction to develop. The configuration used for this purpose is quite simple. It consists of a single buffer line into which the sample containing luminol and the copper complex in a strong basic medium (0.5 M KOH) is injected. An additional channel drives the enzyme in potassium hydrogen phthalate buffer of pH 5 to the detector compartment (capacity, 1 ml), which is in contact with the pressurized membrane. The Celgard 5511 membrane used is made of polypropylene coated with a non-ionic surfactant to ensure wettability; it has a pore size of 0.04 m and an effective molecular weight cut-off of 200 000. The reagent reservoir is pressurized to about 6 p.s.i. by means of a regulated tank of compressed air. The flow-rate of the enzyme solution through the membrane is typically 3.5 μl min^{-1} and depends on the applied pressure. The membrane area is 0.92 cm^2. The reaction between the analyte and the reagent takes place on the analyte side of the membrane close to the membrane surface.[61] Since the reagent flow-rate across the membrane is only of a few μl min^{-1} and the analyte flow-rate is of about 10 ml min^{-1}, there are effectively three orders of magnitude dilution of the reagent into the analyte flow stream. This approach offers several analytical advantages for application in continuous flow systems, namely: (*a*) the sample integrity is preserved and contamination of the sample by the reagent is minimal; (*b*) the reagent volume used is only a few μl per assay, which is of utmost importance when using enzyme solutions (the amount of enzyme required for each analysis is 25 times smaller than with earlier methods); (*c*) light is emitted in the vicinity of the microporous membrane;[60] (*d*) a pH gradient perpendicular to the membrane can be readily obtained in order that reactions requiring different pH conditions may take place simultaneously in the same flow-cell; (*e*) the reagent and the analyte are mixed quite rapidly under controlled conditions without the need to stir the solution; and (*f*) the instrumentation required is fairly simple and inexpensive.

3.2.5 In-line Dialysis

Intracerebral dialysis, customarily used for the analysis of extracellular fluid[62,63] since its introduction by Tossman *et al.* in 1981[64] overcomes many of the problems posed by push-pull perfusion. This type of in-line dialysis, more commonly referred to as 'microdialysis', allows chemical substances to be introduced into the human body without separating the fluid.

The principle behind microdialysis is an artificial blood vessel surgically inserted into the tissue. Chemical substances will diffuse in the direction of lowest concentration. In this way, substances can be recovered from an organ or added to it depending on their relative concentration in the perfusing fluid. There will be two-directional molecular and ionic traffic between the inside of the microdialysis probe and the surrounding tissue. This provided the unique possibility of performing an entire pharmacological experiment within less than 1 mm^3 of tissue. Physiological processes can be closely monitored in animals, whether anaesthetized or awake.

The analytical chemistry required to determine substances in the perfusate posed rather a serious problem in the early years of microdialysis. However, the rapid development of liquid chromatographic techniques has facilitated the determination of a number of physiologically significant substances in the perfusate. Microdialysis can be regarded as a biosensor system, with the sole difference that substances are determined outside rather than inside the body by means of an inserted sensor. The number of substances that can be assayed in this way is much larger than one would expect from a biosensor, at least so far. In addition, microdialysis is applicable to experiments of different complexity. Different substances can be introduced into a tissue in order to interfere with its normal physiology and reveal non-apparent changes in basal level release. Total metabolic experiments can be carried out by radioisotope techniques in conjunction with conventional analytical chemistry. Also, precursors can be introduced into a tissue and the products of their enzymatic conversions be collected in the perfusate.

One major feature of microdialysis is the possibility of comparing *in vivo* and *in vitro* experiments. The latter can be performed on the solute held in an ordinary vessel. Recovery of individual substances can be studied by determining the relative concentration in the perfusate as compared to that in the outside medium. The concentration will depend on the membrane properties (molecular cut-off and thickness), the perfusion rate, and the initial concentration of the compound in the perfusate. By using an adequate membrane thickness and a sufficiently slow perfusion rate, recoveries of up to 100% can be achieved and the concentration of several components of the extracellular space be reliably determined.

As stated above, *in vivo* and *in vitro* conditions can be compared by employing appropriate membranes, use of which, however, poses some potential problems. Thus, a small pore size will exclude some substances, which can be an advantage when enzymes potentially causing the species of interest to break down are to be excluded. The membrane should act as a hurdle to diffusion and slow down substance recovery. Studies made on extracellular brain fluid showed the rate of transfer across the contorted space of the brain to be lower than through most membranes, so this is clearly not the limiting factor. On the other hand, the binding of some compounds to the membrane can indeed pose major problems. In most cases, this pitfall can be circumvented by allowing the system to equilibrate (*i.e.* non-specific binding sites to become saturated).

3.2.5.1 Features of Microdialysis

Microdialysis is closely related to the push-pull technique. The probes used are quite similar except for the membrane and their features determine whether or not they can be used for a particular purpose. This technique offers a number of major advantages, namely:

(*a*) The perfusion system used is quite straightforward. The presence of the membrane enables the use of a syringe pump to push the fluid through the probe. On the other hand, the absence of a pull facilitates collection of samples in the vicinity of the probe.

(*b*) Tissue irritation is minimized thanks to the absence of a circulating fluid in direct contact with the tissue. The risk of tissue being unintentionally aspirated into the probe is virtually nil. Sterile conditions are quite easy to establish and maintain even if the perfusion medium used is not sterile as no micro-organisms ever penetrate the membrane. This avoids cumbersome preparation of sterile drug solutions.

(*c*) Microdialysis offers the unique opportunity to control the target area by using membranes of different lengths and diameters.

(*d*) The possibility of comparing *in vitro* and *in vivo* microdialysis perfusions allows concentrations to be extrapolated to those probably occurring in the extracellular space.

(*e*) The fact that substances penetrate the organism by diffusion allows them to be infused into such a closed space as the brain for long periods of time with no change of volume or pressure changes. This is of use in acute experimental or chronic situations, where the microdialysis probe can be connected to a subcutaneous minipump.

(*f*) Interferences are isolated immediately upon entering the perfusion medium, which makes automated sample handling methods easier to implement. An additional function of the membrane is reducing alteration of the sample by enzymes. As molecules with molecular weights above 5000 are excluded by the membrane, the sample will be accurately representative of the extracellular composition at the time the material crosses the membrane. In relation to the extracellular composition, insofar as only those molecules with a molecular weight smaller than a preset value dictated by the membrane features are separated from the extracellular fluid, the integrity of the extracellular matrix is better preserved than with unrestricted removal. Hence the membrane improves the quality of both the remaining extracellular fluid and the sample. Membranes developed in the future may be more selective to given types of molecules.

3.2.5.2 Components of an In-line Dialysis System

The basic set-up for implementation of in-line microdialysis consists of a micro-infusion pump, a stereotactic instrument, a probe holder, the probe, and a stand for small test tubes where the perfusate is collected in batchwise analyses, or a connector to the injection unit of a continuous configuration. Despite their simplicity, they are always in great demand.

Microdialysis probes can be designed in a number of ways. Thus, bundles of fibres can be implanted across the brain, entering a hole in the skull on one side, crossing the brain and leaving it through a hole on the other side. Alternatively, a loop can be implanted from the dorsal surface of the brain. Figure 3.20A depicts a commercially available probe consisting of two concentric steel cannulae whose ends are covered with a dialysis membrane. The physiological fluid is introduced through the inner cannula, flows across the membrane and leaves by way of the outer cannula. This probe can be used to recover any substances of low molecular weight present in the extracellular space. Molecular cut-off ranges between 5000 and 50 000, depending on the membrane used. The probe is manufactured in different sizes, the length and diameter of the steel shaft being adapted to the animal species and structure to be tested. The probe can be sterilized and the membrane that separates the tissue from the perfusing fluid is a further guarantee for the maintenance of sterile conditions.

Figures 3.20B and C show two laboratory-made microdialysis probes. The former probe was designed by Justice Jr. *et al.*[65] and consists of two lengths of fused silica (0.150 m × 0.025 mm ID) which are inserted into a dialysis tube (7.0 mm long, 320 μm OD) of Cuprophan hollow fibres that are sealed at one end with polyamide sealing resin. The distance between the end of the fused silica tubes is 4 mm and they make the active length of the dialysis probe. The membrane is sealed at the top with polyamide resin and the probe is inserted into a threaded cannula base and secured with epoxy.

Figure 3.20C shows a loop design. The dialysis tube (150 μm ID, 9 μm wall thickness) is made of regenerated cellulose and features a molecular weight cut-off of 9000. The ends of a short piece of dialysis tubing are inserted into 26-gauge stainless steel cannulas and sealed with hot melt glue. The probe size can be varied by using different lengths of dialysis tubing, but is usually 0.5 cm.

Microdialysis pumps are typically precision syringes capable of providing pulse-free flows between 0.1 and 20 μl min^{-1}. The flow should lend itself to immediate starting and stopping.

A liquid switch is of great significance as many microdialysis experiments involve perfusion fluid changeovers (*e.g.* from artificial spinal brain fluid to a solution containing some drug or isotope). Such changeovers cannot be accomplished by moving the tube from one syringe to the other as this often results in the aspiration of air into the system and the

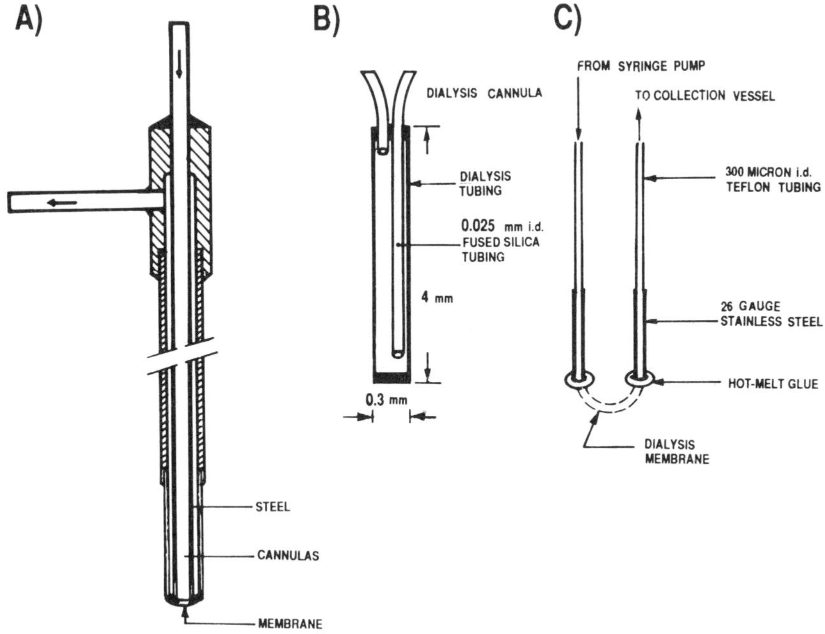

Figure 3.20 *Different types of in-line dialysis probes.* (A) *Commercially available,* (B, C) *laboratory-built. (For details, see text)*

consequent perturbation of the liquid flow and settling of bubbles on the membrane, the dialysis surface of which is thus reduced.

The CAM/200 is the only commercially available refrigerated *fraction collector* specially designed for microdialysis. Thanks to its small size, it can be positioned close to the animal, thereby minimizing the length of the connecting tube required. It is refrigerated to prevent breakdown of the chemical compounds recovered by the microdialysis probe and is able to collect fractions smaller than a falling droplet. This is essential because a microdialysis fraction may be only a few microlitres. The fractions can be collected in capped vials as the cannula will penetrate the membrane of the cap. This unit is also a microsampler that can be converted to an autoinjector for liquid chromatography by simply fitting a CMA/240 sample injector. This system handles extremely small volumes and the samples of the fraction collector can be directly transferred from the fraction collector to the injector system by simply lifting the carousel from one instrument to the other.

3.2.5.3 Applications

Most of the applications of microdialysis are related to the sampling of extracellular fluids, particularly from the brain; however, the operational principle is applicable to many other organs and even plants. Microdialysis units are most often coupled on-line to a liquid chromatograph for the subsequent determination of the sample obtained, which is normally still complex enough to call for a determinative technique of higher resolution. The small volume of liquid usually drawn makes microbore liquid chromatography (MBLC) rather suitable for this purpose.

Justice Jr. *et al.* developed a number of applications of microdialysis coupled on-line to MBLC for the determination of various species (generally drugs) in extracellular brain fluids. Thus, these authors carried out comprehensive work on the kinetics of dopamine metabolism.[65] They used a probe like that depicted in Figure 30.20B, constructed from 300 μm OD hollow fibre dialysis tubing with a molecular cut-off of 5000. The components of the system used for sampling and the determination are depicted in Figure 3.21A. There are two streams that converge on the injection valve of the HPLC instrument. The system uses a Rheodyne 7413 injection valve that enables perfusion flow in the sampling lines in its injection position and hence avoids increased pressure in the brain. The other components of the perfusion system are an infusion pump altered for simultaneous push-pull perfusing two 2.5 ml gas-tight syringes and 25 μm ID fused silica tubing between the injection valve, the dialysis cannula, and the push-pull syringes. The perfusion flow-rate used is 0.4 μl min^{-1} and 5 μl volumes are injected at 15 minute intervals. For work on behaving animals, a six-pin female connector including a 20-gauge, 15.3 mm long stainless steel cannula is implanted in the brain and extends just below the dura. Before the dialysis cannula is placed in the brain, the animal is slightly anaesthetized with ether. The cannula is then inserted and locked in place by the six-pin connector. The perfusion pump is started and sampling begun. After the experiment, the cannula is removed and the animal is returned to its home cage. The dialysed sampling system has been characterized in a series of experiments. One set examined the effect of the flow-rate and variations in the cannula design and pumping technique on recovery. The presence of cerebrospinal fluid and the occurrence of diffusional barriers were also examined. Recovery increases as the flow-rate decreases, while the amount of sample obtained per minute goes through a maximum. The maximum amount of material per minute is recovered in the range 2–4 μl min^{-1}. However, from an analytical point of view, the critical parameter is the amount detectable rather than the amount recovered. Thus, the use of MBLC results in improved detectability even though less material may be used in the analysis.[66] The chromatographic column to be used depends on the demands posed by the experiment. For monitoring metabolites, a 25 cm × 4.6 mm reverse-phase C_{18} column is quite satisfactory. However, for short sampling intervals, the column should be one enabling relatively rapid analyses. For monitoring dopamine itself, 10 cm × 1 mm ID microbore columns yielded better results through improved detection limits. Effective application of voltammetry *in vivo* requires identification of the neurochemicals responsible for the oxidation currents detected for the different stimuli, which is critical in chronoamperometry. Studies in this context pro-

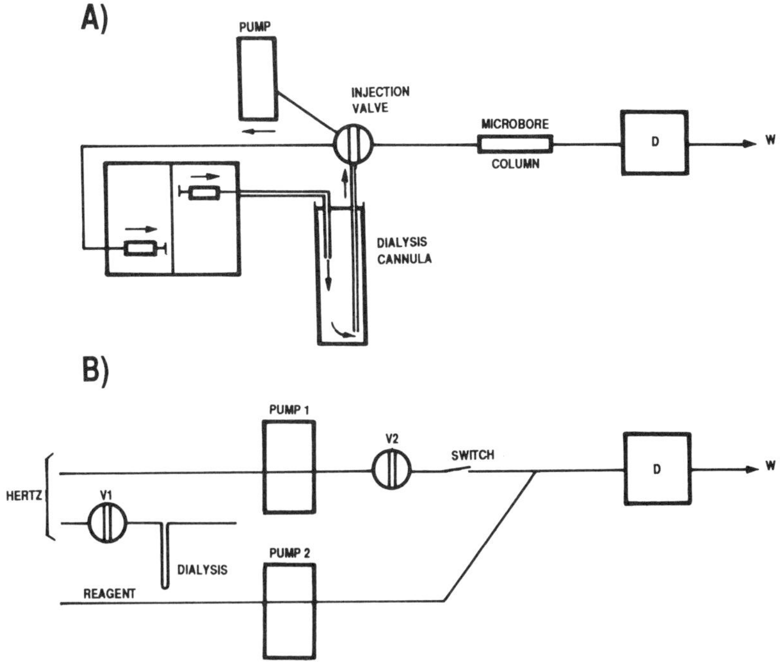

Figure 3.21 (A) *In-line coupling of dialysis and HPLC. The microbore column is fitted directly into the injection valve and the stainless steel block of the amperometric detector. The distance from the dialysis cannula to the injection valve is 40 cm. A Rheodyne 7413 injection valve is used to maintain a constant dialysis flow while the valve is in the inject position.* (B) *In-line coupling of dialysis/FIA with enzymatic fluorescent detection.* V_1 *and* V_2, *valves; D, detector; W, waste*
(Adapted from reference 65(A) and 69(B) with permission of Elsevier Science Publishers)

vided interesting conclusions; the sampling method used was particularly useful to clarify the chemical source of the voltammetric signals detected *in vivo*. The final results show voltammetry to be extremely useful for studying metabolic kinetics on account of the high sampling rate it affords. The same instrumental set-up was also used by these authors to characterize neurotransmitter release in deep structures of the brain of freely moving animals.[66] The inner volume of the system was somewhat smaller, so the perfusion rates were also lower. Work in this respect included the determination of compounds occurring at concentrations well above the detection limit of the system (metabolites), as well as compounds at relatively low extracellular concentrations (neurotransmitters), which required the detection limit to be lowered even further. Four categories of analyses can be established according to whether they involve high or low concentrations and/or sampling rates. All four were studied by monitoring the dopamine neurotransmitter and its metabolites. The work was subsequently extended by using smaller injected volumes (0.5 μl) and evaluating the performance of the system, both *in vitro* and *in vivo*.[67]

The microdialysis probe shown in Figure 3.20C was implanted in the liver of rats to study the hepatic metabolism of phenol and conjugation by glutathione. The results were compared with those obtained *in vitro* using liver homogenate and liver microsomal protein.[68] In this instance, the probe was not coupled to a liquid chromatograph with electrochemical detection, but the samples were collected in 1.5 ml centrifuge tubes and injected manually into the chromatograph. This work is proof of the great potential of this technique for studying hepatic metabolism. Insofar as the integrity of the biological system is preserved and endogenous levels of cofactors are used, these experiments accurately reflect actual metabolic reactions from exposure to xenobiotics.

Flow injection analysis has also been used in conjunction with this sampling technique for the enzymatic determination of lactic acid in extracellular brain fluid, thereby overcoming the most serious problem posed by the microdialysis/HPLC association, namely the infeasibility of near-real time measurements, which are of great significance in many cases. The scheme of this combined set-up is shown in Figure 3.21B. The perfusion dialysis system consists of two matched flow lines, one of which is used to perfuse the dialysis cannula. This line includes an injector with a 10 μl loop for direct introduction of drug solutions used to investigate the local action of pharmacological agents. The other line is used to obtain the blank signal or inject lactate standards via an injector with a 100 μl loop. Either line can be connected to the detection system, so both the blank and the standard solutions can be rapidly assayed. Experiments were carried out on anaesthetized rats placed in the stereotactic apparatus. U-shaped cellulose tubes were implanted bilaterally in the rat stratium. The stratium of freely-moving conscious animals was perfused with a Hertz medium equilibrated with 95% O_2 and 5% CO_2 by changing the dialysis rate between 0.8 and 20 μl min^{-1} to evaluate the dependency of the lactic acid concentration in the dialysate on the flow-rate. One or two days after surgery, the rats were shocked with an Elther Electroshock apparatus and perfused continuously before, during, and after the shock was applied. The rapid response of the FIA/microdialysis system (40 s) enabled near-real measurements, so transient changes in the lactate level upon electrical stimulation were readily and precisely monitored.[69]

Microdialysis has been used in conjunction with thermospray tandem mass spectrometry to follow the time course of a dopamine uptake blocker (GBR-12909) *in vivo*. The detection limit achieved was 100 pg. The GBR-12909 concentration in the extracellular fluid was found to be much lower than that of cocaine for equivalent intraperitoneal doses, which accounts for the divergence between the relative *in vivo* and *in vitro* power of the two drugs.[70]

3.3 Liquid–Liquid Extraction

Mass transfer between two liquid phases previously brought into contact by a variety of procedures is one of the earliest analytical methods developed and has experienced a revival in the last few decades thanks to its inherent major advantages. In fact, liquid–liquid extraction as a separation step is rather commonly applied in analytical processes. As a rule, this technique is rather an appropriate nexus between the real sample and the analytical determinative technique to be applied as it serves or aids in major functions such as matrix effect elimination, preconcentration, and solvent changeover, facilitates implementation of the determinative technique or preserves the integrity of the analytical system.

With the exception of a few special modes, most alternatives to liquid–liquid extraction with analytical purposes are reflected schematically in Figure 3.22. In fact, there are five general modes, which are classified in the Figure according to virtually coincidental criteria, namely:

(1) Liquid–liquid extraction as performed in separatory funnels that are shaken manually or mechanically, where a single partitioning equilibrium is reached in each operation.[71]

(2) The use of a distillation/condensation system which allows the organic phase to circulate through the aqueous phase by dripping. The configuration to be used will depend on the density ratio between the two phases. In any case, it is quite an impractical alternative.[71]

(3) The use of a system made up of hundreds of tubes where equilibrations take place simultaneously; the tubes contain a fixed volume of one phase and allow circulation of volumes of the other, lighter phase in a complex assembly called 'countercurrent extractor' which was originally designed by Craig[72] and is of greater pedagogical than practical significance.

(4) The establishment of a flowing stream consisting of segments of the two immiscible phases; mass transfer takes place through the multiple interfaces established, as well as through the film formed inside the tube.[73, 74]

(5) The use of a liquid–liquid chromatographic partitioning system in which the liquid (stationary) phase is bound (by adsorption or chemical bonding) to a solid support through which an immiscible (mobile) phase transporting the solute is circulated. Equilibrium is reached many times (10^3–10^6) in a zone called 'theoretical plate'.[75] Countercurrent chromatography, a newly developed mode, resembles closely liquid–liquid extraction because of the use of no solid support.[76, 77]

From an *operational point of view*, liquid–liquid extraction modes can be classified into two broad groups, namely: (*a*) batch methodologies, which are implemented in separatory funnels and where phases are

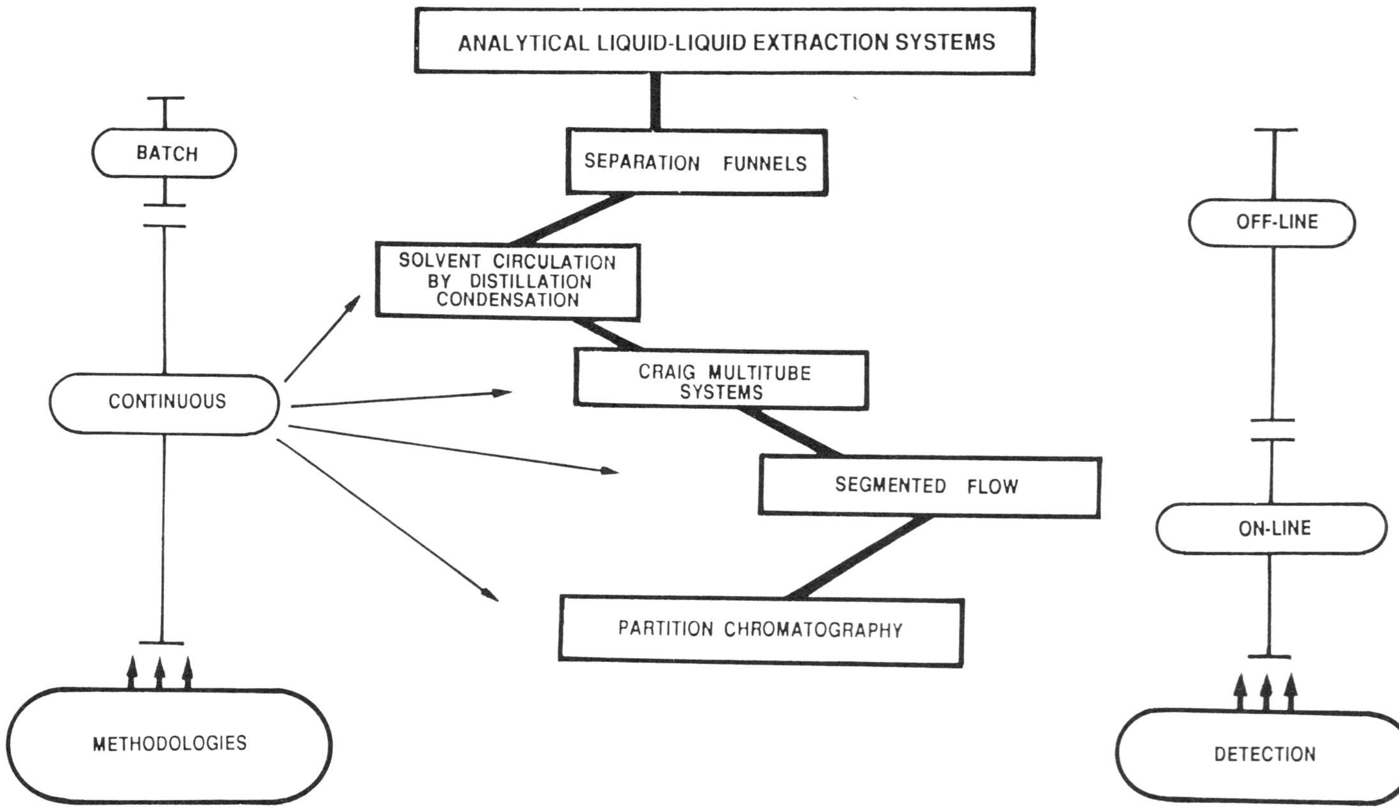

Figure 3.22 *Classification of analytical liquid–liquid extraction methodologies in increasing order of dynamism and efficiency. (For details, see text)*

brought into contact only once; and (*b*) continuous methodologies, where one phase (distillation/condensation, Craig countercurrent and partition chromatography) or both (segmented-flow systems) are kept in continuous motion. Emulsions are only formed and broken down in separatory funnels and Craig multitube systems.

The significance of *chemical and physico-chemical kinetic aspects* increases in descending order in the scheme in Figure 3.22. In fact, with separatory funnels, shaking for an optimal time (0.5–5 min) is usually sufficient to ensure equilibrium is reached, as is shaking Craig tubes following each transfer. The lack of such broad interfaces in segmented-flow systems resulting from the formation of emulsions makes the time a critical variable insofar as, in general, the extraction equilibrium is never completely reached. In partition chromatography the flow-rate of the stationary phase is a key to the equivalent height of a theoretical plate, as are other physical kinetic aspects related to diffusion.

Detection in these extraction systems is usually carried out off-line in the less dynamic modes, *i.e.* those based on the use of separatory funnels, solvent circulation by distillation–condensation, and Craig multitube systems. In the more dynamic modes, detection is done on-line by using a generally non-destructive detector that continuously monitors the flow of one of the phases that cross it by means of a suitable flow-cell.

The *separation efficiency* also increases with the dynamic nature of the liquid–liquid extraction mode and is thus maximal for chromatographic methodologies. Obviously, multi-determinations will be more efficiently accomplished with Craig's system—now outdated—than with segmented-flow systems, despite their position in Figure 3.22.

The *reduction* of human involvement in the analytical process, *i.e.* automation, also increases with the dynamic nature and efficiency of the liquid–liquid extraction mode, which is consistent with the way in which analytical signals are measured (off or on-line). Analytical processes that involve mass transfer between two immiscible liquids can be automated in a number of ways encompassing a variety of elements from manually shaken separatory funnels to liquid chromatographs with automatic injectors.

The second part of this chapter deals with analytical systems based on continuous liquid–liquid extraction as accomplished by establishing a stream of alternate segments of the two immiscible phases. Their implementation rests on the principles of air-segmented[78] or unsegmented analysers,[79, 80] which are behind the use of the so-called 'liquid–liquid extractor', the elements and operational foundation of which are described first. Then, the different types of configurations and chemical reactions involved are discussed, with special emphasis on their ability to solve real problems. Finally, some special continuous liquid–liquid extraction systems using no phase separator and the combination of a liquid or gas chromatograph, and a continuous extractor are commented on.

The term 'segmented flow' as used to refer to these liquid–liquid extraction systems is strictly appropriate judging by their actual nature and that of the process involved; however, it may be confusing and ambiguous as it has also been used for over 30 years to denote automatic continuous analysers using air bubbles to preserve the integrity of the sequentially introduced samples. In this case, segmentation takes place between a liquid and a gas, whereas in extractors it involves two liquids. In order to avoid confusion with solid–liquid extraction, dealt with in Chapter 4 (On-line Leaching), we shall stick as far as possible to the term 'liquid–liquid extraction' rather than 'extraction' alone, following IUPAC's recommendations.[81]

3.3.1 The Continuous Liquid–Liquid Extractor

A continuous liquid–liquid extractor receives the separate streams of two immiscible phases and gives rise to a single segmented flow within. Mass transfer (usually from the aqueous to the organic phase) then takes place through the multiple interfaces established and a film formed on the tube walls and, finally two separate streams of the two phases emerge from the separator. An on-line detector is used to monitor the contents of one of the phases. The liquid sample is previously introduced into the first miscible phase by aspiration or injection. Derivatization reactions are frequently implemented prior to the detectors in order to facilitate mass transfer or after it to aid detection.

With a few exceptions commented on at the end of this section, the conventional liquid–liquid extractors typically used in continuous segmented or unsegmented analysers consist of three basic components that are connected on-line in the following sequence (Figure 3.23): a solvent segmenter (SS), an extraction coil (EC) and a phase separator (PS). The features and operation of each are described in detail below.

3.3.1.1 Operational Principles

Although each component of a liquid–liquid extractor has its own peculiar design and function, the operation of all three relies on two fundamental principles, namely:

(*a*) The density difference between the immiscible phases. This is common to other liquid–liquid extraction systems depicted in Figure 3.22.

(*b*) The selective wetting of the inner component surfaces by both organic and aqueous phases. As a rule, organic solvents wet Teflon and other plastic surfaces, while aqueous phases wet glass surfaces; conversely, organic solvents are repelled by glass surfaces and aqueous phases are rejected by Teflon surfaces.

The latter principle is of greater significance than the former in describing the operation of a continuous liquid–liquid extractor based on segmentation of immiscible phases.

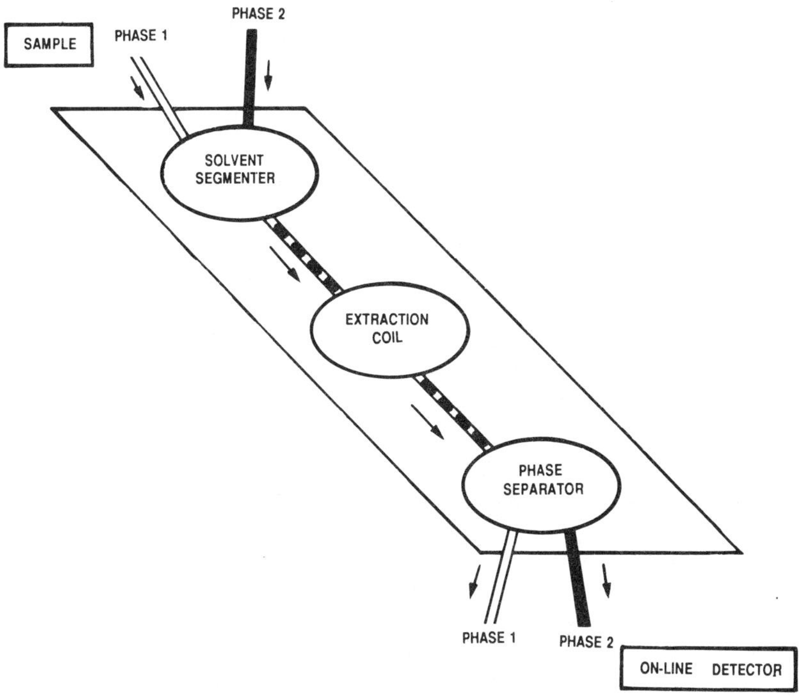

Figure 3.23 *Scheme of the principal components of a continuous liquid–liquid extractor based on phase segmentation*

3.3.1.2 The Solvent Segmenter

This component receives the two streams of the immiscible phases, which converge on a minichamber or confluence point, and establishes an emerging flow consisting of alternate, regular segments of the two phases that is driven to the extraction coil. It should ideally meet two requirements, namely: (*a*) smooth operation in order to avoid irreproducibility in the results and (*b*) allow one to select the length of the segments of the two phases, which can be equal or not.

The literature abounds with reports on a number of segmenter designs; however, current trends point to simplification of this element of the liquid–liquid extractor as far as possible.

The earliest segmenter was a mere modification of connectors A-8 and A-10 from Technicon. It consisted of a tube with two inlets: one in the front in the form of a glass capillary for penetration of the aqueous phase, and the other perpendicular to this in the form of a platinum capillary intended to allow the organic phase in. Two adjustable concentric Teflon tubes shut off the tube at the other opening in the same direction as the aqueous phase; the segmented stream emerged from the central tube. The length of the emerging segments depended on: (*a*) the height of the central Teflon tube, (*b*) the inner volume of the mixing chamber, and (*c*) the ratio between the flow-rates of the two phases (q_o/q_a), which is equivalent to the volume ratio in batch liquid–liquid extraction and determines the preconcentration factor. Most commonly, segments were identical (about 5 mm in length) and the flow-ratio used was roughly unity.[73, 74]

Figure 3.24 shows nine configurations based on three different designs (T, W, and Y) or triple confluence points used to implement the segmentation-mixing process. Most of the segmenters used at present are based on such a simple configuration. The connectors are typically made of Teflon, Daiflon, polypropylene, and Kel-F, so their inner surface repels the aqueous phase and is wet by the organic phase. The organic and aqueous phase streams can merge frontally or laterally at different angles, and the segmented phase can emerge from the central or one of the side openings. Which type of segmenting configuration is used will depend on the density ratio between the two phases. Thus, Motomizu *et al.*[82] studied the influence of the different configurations on the height and width of the FIA peaks obtained in the ion-pair formation reaction between molybdophosphate anion and Malachite Green, which was extracted into benzene/4-methylpentan-2-one. They found a T-segmenter in which the aqueous phase flowed straight through and the organic phase flowed at right angles (type A) to be the most efficient, even though the differences were not too large. Kawase[83] studied the influence of three configurations with capillaries of an inner diameter of 0.8 mm (90–90 T, 30–30 Y, and 45–45 W) and extraction coils of a larger ID (1 mm). All three configurations yielded identical segmentation profiles; however, at smaller diameters (0.5 mm), T and Y segmenters produced larger segments than W segmenters did.

By using a T segmenter (type A) like that shown in Figure 3.24, Cantwell and Sweileh[84] studied the hydrodynamic and interfacial origin of phase segmentation. They developed a semi-quantitative model for the process whereby alternating segments of aqueous and organic phases were obtained upon merging of the two liquids. A growing 'drop' of the organic phase was dislodged to produce a segment when the hydrodynamic force exerted on it by the aqueous phase stream equalled the interfacial force holding it in place. Hydrodynamic forces were expressed by Poiseuille's and Bernouilli's laws, while the interfacial force was expressed in a form of the Tate equation in terms of liquid–liquid interfacial tension and solid–liquid–solid contact angle. Mathematical equations derived by these authors allow one to predict the influence of experimental variables on the volumes involved and hence on the lengths of the organic and aqueous phase segments; they also allow the height and area of the FIA peaks to be predicted.

Recently, Burns *et al.*[85] reported an atypical Teflon segmenter with four openings designed for the determination of cobalt(II) based on the formation of an ion-pair between tetracyanocobaltate(II) and the reagent

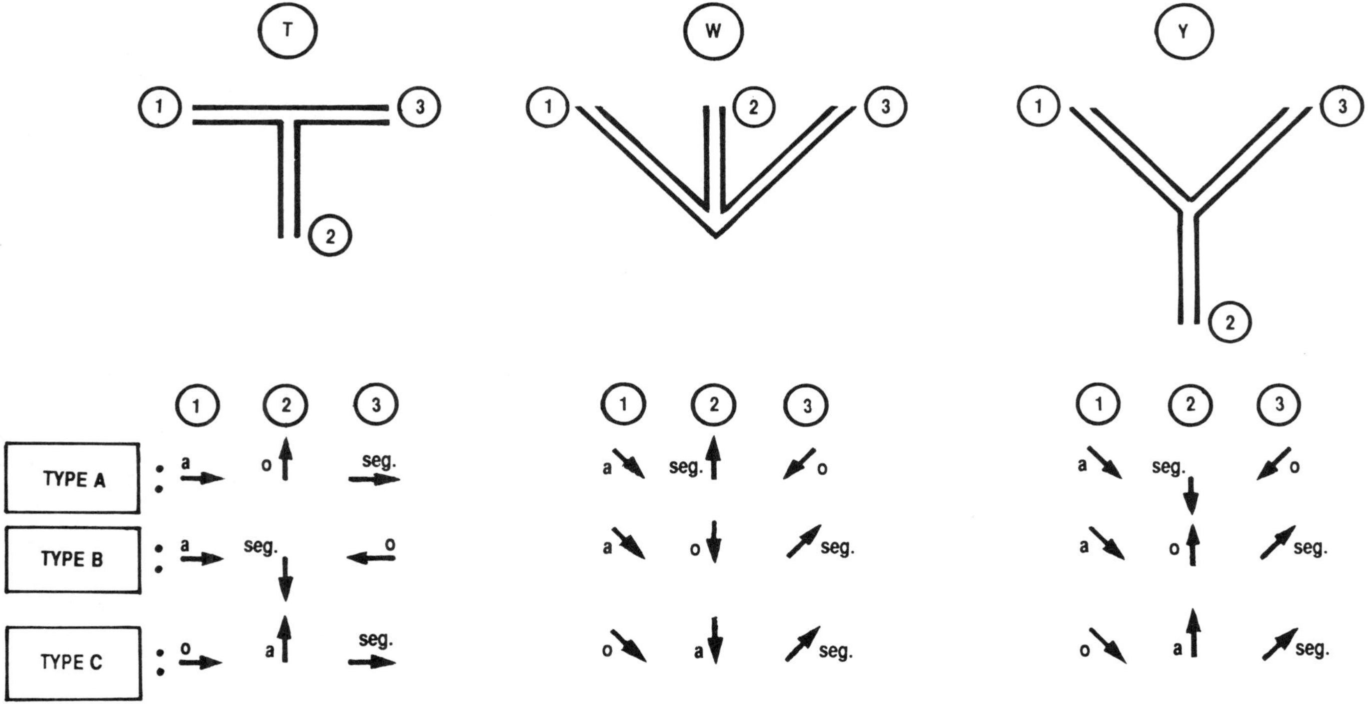

Figure 3.24 *Different configurations* (A, B, and C) *of triple merging points* (T, W, and Y) *used as solvent segmenters. 1, 2, and 3 denote the inlets (two) and outlets (one) for the aqueous phase (a), the organic phase (o), and the segmented flow (seg), respectively*

cation bis(triphenylphosphonium). In addition to an inlet for a chloroform stream, there are another two for aqueous phase streams, one which contains $[Co(SCN)_4]^{2-}$ and the other the counter-cation reagent. From the cylindrical minichamber (10 cm ID, 5 mm height) emerges a segmented stream. A major novelty in this context is the use of a magnetic stirrer (9 mm × 2 mm ID bar) to ensure adequate mixing in transferring the ion-pair formed in the minichamber to the organic phase, thereby avoiding its precipitation in the aqueous phase and the risk of clogging the corresponding tubes.

3.3.1.3 The Extraction Coil

This is usually a helically coiled tube that receives the segmented flow from the solvent segmenter. For a given time determined by its length and the flow-rate, it is crossed by the segmented flow, which undergoes mass transfer between the immiscible phases. Unlike other technical alternatives, the extraction yield ranges between 80–99% in most cases; this calls for strict control of the experimental variables in order to achieve reproducible results. Like other FIA modes, those involved here are fixed-time kinetic variants. Air-segmented continuous systems provide virtually complete extractions thanks to their efficient mixing and longer residence times.

The material from which the tube is constructed markedly influences the profile of the segmented flow that circulates through it because of the different wettability of the tube walls. As can be seen in Figure 3.25, glass coils are wetted by the aqueous phase, while the organic phase is allowed to circulate freely as elongated bubbles. On the other hand, Teflon coils give rise to bubbles of the aqueous phase. Therefore, the phase initially containing the analyte (or its reaction product) should be that circulating as bubbles[86] for a number of reasons, namely: (*a*) carry-over in the aqueous phase between successively introduced samples or standards is reduced; (*b*) the distance to be travelled by the solute from any point in the aqueous phase to the interface is shortened, which decreases diffusion in this phase and accelerates the process; and (*c*) the interface area/solute volume is increased, and so is the rate of mass transfer accordingly. Since the initial phase is usually the aqueous phase, Teflon coils are by far the most commonly used.

The inner walls of glass coils can be made hydrophobic by silanizing them; this involves heating the tube filled with the silanizing agent.[176]

By using an ingenious photographic technique, Nord and Karlberg[87] succeeded in viewing flow patterns, both in the aqueous and in the organic phase in an extraction coil (see Figure 3.26). In addition to gravitational mixing, effective mass transfer was found to result from plug flow in the opposite direction occurring in the segments and also from the secondary flow patterns formed in a coiled tube. Therefore, tubes of small diameters (0.5–1 mm) are recommended.

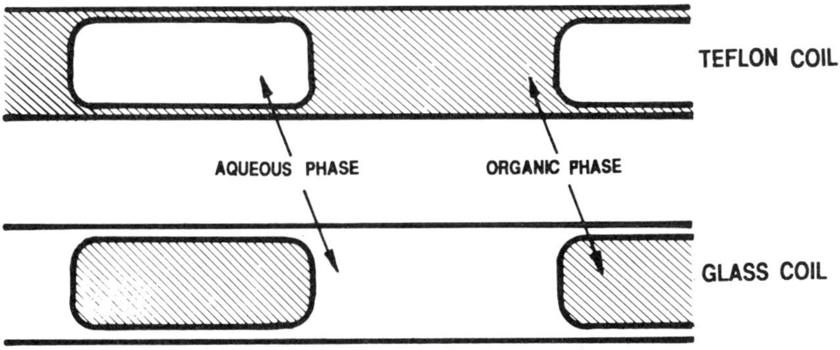

Figure 3.25 *Segmented flow configurations classified according to the material of which the extraction coil is made*

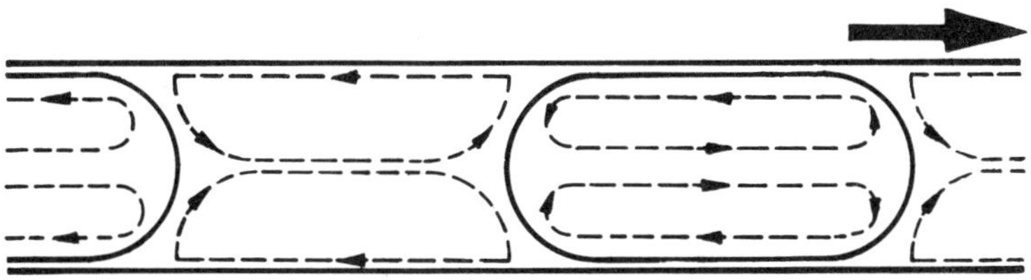

Figure 3.26 *Streamlines in the segmented flow, both in organic and in aqueous phases, that circulates through the extraction coil*
(Reproduced from reference 87 with permission of Elsevier Science Publishers)

The liquid–liquid interfaces depicted in Figure 3.27A cannot fully account theoretically for the high extraction yields provided by these coils. This prompted Nord and Karlberg[87] to carry out an interesting study; with a series of experiments, they provided evidence for the formation of a film (0.01–0.05 mm thick) in the phase wetting the tube walls which was indispensable to account for the foundation of mass transfer in a segmented flow, as well as for the dispersion or dilution of the analyte in the dynamic system, which is reflected in the shape of the transient signals obtained. The occurrence of such a film substantially increases the contact area between phases (see Figure 3.27B) and contributes to the dispersion of the analyte in the system. The film thickness is a linear function of the flow velocity: the higher the flow-rate is, the thicker the film will be. In any case, it is not formed at zero flow-rate.

The same authors derived an equation relating the film thickness (d_{film}, cm) to the inner diameter of the tube (R, cm), the flow velocity (u, cm s^{-1}), the viscosity (η, poise) and the surface tension of the system (γ, dyn cm^{-2}):

$$d_{\mathrm{film}} = \mathrm{cons} \times R\ (u\eta/\gamma)^{K}$$

where K is an empirical constant equal to 1/2 or 2/3.

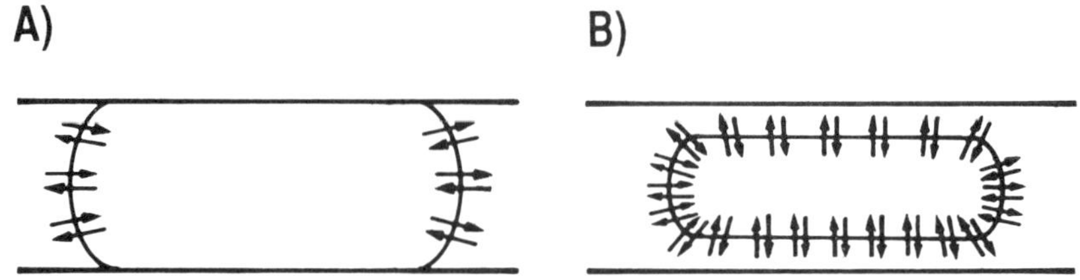

Figure 3.27 *Active interfacial area in relation to mass transfer on the segmented flow that circulates through the extraction coil in the absence* (A) *and presence* (B) *of a static thin film of liquid wetting the tube walls*

Both the viscosity of the film-forming phase and the interfacial tension influence film thickness in such a way that high viscosity in combination with low interfacial tension results in a thick film. Thus, the thickness of the film formed on the inner walls of a Teflon tube of 0.7 mm ID by a phase stream flowing at a velocity of 5 cm s^{-1} is 0.03 mm for a pentanol-water system and 0.004 mm for chloroform–water, which is consistent with the lower viscosity and higher surface tension of chloroform compared to pentanol.

The analyte dispersion is increased by the formation of the film, which can be regarded as static because of the prevailing laminar flow conditions (Reynolds no. < 2000), under which the flow velocity at the tube walls is virtually zero. Figure 3.28 is illustrative of this phenomenon.[88] As a 'bubble' of the aqueous phase containing the analyte reaches a given zone of the tube the partial extraction of the analyte takes place, which remains in the film. When the next plug of organic phase reaches the same point, the extracted analyte should diffuse and virtually disappear from the static zone. Because of the short time elapsed and the prevailing laminar flow regime (absence of whirls), diffusion is scarce, so part of the extracted analyte lags behind and meets the following aqueous bubble. The final result is that the analyte is diluted in the organic phase of the segmented flow. This effect increases as the flow-rate is raised as a result of the increased film thickness and viscosity of the organic phase and the decreased surface tension of the latter. As the extraction radius of the film decreases, so does dispersion through decreased film thickness, and diffusion of the analyte in the film is favoured through a decrease in the distances to be travelled. In this respect, dispersion has also been shown to decrease with increasing segmentation (*i.e.* with decreasing segment length).

Nord *et al.*[89] continued the above studies by dealing with the kinetic aspects of liquid–liquid extraction as implemented on continuous segmented-flow analysers. They found the kinetics of the overall process to be decisively influenced by two factors: (*a*) mass transport in the segments to and from the interface, and (*b*) the interfacial area-to-volume ratio. They provided some guidelines for improving the extraction rate, namely:

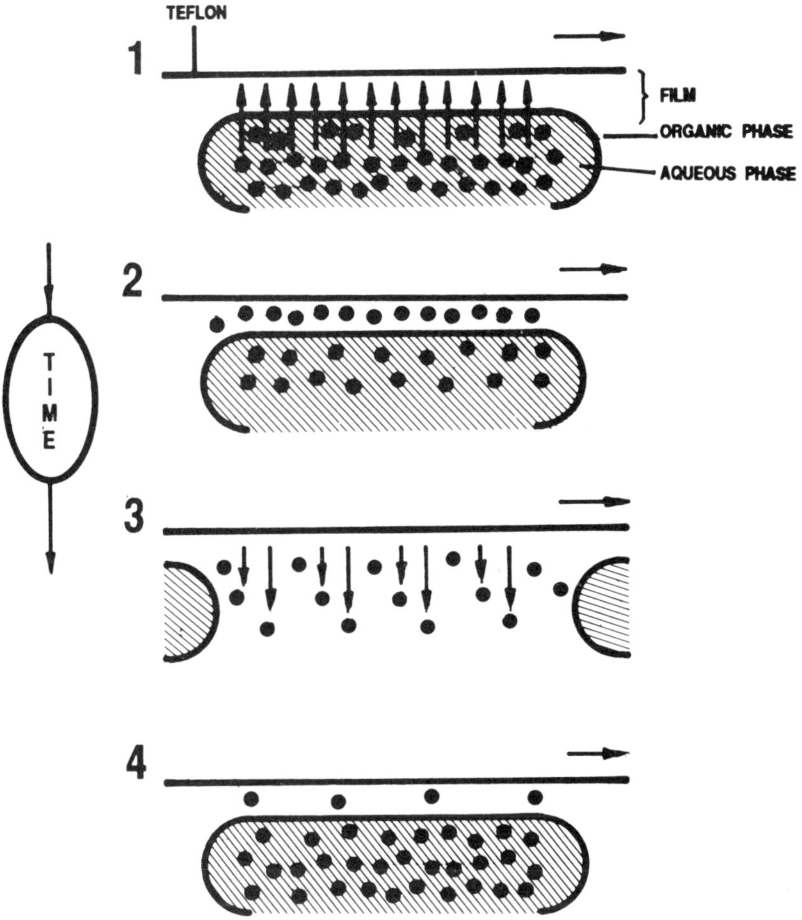

Figure 3.28 *Scheme of the analyte (•) dispersion resulting from the formation of a film of organic phase in a Teflon tube. Only the time evolution (1 to 4) of one of the walls in a given zone of the tube is represented. (For details, see text)*

(*a*) The extraction coil diameter should be quite small in order to obtain a high interfacial area/segment volume ratio.
(*b*) Shorter segments lead to more efficient mixing and hence to increased extraction rate.
(*c*) Increased flow velocities result in increased extraction rates through decreased residence times.
(*d*) Decreased flow velocities also increase the extraction rate in short coils.

The chief constraint to the above considerations is the difficulty involved in applying them in a generalized form; in fact, many systems have some peculiarities that prevent practical applications of the conclusions drawn by Nord *et al.* on the basis of one or two specific systems. As pointed out by Ruzicka and Hansen,[79] it is still adventurous to provide general guidelines for an FIA-extraction system, even though the formation of the film on the tube walls has an admittedly decisive influence.

The functioning of continuous segmented-flow analysers is decisively altered by other extraneous factors in some instances. Such is the case with the determination of terodiline in blood serum. This analyte is strongly adsorbed (60–80% loss) on Teflon or polypropylene walls when included in the aqueous phase. The extent of adsorption is much smaller in glass, nickel or stainless steel tubes. On the other hand, terodiline is not adsorbed at all when included in the organic phase. Therefore, the manifold used for the extraction of this drug from the biological fluid should be suited to these peculiarities.

The use of such auxiliary processes as the incorporation of beads into the coil or subjecting it to high temperatures, vibrations, or ultrasound in order to improve the extraction efficiency is not generally advisable as it usually causes the formation of emulsions that disrupt the functioning of the phase separator.

3.3.1.4 The Phase Separator

This is the last essential element of a liquid–liquid extractor and probably the most relevant to appropriate functioning. It receives the segmented liquid–liquid flow from the extraction coil and separates the segments in such a way that two independent streams of the two phases are obtained at its exit. Indeed, complete phase separation is accomplished only rarely and provided special requirements are met. Thus, the actual purpose of this element is providing a flow of one of the phases containing none of the other; the phase in question is usually that containing the analyte or its reaction product, which is also that subsequently driven to the continuous detector connected on-line with the extractor.

This element must work highly efficiently and rapidly in order to detract from the overall sampling frequency. In addition, it should not add to the dispersion or dilution of the analyte, so its inner volume should be as small as possible. Also, its operation should be smooth and consistent throughout in order to obtain reproducible results. This requires controlling the flow-rate of the emerging phase of interest or of both via a peristaltic pump, rather than sending them to waste. Some workers use restrictor tubes to control the unmonitored phase in order to regulate the differential pressure in the phase separator and hence increase its separation efficiency.[91]

There is a large variety of phase separators. Virtually every research group working in this field develops their own unit. However, they can be classified into three broad categories according to their operational principle, which also dictates their internal shape.

Separators based on density differences. This is the simplest type of separator. A minichamber receives the segmented flow from the top or through one of its sides. A single interface is established as a result of density differences, The flow of the heavier phase leaves the minichamber from the bottom, while that of the lighter phase emerges from the top or one side. This type of separator is formally resemblant of a separatory funnel, with the obvious differences of miniaturization, a marked difference in the time required to accomplish the separation according to whether or not an emulsion is formed and mass transfer takes place in the same location.

Figure 3.29 depicts various models of phase separator relying solely on gravity. The first (Figure 3.29A) is an open design as, in addition to separating the segmented flow, it must act as a debubbler; phase separation is accomplished by means of an overflow system that continuously removes the lighter (aqueous) phase while the heavier (organic) phase is driven to the continuous detector.[92] A similar design constructed from glass was reported by Wallace.[93] The other two designs in the figure are of the closed type and have been used with liquid–liquid extractors incorporated into FIA manifolds. In that depicted in Figure 3.29B, the lighter (aqueous) phase emerges from the same port through which the segmented flow is originally introduced;[94] on the other hand, in the design shown in Figure 3.29C, the segmented flow enters and the heavier phase leaves the separator through one side, while the lighter phase emerges from the top.[95]

The fact that the last two separators do not exploit differential wettability may account for their poorer results in terms of reproducibility. In addition, dispersion is rather high as a result of the larger volume of the minichamber. Because of these two shortcomings, this type of separator is currently outdated.

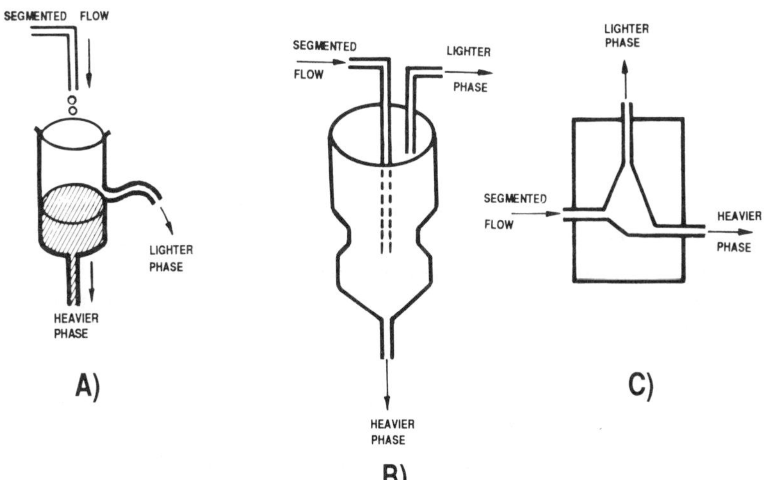

Figure 3.29 *Different types of gravity-based separators. (For details, see text)*

T-type separators. In this type of separator, the segmented flow enters through one side and the phases are continuously separated at their confluence point: the heavier phase emerges from the bottom, while the lighter leaves the chamber at the top. This type of separator is constructed from A-4 and A-10 Technicon connectors, the efficiency of which is increased by inserting small strips of Teflon or hydrophobic paper into the tube end through which the organic phase emerges, thereby increasing process efficiency by wettability in addition to gravity differences.[96, 97] Figure 3.30 shows the scheme of a T-type separator; in this instance, the organic phase is the lighter, so the Teflon insert points to the top of the T. This type of separator offers some advantages over the previous model; however, sandwich separators, which are described below, are much more frequently used at present.

The incorporation of a liquid–liquid extraction system into an automatic continuous segmented-flow analyser requires provision of a triply segmented flow as both gas–liquid and liquid–liquid interfaces are involved. Phase separators designed for this purpose should meet this requirement: thus, they should remove air from the triply segmented flow and introduce it into the outgoing organic phase in order to preserve the characteristic flow profile. Figure 3.31 shows two separators of this type marketed by Technicon and applied according to the relative density of the two phases. They resemble T-type separators in that they are also made from glass; however, they include more than one right-angle connector. A Teflon piece inside aids the separation through wetting by the organic phase. The separator in Figure 3.31A is used when the organic phase is lighter than the aqueous phase, while that in Figure 3.31B is employed in the opposite case. De-aeration is effected immediately upon entry of the triply segmented flow with some organic phase (model A) or of the aqueous phase with some organic phase (model B). The Teflon piece can be straight or curved at an angle, depending on the model. When the organic phase is the lighter, the design to be used is somewhat more complex: it requires an additional (downward) exit to separate the aqueous phase.

Sandwich separators. This type of segmented liquid–liquid flow separator is formally similar to most gas-diffusion and dialysis units as it is based on the use of a semi-permeable membrane that is wetted selectively by one phase. Depending on how the outgoing flows of the two phases are arranged, density differences can also be exploited. The chief difference between this type of separator and dialysers and gas diffusers is that it separates whole phases rather than analytes (or their reaction products) alone.

Sandwich phase separators typically consist of two parallelepipedal, cylindrical, or round blocks of Kel-F, Teflon, Daiflon, Perspex, or PCTFE with holes that allow entry of the segmented flow and exit of the organic and aqueous phase flows. As a rule, the blocks also have threaded holes for connections to be made within. Their central zone, on

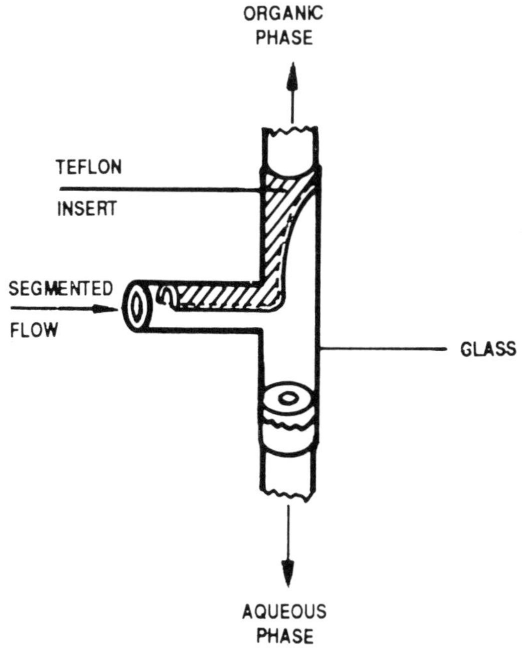

Figure 3.30 *Scheme of T-type phase separator*

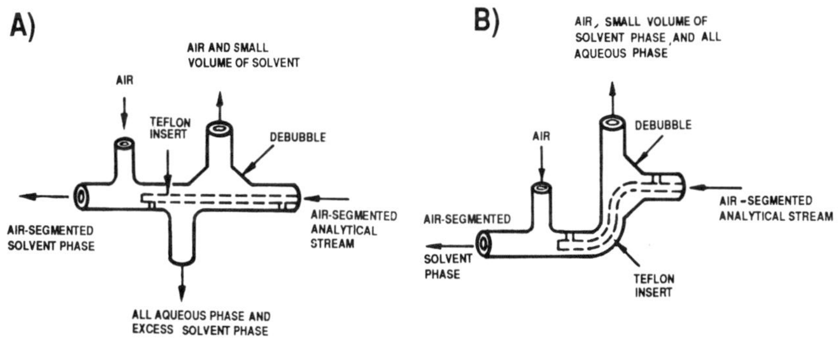

Figure 3.31 *Phase separators used in continuous air-segmented analysers. Designs used when the organic phase is the lighter* (A) *and heavier* (B) (Reproduced by courtesy of Technicon)

which the three conduits converge, is constructed in such a way that it provides a minichamber of 30–100 μl inner volume on assembly. The membrane is placed between the two blocks, thereby establishing two parts (equal or not). In order to avoid leakage, the two blocks are tightly squeezed together by means of four to eight screws that cross them at the ends—occasionally, two metal plates are also applied to the sides of the blocks. Greater caution should even be exercised if working pressures above 1 atm are used. The central part should also be carefully machined in order to avoid leakage. Also, the two blocks should be joined in such a way that the minichamber preserves its shape and dimensions throughout, which is normally accomplished by using metal slides fixed to the blocks.

This type of separator normally uses hydrophobic membranes which are only permeable to the organic solvent—the organic phase is also normally that of interest and should therefore be free from aqueous phase for proper monitoring. The membranes most frequently used in this respect are made of polyethylene-backed Teflon (PTFE) and Fluoropore filters with a pore diameter between 0.2 and 1 μm. Some firms (*e.g* Millipore, Goro-Tex) market a wide variety of membranes of this type.

The ability of a sandwich phase separator to exclude water can be expressed by the water intrusion pressure. This in turn depends on: (*a*) the pore size of the membrane (*e.g.* 2.8 bar at 0.2 μm and 0.7 bar at 1.0 μm); (*b*) the hydrophobicity of the organic phase (the higher the hydrophobicity is, the higher the pressure will be); and (*c*) the presence of substances that can be adsorbed by the membrane and thus decrease the pressure. Hydrophilic membranes are much less often used for this purpose. They consist of one or several layers of filter paper (*e.g.* Whatman no. 5, Whatman 1PS).

In order to avoid problems arising from the physical instability of the membrane zone which is not supported (*i.e.* the active zone in phase separation), an inert Teflon support is increasingly being used. This can be a 1 mm thick sheet of PTFE, a perforated Teflon strip, or a commercially available (Millipore) Teflon-coated stainless steel filter. In this last case, the filter thickness should be consistent with the design of the separator blocks. The support is virtually indispensable when large active membrane zones are involved, as is the case with cylindrical separators or when high pressures are used (*e.g.* use of a liquid–liquid extractor as a post-column reactor in HPLC).

The efficiency of a sandwich phase separator can also be further increased by slightly increasing the overpressure on the channel through which the phase that is not completely separated emerges—this channel should be connected to the segmented entry flow. Thus, with hydrophobic membranes, the exit channel for the aqueous phase is provided with a restrictor to achieve the said overpressure and facilitate passage of the organic phase across the membrane. Alternatively, the exit flow-rate can also be controlled by additional pumping.

Because of their ease of construction, virtually every group working on this topic have designed their own sandwich separator. A comprehensive description would thus be tedious and reiterative, so only the two models depicted in Figure 3.32 will be described in some detail. Insofar as the organic phase is usually that which must be free from the other, a hydrophobic membrane is used and the flow that crosses it leaves through the

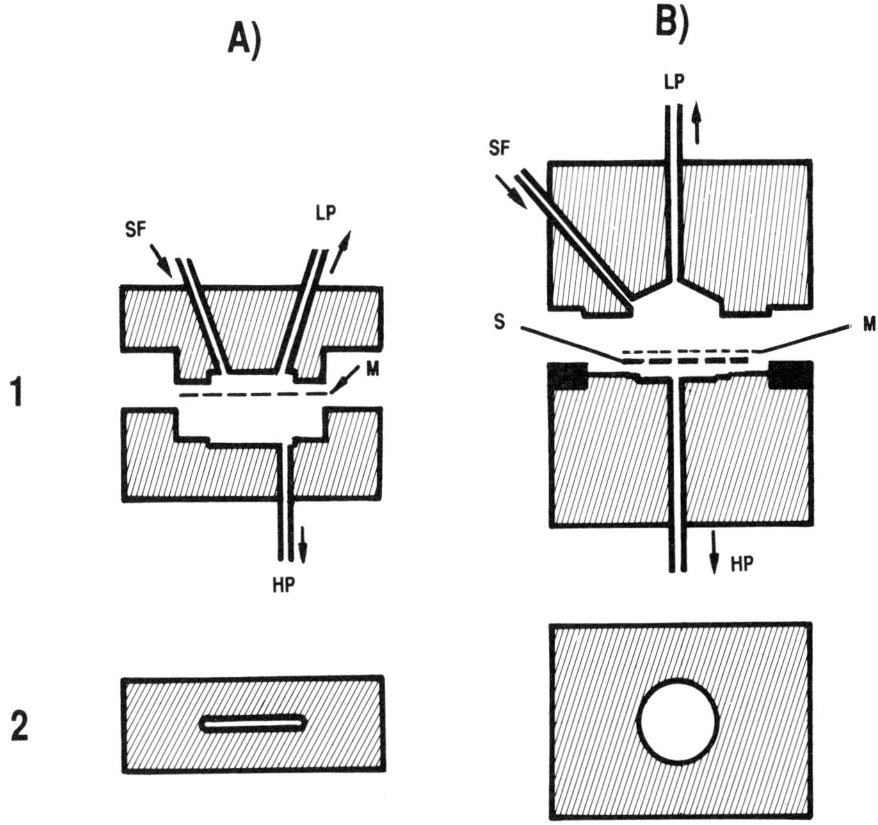

Figure 3.32 *Sandwich membrane phase separators.* (A) *Grooved.* (B) *Cylindrical. Side* (1) *and top* (2) *views. SF, segmented flow from the extraction coil; LP, lighter phase; HP, heavier phase; M, membrane; S, membrane support*

opposite direction to the segmented flow entry. In models A and B in Figure 3.32, the organic phase is the heavier; otherwise, the sandwich separator must be reversed in order to take advantage of this density difference in conjunction with the separating ability of the membrane.

Grooved sandwich separators (Figure 3.32A) are the simplest and earliest used in FIA.[98, 99, 106] The two plastic blocks fit through several profiles and squeeze the membrane in between. Both blocks have central grooves (20–120 mm long, 0.4–1.5 mm wide, 0.2–1.5 mm deep) that give rise to minichambers with inner volumes between 3.5×10^{-3} and 0.5 ml. Like dialysers, there are also some separators with widening grooves shaped like Roman frets,[100] asymmetric conical cavities[82] and flat coils (commercially available Tecator model). One should aim at obtaining a large active separation zone in order to increase the efficiency of the process; however, the inner volume should be as small as possible in order to avoid dispersion or dilution of the analyte and hence poorer transient signals. Normally a compromise must be made between these two aims and with an appropriate minichamber design. The influence of the shape and volume of each of the minichambers separated by the membrane on dispersion was found to be rather different;[101] this prompted workers to design sandwich separators with asymmetric chambers. The best results are obtained by altering and enlarging the zone through which the segmented flow enters and the unmonitored phase leaves. The way in which the entry and exit channels are arranged also influences the separation yield; however, the difference between lateral, central and angled entry or exit is usually insignificant. In any case, the arrangement in Figure 3.32A is one of the most efficient in this respect as it provides smooth, long-term, and efficient (100%) phase separations.[82]

The active zone of *cylindrical cavity-type phase separators* (Figure 3.32B) is circular (6–15 mm in diameter) rather than parallelepipedal. This larger separation surface results in much increased efficiency. Problems arising from the mechanical stability of the membrane are avoided by using an appropriate inert support. The zones on each side of the membrane are normally different. That receiving the segmented flow and yielding the outgoing aqueous phase can be parallelepipedal—the two flows can run parallel or at an angle of 45 or 90°[102]— or conical (*e.g.* that in Figure 3.32B). This also applies to the other zone, which yields the outgoing organic phase. A detailed study of all the possible arrangements revealed an angle of 45° and the use of a conical cavity to yield the best separation efficiency (*ca.* 100%). Experiments with different inner volumes and flow-rate ratios (q_a/q_o between 0.05:1 and 30:1) allowed the following conclusions to be drawn:[31] (*a*) yields of about 99% can be obtained over a wide range of flow-rate ratios; (*b*) the use of an inert rigid support is essential to applying the pressure drop across the membrane required to accomplish an almost complete separation; (*c*) the contribution of this unit to the overall dispersion in the system is virtually negligible; (*d*) when high overall flow-rates ($q_a + q_o$) are used, the microvolume of the cavity receiving the segmented flow should be increased by an appropriate factor. A recent modified design of this type provides slight improvements at a lower cost.[104]

Whenever the two phases are to be monitored after the liquid–liquid extraction, the separator should provide two completely 'clean' fluids. This can be accomplished by using a dual-membrane sandwich phase separator (Figure 3.33).[105, 106] The segmented flow enters the central minichamber laterally; this contains a hydrophilic and hydrophobic membrane which separate the cavities through which the aqueous and organic phases leave. Insofar as extraction may be incomplete, there is an exit flow to waste. Thus, unlike other phase separators, this unit has one entry and three exits.

In 1987, Frei *et al.*[107] reported a sandwich phase separator using no membrane (Figure 3.34). Its functioning relies on two basic principles

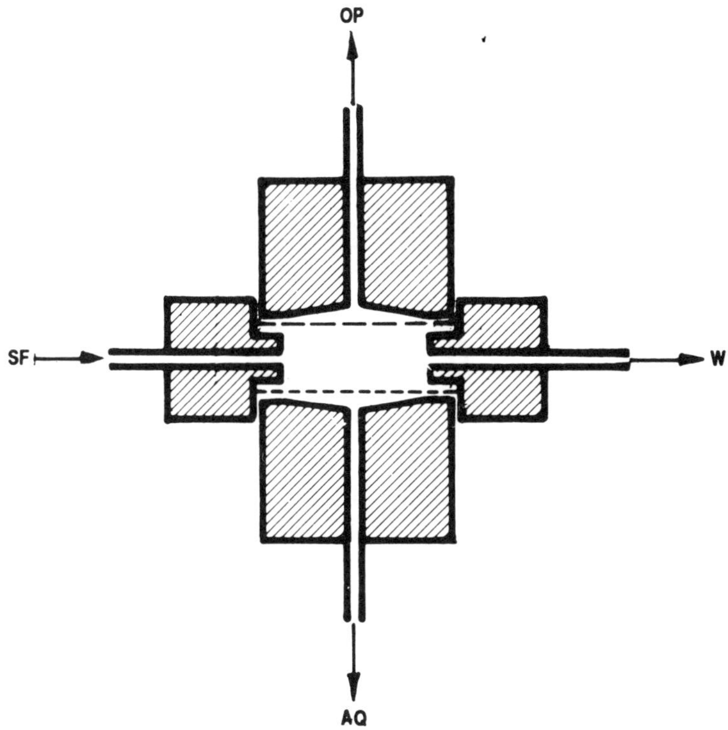

Figure 3.33 *Phase separator furnished with two membranes: hydrophobic (— — —) and hydrophilic (- - -). SF, segmented flow; AQ, aqueous phase; OP, organic phase; W, waste*

named above: selective wetting and phase density difference. Two stainless steel blocks squeeze a Teflon disk. One of the blocks (A in Figure 3.34) and the central Teflon disk have engraved grooves that bind the separation minichamber (8–43 μl inner volume). This has three connections: one for entry, which is located on one end and receives the segmented flow, and two for exit of the aqueous (lighter) and organic (heavier) phase. The use of no membrane calls for non-periodical replacement and hence avoids stopping the continuous analyser from time to time. This separator has only been applied to a given type of liquid–liquid interface made up by acetonitrile/water or methanol/water as aqueous phase and by 1,2-dichloroethane or n-heptane as organic phase. Hence, its full potential is yet to be exploited. Formally, this type of separator is similar to a sandwich separator, whereas operationally, it is more closely resemblant of a T-type separator.

Sandwich phase separators have also been used to implement mixed dialysis/liquid–liquid extraction processes by using special membranes to separate streams of two immiscible liquids continuously and directly reaching the separation unit. It is the analytes rather that the liquid

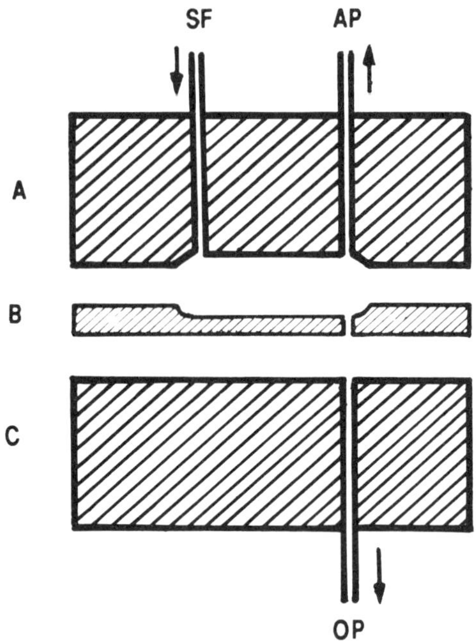

Figure 3.34 *Membraneless sandwich phase separator.* A *and* C *denote grooved and ungrooved stainless steel blocks, respectively, and* B *a grooved PTFE piece* (Adapted from reference 107 with permission of Elsevier Science Publishers)

phases that are transferred across the membrane in this case.[106, 108] They are commented on in greater detail later on.

Compared to other types of phase separator, sandwich membrane separators feature a number of advantages, namely: (*a*) higher efficiency in phase separation (typically between 90 and 95%, but even 100% if some technical modifications are introduced); (*b*) they are applicable to a variety of organic immiscible solvents as no substantial density difference with the aqueous phase is required; (*c*) they result in smaller dispersion and dilution of the analyte as a result of the separation chamber being two to five times smaller; (*d*) they can be used over a much wider range of flow-rate ratios, which results in increased preconcentration capacity; and (*e*) they can be used with high overall (organic plus aqueous phase) flow-rates, which results in increased sample throughput.

3.3.2 Continuous Configurations

Continuous analysers—whether segmented or not—that use a liquid–liquid extractor are not substantially different overall from other analysers of the same type. There are obviously some specific features such as the need to control the outgoing flows in order to regulate the extractor functioning, pumping of organic solvents, *etc.* This section places special emphasis on these aspects rather than on such common elements as tubes, connections, *etc.*

3.3.2.1 Components

Apart from the typical elements making up a continuous liquid–liquid extractor described above, all other modular components of an automatic continuous configuration used for liquid–liquid separations are those normally employed for other purposes, which are occasionally entrusted other functions.

Propulsion systems. Peristaltic pumps establishing several flows of the two phases that are controlled on exit represent the most common situation in this context. However, pumping the organic solvent through ordinary pump tubing causes major technical problems arising from internal corrosion that in turn results in serious fluctuations in the flow-rate. This compels one to adopt one of the technical solutions illustrated in Figure 3.35. The use of flexible tubing resistant to organic solvents is the simplest alternative in this respect (Figure 3.35A). Traditional Tygon tubes (suitable for aqueous solutions) are replaced with others of a special material such as Solvaflex (for alcoholic solvents like iso-octane, cyclohexane) or Acidiflex (for chloroform, toluene, or dichloromethane), Viton (chloroform), among others. Apart from their high price, these materials are not comparable in mechanical resistance to that of conventional pump tubing for aqueous solutions. This compels frequent checking of the flow-rate. These shortcomings are normally circumvented by using displacement bottles (0.5–2 l), which are schematically depicted in Figures 3.35B and C. These are glass bottles with a special stopper that allows entry and exit of two streams (one of water and the other of organic phase). The water flow established by the peristaltic pump enters through the bottom (B) or top (C) of the bottle and displaces the organic phase, which leaves the bottle through the top (B) or bottom (C) depending on the relative density of the

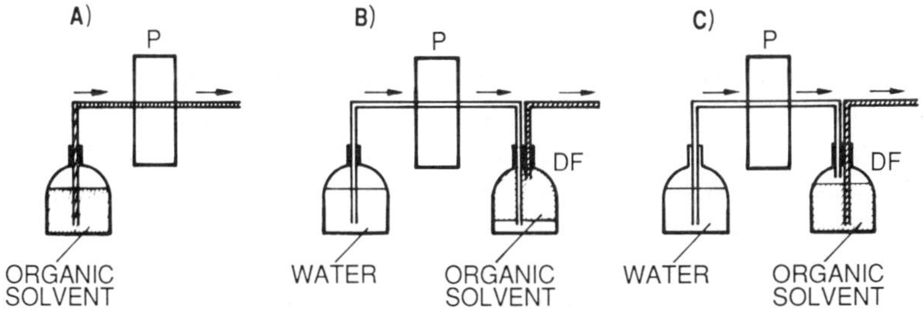

Figure 3.35 *Use of peristaltic pumps (P) to establish flows of organic streams by direct pumping or by use of a displacement flask (DF) for organic solvents lighter (B) or heavier (C) than water*

two phases. The flow-rates thus obtained are perfectly constant and controllable. The only problem in this respect is that their operational performance is limited by the bottle volume; however, by working at ordinary flow-rates (1–3 ml min^{-1}), this poses no special problems.

The problems arising from the establishment of a stream of immiscible organic phase can also be circumvented by using special propulsion systems based on principles other than those of peristaltic pumps. Such is the case with gas pressure-based systems[109] or special pumps such as double-plunger micropumps.[110, 111] Both succeed in overcoming the aforementioned pitfalls, though at the expense of increased analyser complexity.

Valves. Valves used with this type of system serve their usual functions. On the one hand, they are used to inject microvolumes of sample and of one of the phases after the liquid–liquid extraction has finished. On the other, they can also be used as channel (sample, blank) switches by placing them before or after the propulsion unit. Also, switch valves are used for specific purposes in close-loop solvent extraction systems,[112, 113] and on-off valves are used to control exit flow-rates. The roles of valves are commented on below in describing the different types of configuration.

Detectors. Over 70% of all the applications of liquid–liquid separation reported so far uses photometric detectors furnished with flow-cells through which the organic phase was passed and subsequently driven to waste by means of additional pumping when necessary. In some instances (see Figures 3.37.5 and 6), two detectors are used to monitor simultaneously the two flows emerging from the continuous liquid–liquid extractor. Fluorimetric monitoring has also occasionally been used to increase sensitivity and selectivity in this type of application. The direct determination of metals and indirect determination of either species has been addressed by using an atomic absorption spectrophotometer coupled on-line to a liquid–liquid extractor; coupling can be accomplished directly or via an injection valve. A detailed study of the performance of this association was recently reported.[114] On the other hand, electroanalytical detection (both potentiometric and voltammetric) has scarcely been used in this context.

A wave-guide capillary cell (0.2–2 mm long, 0.5 mm ID) can be used as a spectrophotometric detector to continuously monitor the outgoing organic phase. Because carbon disulphide has a refractive index which is sufficiently higher than that of the fused silica from which the capillary cell is made, the light from the source propagates inside the capillary cell via total reflection at the inner cell wall as in an optical fibre.[115]

Occasionally, detection is carried out batchwise by using a sampler as an interface between the manifold including the continuous liquid–liquid extractor and the sensing instrument. In this way, metal traces have been determined by electrothermal-vaporization (graphite chamber) atomic absorption spectroscopy;[116] the inherent batchwise operation of the detector prevented direct on-line connection. Also, a liquid–liquid extractor has been coupled off-line to a gas chromatograph.[90]

3.3.2.2 Types of Configuration

Liquid–liquid extractors have so far been incorporated into an automatic continuous analyser in a number of ways which have resulted in the development of a variety of configurations intended to overcome successively encountered problems or meet increasingly demanding requirements.

These configurations can be classified according to various criteria. Thus, a first classification could be based on whether or not air bubbles are used to avoid carry-over between samples, *i.e.* according to the essential difference between automatic continuous analysers. Triply segmented systems involving two types of interface (gas–liquid and liquid–liquid) were introduced over 20 years ago as commercially available instruments (Technicon, Skalar) and are characterized by their complexity: air must first be removed and then inserted into the phase separator (see Figure 3.31), which decreases reproducibility and increases start-up time. Because of these and other technical reasons that make segmented analysers in general more complicated than their unsegmented counterparts, triply segmented systems are now well outdated. Conversely, continuous unsegmented flow analysers including a liquid–liquid extractor are arousing increasing interest, so much so that some manufacturers have already started commercial production of these systems.

Figure 3.36 shows two generic schemes of continuous analysers including a liquid–liquid extractor; they differ in the relative positions of the injection valve and the liquid–liquid extractor. In the module shown in Figure 3.36A, the sample is injected into an aqueous carrier that is continuously merged with one or several reagent streams. This and another stream carrying an organic solvent, converge on the extractor. Finally, the outgoing organic phase stream is continuously monitored. In the configuration depicted in Figure 3.26B, the sample (or the blank, depending on the position of the selecting valve) is continuously introduced by aspiration and mixed with the reagents before reaching the solvent segmenter of the continuous extractor. The outgoing organic phase fills the loop of an injection valve through which a carrier continuously reaching the detector is circulated. In the inject position, the organic content of the valve loop is directly transported to the detector. The final carrier can be aqueous, which enables undiluted transfer, of special use in using an atomic absorption spectrophotometer as detector. A channel carrying the organic solvent itself can also be used with photometric detectors; however, this results in dispersion or dilution of the analyte or its reaction product. We should note that the two designs in Figure 3.36 are not representative of all the possibilities in this context. Thus, the system in Figure 3.36A uses a displacement bottle, while that in Figure 3.36B utilizes flexible peristaltic pump tubing. In many cases, the outgoing streams are propelled by additional pumps which have also been omitted from the figure.

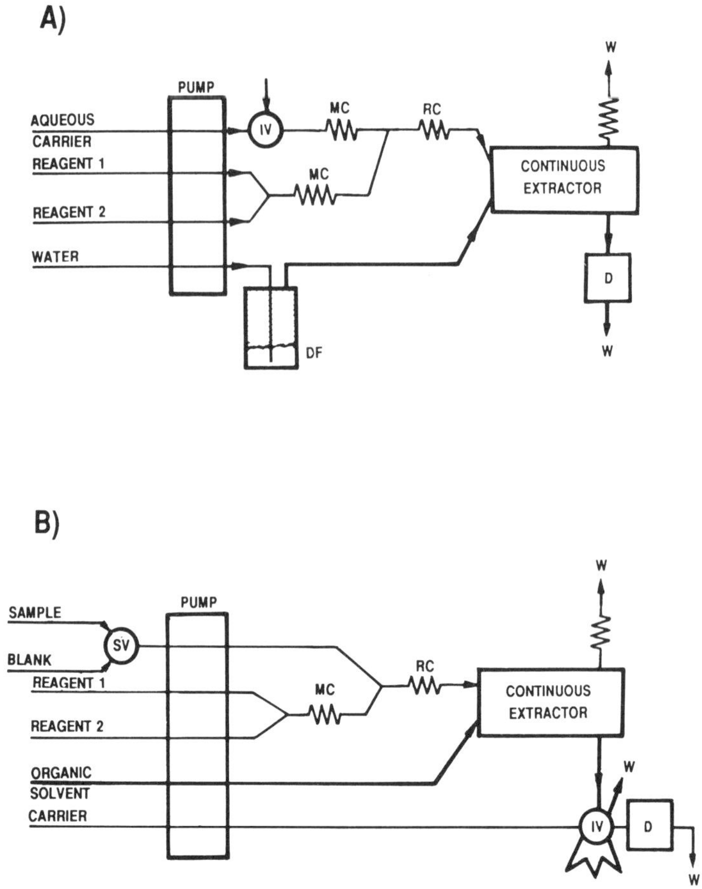

Figure 3.36 *Different types of continuous analysers involving liquid–liquid extraction and injection prior to* (A) *or after the extractor* (B). *In* (A), *the sample is injected; in* (B), *it is aspirated into the system. SV, selecting valve; IV, injection valve; MC and REC, mixing and reaction coil; D, detector; W, waste. Model* (A) *uses a displacement flask* (DF), *while model* (B) *uses solvent-proof pump tubing*

Recently, Toei[100] reported a system in which the sample is directly injected into the segmented flow. The injection valve is located between the solvent segmenter and the extraction coil. This, according to Toei, diminishes dispersion and dilution of the analyte, thereby increasing sensitivity and sample throughput.

The way in which the outgoing flows emerge from the liquid–liquid extractor is of great practical interest as it determines its smooth functioning. The alternatives most commonly used in this respect are schematically illustrated in Figure 3.37. In broad terms, systems can be classified according to whether (models 2, 4, and 6 in Figure 3.37B) or not (models

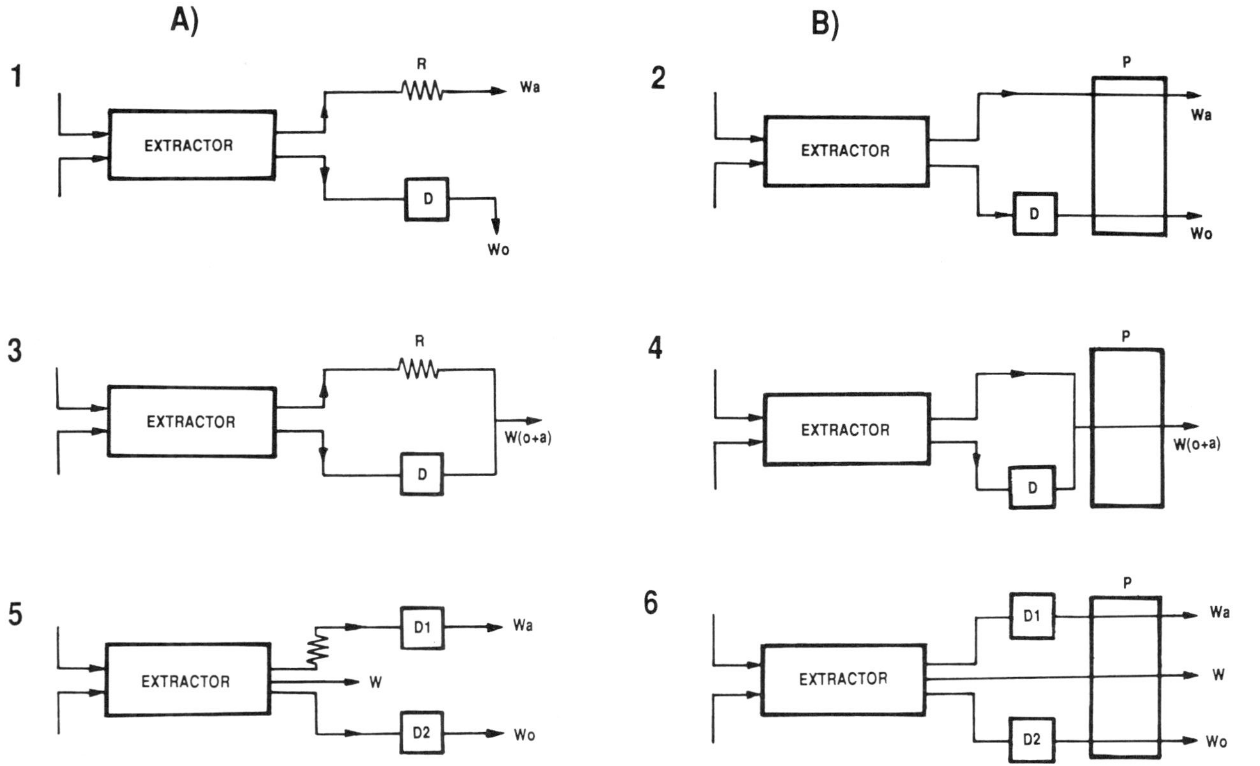

Figure 3.37 *Different alternatives to the exit from the continuous liquid–liquid extractor.* (A) *Without and* (B) *with additional pumping: (1, 2) independent outlets; (3, 4) convergent outlets; (5, 6) use of two detectors. P, pump; R, restrictor; D, detectors;* W_a, W_o *and* W_{a+o} *wastes*

1, 3, and 5 in Figure 3.37A) additional pumping, which can be provided by the same peristaltic pump used to propel the entry flows (see Figure 3.38), or by another pump, is used. The organic solvent can be pumped directly or via a displacement bottle. In the absence of additional pumping, it is advisable to use a restrictor at the aqueous phase exit, as noted earlier. The two immiscible phase streams can leave the unit independently (designs 1 and 2) or be merged after the detector and be subsequently led to waste as a single stream (designs 3 and 4). The use of two continuous detectors to continuously monitor the two phases obviously calls for an additional waste line (see Figure 3.33); thus, all three streams can be sent directly to waste or, alternatively, their flow-rates can be controlled by means of a peristaltic pump (designs 5 and 6).

The schemes in Figures 3.36 and 3.37 are representative of most of the configurations reported so far. There are, however, some exceptions worth mentioning on account of their future potential and ingenious design. Such is the case with *multi-extraction systems* based on the use of two liquid–liquid extractors connected on-line (Figure 3.38). The sample is injected into the stream that carries the first phase. Mass transfer into the second phase takes place at extractor I; the stream carrying such a phase continuously reaches extractor II, where a stream of the first phase (usually containing an additional dissolved reagent and/or at a pH other than that of the initial phase) collects the analyte or its reaction product and drives it to the continuous detector, where it is monitored, or, alternatively, collects fractions in a sampler. There are thus three waste lines: two for the first phase and one for the second. Depending on their nature, it is advisable to use extraction coils of different materials. Thus, if the first phase is aqueous and the second is organic, extractor I should be made of Teflon, while extractor II should be glass so that mass transfer is accomplished as expected. These continuous multi-extraction systems have two basic aims: preconcentration, which requires controlling the flow-rates of the incoming and outgoing phases; and removal of interferents with similar features to those of the analytes, which would not be eliminated by a conventional extraction procedure. Their most serious shortcoming is the difficulty involved in accomplishing smooth functioning from the analyser, which is complicated by the incorporation of the extra elements required. Only two of the relatively few systems of this type reported so far are described below.

The manifold depicted in Figure 3.38 has been used for the determination of *polynuclear aromatic compounds* (PNAs); the system includes two T-type separators (see Figure 3.30) and the first phase is organic whereas the second is aqueous. A crude oil ash sample residue dissolved in cyclohexane is injected into a fresh cyclohexane stream that is merged with a dimethylsulphoxide (DMSO) stream in the first solvent segmented connected to a glass extraction coil. The organic phase is sent to waste, while the DMSO stream containing the analytes is merged with a water stream. After passage through a cooling coil, this stream is segmented

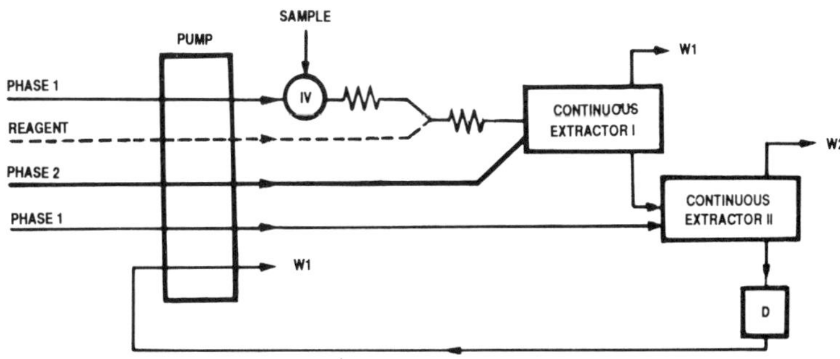

Figure 3.38 *Continuous multi-extraction by means of two serially coupled extractors* (I and II). W_1 *and* W_2 *denote the wastes of phases 1 and 2, respectively. (For details, see text)*

with fresh cyclohexane in the second extractor, which accommodates a Teflon coil. The outgoing organic phase is then passed through the flow-cell of a videofluorimeter, while the DMSO–water phase is driven to waste.[97] By using the same configuration but introducing the phases in the opposite order (aqueous first and organic second), Bengtsson *et al.*[116] and Backstrom *et al.*[103] described continuous liquid–liquid extraction systems for preconcentration and clean-up of metal species in two stages: extraction of metal carbamates into Freon 113 and subsequent back-extraction into an aqueous carrier containing Hg^{II} as displacing reagent. The resulting aqueous phase was collected into sample cups and batch detection was carried out by graphite-furnace atomic absorption spectrometry. The two coils used in this instance were made of Teflon and the phase separators were of sandwich type.

Christian *et al.*[112, 113] recently reported quite a complex *open–closed loop system* including a liquid–liquid extractor (Figure 3.39). The loop is located in the central zone and is open or closed according to the position of the selecting valve. In the load position, the loop is filled with organic phase that is circulated through it. Then, an aqueous blank is pumped through the system and the loop is closed. A small make-up organic stream (syringe pump 3) is continuously added to replenish any solvent that is lost. The system is allowed to run for a few minutes until a stable background signal is obtained. By switching the sample selecting valve, the sample is loaded (3 to 20 minutes, depending on the preconcentration factor required) into the system. In the closed-loop arrangement, the sample flows through the extraction coil, is isolated from the organic phase in the membrane phase separator, and finally driven to waste. At the end of the loading time, the sample selecting valve is switched again to deliver a blank solution. Solvent for a new cycle is then generated by switching the loop selecting valve to allow delivery of fresh organic solvent to replace the old organic solvent in the loop. A check valve is attached to the organic waste stream to prevent back-flow. A syringe fitted to one of the valve inlets is used to wash the sample tubing in order to avoid carry-over

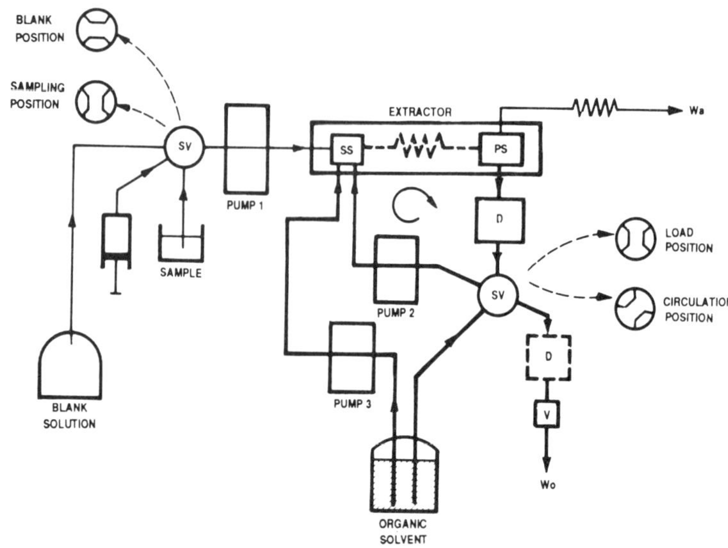

Figure 3.39 *Open-closed loop system for continuous liquid–liquid extraction: SV, selecting valves; V, check valve; SS, solvent segmenter; PS, phase separator; D, detector. (For details, see text)*
(Adapted from reference 112 with permission of the American Chemical Society)

between successive samples. Continuous monitoring is accomplished by placing a flow-cell in the open-closed circuit. The signal obtained is a continuous trace that increases gradually with the amount of sample introduced and returns to the baseline as the blank is inserted. Placing the continuous detector between the selecting valve and the check valve results in a typical FIA transient signal when the circulating organic phase is removed. The chief advantage of this open-closed configuration is the possibility of carrying out preconcentration over a wide range of V_a/V_o ratios, which cannot be achieved by simply changing the flow-rates of the organic and aqueous phase in conventional continuous liquid–liquid extractors.

Finally, it is interesting to note that some of these configurations involve a rinsing step intended both to condition the separation membrane and to flush the system between successively inserted samples. As a rule, this is done by using a miscible organic solvent (ethanol, methanol, acetone) that sweeps both the organic and the aqueous phase. Figure 3.40 illustrates the two ways in which this intermediate operation can be implemented. One of them (Figure 3.40A) involves introducing the rinsing solution through a selecting valve via the same tubes as the immiscible organic solvent and cleaning all the internal elements of the liquid–liquid extractor. The other (Figure 3.40B) is self-implemented by the model marketed by Tecator and involved using a syringe loaded with the miscible organic solvent, which is connected to the immiscible organic solvent

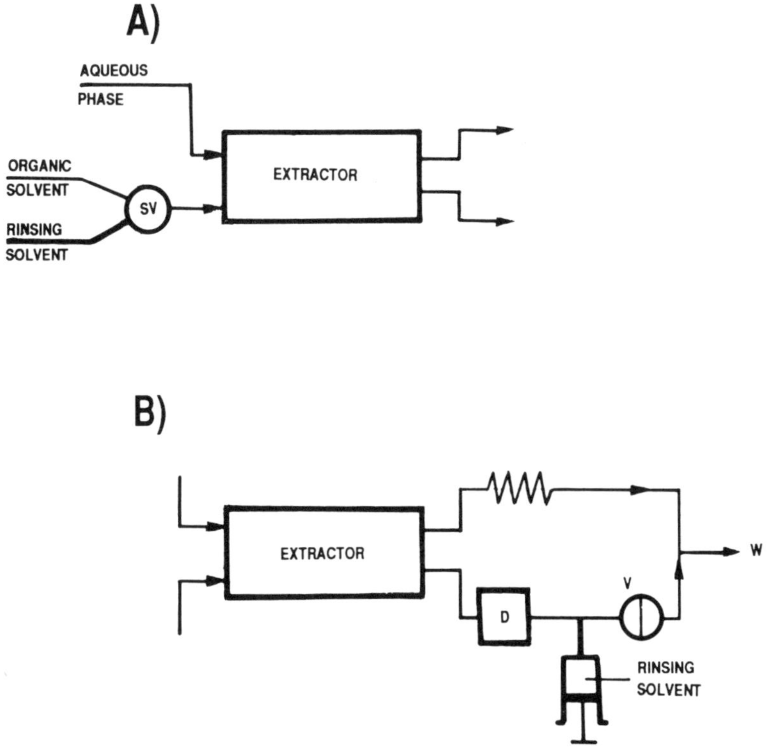

Figure 3.40 *Rinsing systems for liquid–liquid extractors. The rinsing solution can be inserted before* (A) *or after the extractor* (B). *SV, selecting valve; V, open-closed valve; D, detector*

stream after the continuous detector. A closed configuration of the outgoing flows and the use of an open-close valve between the syringe and the waste allows the membrane separator to be cleaned. During the rinsing operation, the peristaltic pump used to propel the organic and aqueous phase streams is halted as the above-mentioned valve is closed. On pushing the syringe plunger, the miscible solvent is forced to circulate through the phase separator and emerge through the aqueous phase line.

3.3.3 Reactions Involved

Chemical reactions can be implemented in continuous analysers including a continuous liquid–liquid extractor for one or two purposes, namely: obtaining a reaction product capable of facilitating mass transfer and/or continuous detection, which in turn result in increased sensitivity and/or selectivity.

Figure 3.41 shows the different possible locations where a derivatizing reaction can be developed within this type of system. Most commonly, such a reaction takes place in a coil after the sample, whether injected or

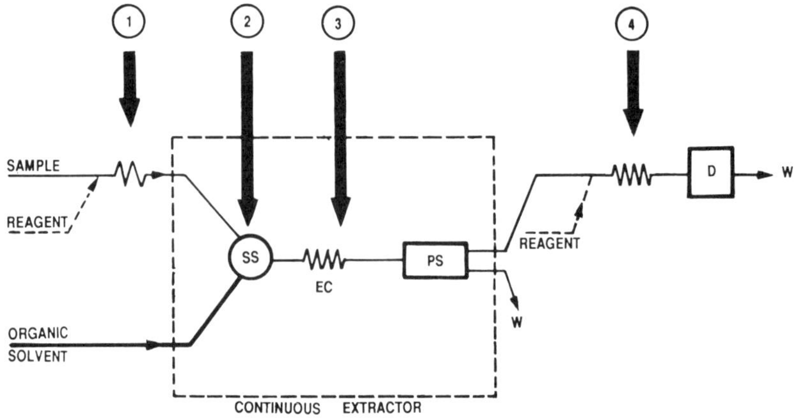

Figure 3.41 *Zones where a derivatizing chemical reaction can be implemented in a continuous liquid–liquid extraction system:* (1) *in a coil located before the extractor,* (2) *in the solvent segmenter (SS),* (3) *in the extraction coil (EC), and* (4) *in a coil located after the extractor. PS denotes the phase separator*

aspirated, is merged with a reagent stream. In order to avoid precipitation of the extractable compound in the aqueous phase before it reaches the continuous extractor, sample and reagents can be mixed at the solvent segmenter, which will thus receive three streams: two of aqueous phases and a third of immiscible organic solvent.[85] Many liquid–liquid extraction processes based on the formation of metal chelates are started with the ligand dissolved in the organic phase; in continuous extractors, the derivatizing reaction takes place virtually simultaneously with the transfer of the chelate in the extraction coil. Finally, in some continuous analysers, the derivatizing reaction takes place after the extractor and is accomplished by using a reagent stream that is merged prior to the detector with the phase containing the analyte.

The derivatizing reactions implemented in analysers–extractors are not substantially different from those used in other technical variants of liquid–liquid extraction. The most common in this respect are ion-pair formation reactions, where the analyte forms the ion-pair directly (*e.g.* determination of anionic surfactants by using bulky organic substances such as Ethyl Violet or Methylene Blue, and the determination of codeine with picric acid and enalapril with Bromothymol Blue) or after reacting with a reagent to form a bulkier, charged species (*e.g.* cobalt tetracyanate, phospho-molybdate, metal ion/macrocyclic ligand, *etc.*). Transition metal ions, though, are more often transformed into extractable neutral chelates by using such common ligands as dithizone, APDC, oxine, *etc.* Oxidation reactions yielding an extractable product such as pyrimidic acid (resulting from the oxidation of felodipine[91]) or thiochrome (an oxidation product of vitamin B[118]) have only occasionally been used for this purpose.

3.3.4 Applications

Good proof of the great analytical potential of continuous analysers furnished with a liquid–liquid extractor is the wide variety of problems addressed and successfully solved with them and their increasing availability as commercial systems.

In the 1960s and 1970s roughly fifty automatic determinations were reported based on the use of *continuous air-segmented analysers* furnished with liquid–liquid extractors similar to those depicted in Figure 3.31, which are described in detail in the scientific publications sponsored by Technicon.[78] Their current scarce use imposed by the limitations mentioned earlier in this chapter warrants only a brief description. The following are probably their most salient applications so far: the determination of (*a*) anionic surfactants in water and in commercial products based on the formation of ion-pairs with cationic dyestuff;[119, 121] (*b*) non-ionic surfactants in water;[52] (*c*) quinizarin in hydrocarbon oils;[123] (*d*) amines[124] and phenmetrazine[125] in pharmaceuticals; and (*e*) drugs in urine.[126, 127]

Continuous extraction processes implemented in unsegmented analysers are particularly relevant to *pharmaceutical analyses*, which quite often include a liquid–liquid extraction. The automation of these processes is thus the chief aim of these applications. The earliest determination based on an FIA-extraction system was that of codeine in acetylsalicylic acid tablets based on the formation and extraction into chloroform of an ion-pair of the analyte with picrate ion at a sampling rate of 60 samples h^{-1} and with an r.s.d. of 1%;[128] by a similar procedure, procyclidine was later determined in tablets.[102] Thiamine (vitamin B) has been determined in pharmaceutical preparations in a continuous extraction system by oxidation to thiochrome.[118] Bromothymol Blue has been used as counter-ion to determine enalapril in pharmaceuticals by extraction into dichloromethane in quite a straightforward continuous system featuring a sample throughput of 80 samples h^{-1} and an r.s.d. between 1.0 and 3.4%; neither the products resulting from the degradation of the analyte nor the excipients interfere with the determination.[129] The bismuth content of pharmaceutical samples has been determined by formation of tetraiodobismuthate and continuous extraction into dichloromethane using the salt of a phosphonium cation as counterion.[130] Also, diphenhydramine and 8-chlorotheophylline have been determined simultaneously in dramamine motion sickness tablets by using a monitoring system consisting of two detectors, the dual phase separator depicted in Figure 3.33 and a configuration similar to that shown in Figure 3.37.6.[105] The dissolution test of felodipine tablets has been monitored by coupling an FIA-extraction system on-line to the dissolution test vessel, with remarkable advantages (rapidity, simplicity, low cost) over the conventional method based on the use of a liquid chromatograph.[91] Active principles in pharmaceuticals can be determined indirectly by atomic absorption spectroscopy by formation and extraction of ion-pairs. In this way bromazepam has been assayed, both in tablets and in urine by formation of a complex with Cu^{II} and subsequent formation of an ion-pair with perchlorate, with remarkably high sensitivity and selectivity.[131] The formation of a CS_2 derivative of amphetamine and subsequent formation of a neutral chelate with copper, nickel, or zinc allowed the development of an automatic determinative method for this drug in the presence of high concentrations of other drugs of abuse.[132]

The elimination of well known matrix effects affecting *biological and clinical samples* makes a fundamental step of the analytical process. Liquid-liquid extraction as implemented on a continuous analyser has proved to be an advantageous alternative in this context. The selective extraction of acid drugs in blood serum makes an excellent example of interference removal; thus, terolidine has been determined by liquid–liquid extraction after mixing the injected sample with NaOH and subsequently collecting the n-heptane/pentanol extracts (ratio 98:2) batchwise and injecting them into a gas chromatograph.[90] Most of these determinations involved inorganic species, which allows one to classify the methodologies used into two broad groups according to the type of detector (atomic or molecular spectroscopic) used. By using an ingenious manifold involving injection of a complex sample into a buffer carrier merging with an MIBK stream, zinc was determined at concentrations between 20 and 800 ng ml^{-1} in biological samples with a concentration factor of two and an r.s.d. of 1.5%.[133] On the other hand, lead was determined in urine by manual liquid-liquid extraction into the APDC/MIBK system and subsequent injection of 20 μl of the extract into an aqueous carrier transferring the injected plug to the AAS instrument undispersed.[134] Perchlorate traces (0.1–5 mg l^{-1}) can be determined indirectly in urine by AAS on an FIA-extraction set-up by using a cuproine-like chelate as counter-ion, at a sampling frequency of 45 h^{-1}.[135] Also, cadmium traces (concentrations above 0.2 μg l^{-}) have been determined photometrically in human urine by continuous extraction of cadmium dithizonate,[136] and orthophosphate has been also determined photometrically in urine and human seminal plasma by formation of a heteropolyacid with ammonium molybdate in an aqueous medium, followed by extraction into isobutyl acetate in an FIA-extraction manifold with a detection limit of 200 pg P at a sampling rate of 80 samples h^{-1} and with no sample deproteination.[137] Quite an interesting determination for lithium in blood serum and urine is based on the extraction of the complex formed between the cation and a crown ether or a cryptand, which in turn forms an ion-pair with a bulky organic counter-ion; this endows the determination with a high selectivity—sodium concentrations up to 1000 times higher than that of the analyte cause no interfering effect.[138, 139]

The field of *environmental analysis* has benefited to a smaller extent than that of pharmaceutical analysis from the assets of continuous extraction as implemented on continuous unsegmented analysers. In fact, fewer determinations have been reported and analytes assayed in this context. The determination of anionic surfactants in various types of water has been paid the greatest attention in this respect, probably as a result of the corresponding manual method being the most earnestly recommended for this type of determination.[140] Methylene Blue is used as a counter-ion to form the ion-pair, which can be extracted into various solvents (chloroform, 1,2-dichlorobenzene) in an FIA-photometric manifold including a liquid–liquid extractor into which the sample is injected (Figure 3.36A).[11, 141, 151] Aspiration of the sample (Figure 3.36B) results in markedly increased sensitivity.[142] Other bulky coloured cations such as 1-alkyl-4-(4-dialkylaminophenylazo)pyridinium salts have also been used for this purpose.[143] This determination can also be carried out by using an atomic absorption detector, the configuration depicted in Figure 3.36B and the chelate Cu^{II}–1,10-phenanthroline as counter-ion, with rather a high selectivity.[144] Also, cationic surfactants have been determined in water by formation of ion-pairs with bulky anions in an FIA-extraction set-up with photometric detection with Orange II[83] and by atomic absorption spectroscopy using the tetrathiocyanatocobaltate II complex as counter-ion.[145] Potassium has been determined in river water on a manifold such as that depicted in Figure 3.36 by sequential formation of a crown ether and an ion-pair with a bulky organic anion in the aqueous phase prior to reaching the continuous liquid–liquid extractor used.[110, 146] By a similar photometric procedure orthophosphate was determined in water with a detection limit of 0.1 ng ml^{-1} P on the basis of the formation of phosphomolybdate and the formation and extraction of an ion-pair with Malachite Green.[12] The formation of neutral metal chelates has been exploited for the determination of metal traces in water in configurations resembling that in Figure 3.36B;[147] however, uranium has been determined in nuclear waste reprocessing solutions on the open-closed configuration depicted in Figure 3.39 in order to achieve the required preconcentration.[113]

A number of determinations in the field of *food analysis* are constrained by the low concentrations at which the analytes usually occur and by effects from the sample matrix. Liquid–liquid extraction in its automatic continuous variant is thus of great help in this context. Total aliphatic amines have been determined in alcoholic beverages such as beer and wine over the concentration range 1–130 mg l^{-1} by reaction with 1,2-naphthoquinone-4-sulphonate and extraction of the reaction products into chloroform in an FIA-extraction manifold with photometric detection at 460 nm. A manifold such as that in Figure 3.35B was used for the individual[162] and simultaneous determination[163] of nitrate and nitrite in meat based on the formation of ion-pairs with the cuproine chelate, extraction into MIBK and measurement of the metal content of the organic phase by atomic absorption spectroscopy. Auxiliary chemical agents allow discrimination between these two analytes. First, Ce^{IV} is used to oxidize nitrite to nitrate and the overall content is determined; then, sulphonic acid is used to decompose nitrite, so only nitrate is determined. The straight forward FIA-extraction manifold shown in Figure 3.36A was used to determine bittering compounds in beer with clear advantages (rapidity, lower reagent consumption, the need for no blank) over the standard manual method.[149] The automatic extraction of water-soluble compounds in samples with basically organic matrices reduces human participation in the tedious preliminary steps of the analytical process. Such is the case with the determination of free fatty acids in oils implemented on a continuous analyser. The sample (100 μl) is injected into a toluene stream that is then segmented by the aqueous reagent stream, which contains Cu^{II} and pyridine. The copper–fatty acid complex contained in the organic phase is monitored photometrically at 699 nm. This automatic method features improved sensitivity, selectivity, precision, and rapidity, as well as reduced reagent consumption and hazard from toxic fumes.[150]

Automatic continuous extraction systems have also been used to develop interesting applications in the field of industrial analysis. Many of the systems concerned rely on the formation of extractable ion-pairs in which the metal ion (analyte) undergoes a prior transformation.[152] Such is the case with the determinations of cobalt[85] and manganese[153] in steel, in which the metals are previously converted into $[Co(SCN)_4]^{2-}$ and $[MnO_4]^-$, respectively, and with the direct analysis for perchlorate in commercial chlorate samples by formation of an ion-pair with Brilliant Green.[154, 155] Also, indium traces have been determined in synthetic industrial samples by using an FIA-extraction set-up coupled on-line to an atomic absorption detector, using bis(2-ethylhexyl)phosphoric acid dissolved in 4-methylpentan-2-one as extractant.[156] Finally, available zinc has been determined in an automatic fashion in soils by extraction as dithizonate.[157]

Some applications of automatic extraction systems are aimed at determining physico-chemical parameters rather than analytes (*e.g.* the determination of partition ratios,[36] acid constants[158] or extraction constants[159]).

It is also worth mentioning here the selectivity studies carried out in connection with continuous liquid–liquid extraction systems. Thus, the removal of classical interferences with atomic absorption spectroscopy has been investigated by coupling an FIA-extraction manifold on-line with the nebulizer.[160] By using such a classical system as the extraction of aluminium with 8-hydroxyquinoline in chloroform, the interference limits of other cationic and anionic species afforded by batch and continuous methods were compared; by virtue of kinetic discrimination, the selectivity of the automatic method was substantially higher.[161]

3.3.5 Atypical Continuous Liquid–Liquid Extraction Systems

The systems discussed so far are continuous analysers (air-segmented or not) including a liquid–liquid extractor with its three essential elements: segmenter, extraction coil, and phase separator (see Figure 3.23). This section deals with the most salient variants of this basic configuration involving substantial alterations and major gains is some aspects.

The possibility of accomplishing continuous liquid–liquid extraction with none of the essential elements of the extractor was first shown in 1978[164] with the fluorimetric determination of potassium on a single-channel FIA manifold. The aqueous sample, mixed with a fluorescent counter-ion (anilinonaphthalene sulphonate) was injected into a 1,2-dichloroethane carrier containing a macrocyclic compound that formed a cationic complex with the analyte. Mass transfer and the formation of the complex and ion-pairs took place in the mixing coil. The resulting turbid plug yielded an FIA peak at the fluorimetric detector. A blank (no potassium) injection yielded a small peak. Despite its simplicity and apparent high potential, this remains the sole system of this type reported so far, which can be attributed to reproducibility constraints and to the lack of analytical chemical systems meeting the requirements of such a simple liquid–liquid extraction manifold.

The analytical applications of organized molecular assemblies (*e.g.* micelles, reversed micelles, and microemulsions) have opened up previously unsuspected prospects.[165] In addition to altering reaction rates, chemical equilibria, and reaction pathways, these media give rise to solubilization and concentration processes that resemble interfacial liquid–liquid extraction and result in significantly increased sensitivity and selectivity in relation to analytical processes involving a single phase. The use of microemulsions in FIA enables replacement of the liquid–liquid extractor with a suitable surfactant contained in the carrier or a microemulsion acting as the carrier itself. The analytes, usually contained in the aqueous phase, are injected into both types of carrier. If the microemulsion is formed *in situ* in the FIA system, it is advisable to use a reactor packed with inert beads or place the mixing coil within an ultrasound field in order to favour the formation of the microemulsion in the carrier to be directly passed through the flow-cell of a photometric or fluorimetric detector. In this way have been determined C_6–C_{10} primary amines by derivatization with *o*-phthalaldehyde and 2-mercaptoethanol in a heptane/water system containing dioctyl sulphosuccinate as surfactant and fluorimetric detection,[106] as well as traces of copper(II) by derivatization with neocuproine and using the same organized medium.[167]

The use of minicolumns capable of (ir)reversibly retaining one of the phases taking part in the liquid–liquid extraction process also involves a substantial alteration to continuous systems including an extractor. This type of system is schematically depicted in Figure 3.42. Toei[168] recently reported an interesting FIA manifold (Figure 3.42A) for the determination of caffeine. A volume of 10–15 μl of the aqueous sample is injected into a chloroform carrier. The column used is packed with a water sorbent such as Sumicagel base gel or cotton fibres and is preconditioned with a mixture of chloroform and methanol. Mass transfer and the complete retention of the injected microvolume of aqueous phase take place

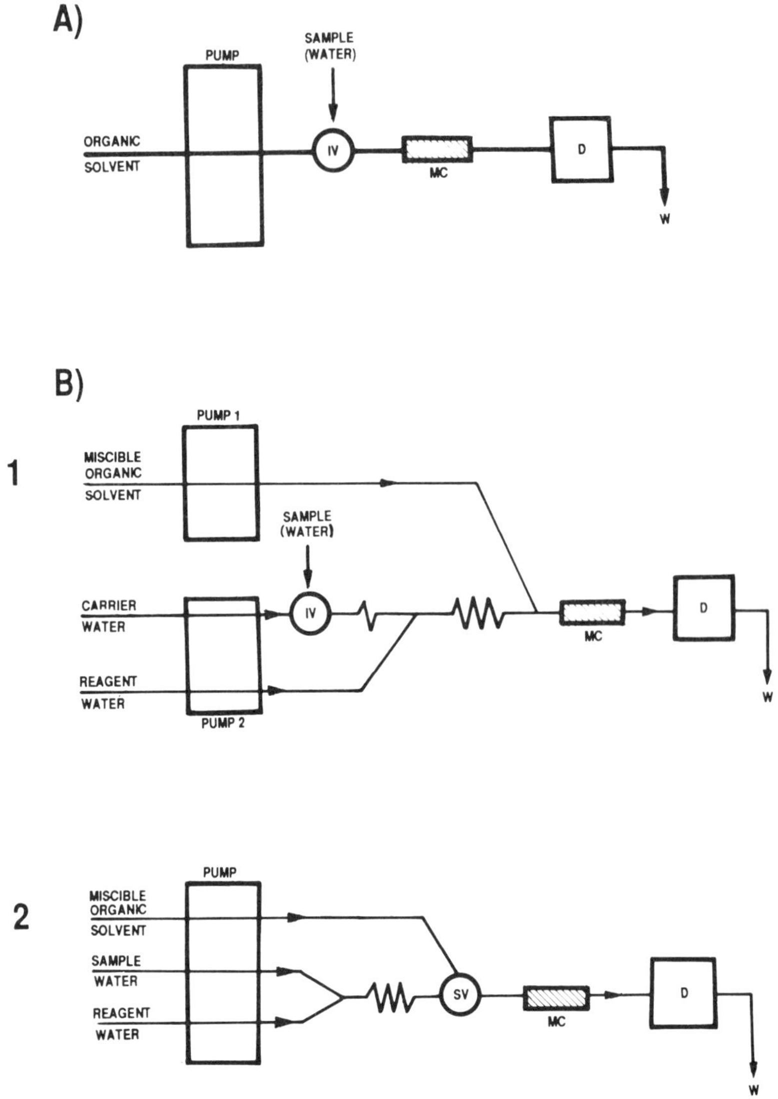

Figure 3.42 *Continuous liquid–liquid extraction based on the use of minicolumns (MC) packed with an active material.* (A) In situ *retention of the aqueous phase.* (B) *Covalent retention of the organic phase. IV, injection valve; SV, selecting valve; D, detector. (For details, see text)*

in it. An organic phase stream continuously reaches the photometric detector (254 nm). The extraction yield thus achieved is close to 100%. Because of the small volume of aqueous phase injected, the column only becomes saturated after rather a large number of injections. Ruzicka and Arndal[169] suggested the use of a minicolumn packed with C_{18} bonded-phase silica gel (159 μm diameter beads) to extract metal chelates previously formed on merging the sample containing metal traces (lead, copper) with ligands forming neutral chelates (DDC, PAR, PAN). First, the chelates are retained by the minicolumn in the bulk of the 'organic phase'; then, they are eluted by a methanol carrier that drives them to the atomic absorption spectrophotometer used. These two sequential operations can be done by intermittent pumping of the sample (Figure 3.42B.1) or by using a selecting valve to aspirate the sample for a preset time (Figure 3.42B.2). This latter alternative also enables preconcentration. This is but a further example of solid–liquid separations that is described in further detail in Chapter 4.

One of the pioneers in the use of continuous separation systems coupled to FIA manifolds reported in 1986 an atypical system which is schematically depicted in Figure 3.43A.[108] A membrane-type sandwich module received and allowed the streams of the two immiscible phases to leave it independently. The analyte or its reaction products were transferred across an inert porous PTFE membrane; the situation was thus virtually identical to dialysis, except for the immiscibility of the phases involved. The chief shortcoming of this system was its low separation yield (1.5–18%), which depended on the injected sample volume and flow-rates used; in fact, its only practical application was the determination of analytes (caffeine, anionic surfactants) occurring at high concentrations. Gluck[36] used a configuration similar to that shown in Figure 3.43A and a Cuprophane cellulose derivative membrane with a cut-off of 10 000 to determine coefficients of partition between the octanol/water system. This system was described in greater detail in the section devoted to dialysis.

The possibility of using thin layers of organic solvents—which may or may not contain dissolved reagents—immobilized on microporous inert supports between two aqueous solutions makes an appealing alternative to liquid–liquid extraction.[170] This type of supported liquid membrane has scarcely been applied so far for analytical purposes. Its continuous

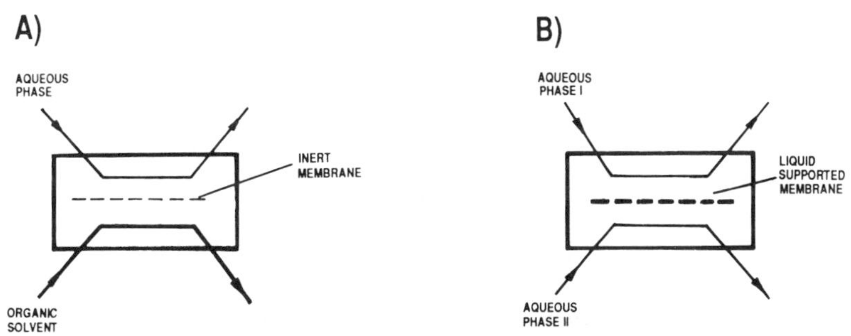

Figure 3.43 *Scheme of continuous liquid–liquid extraction processes as implemented with a membrane sandwich module without subsequent phase segmentation and separation.* (A) *Use of an inert membrane.* (B) *Use of a liquid supported membrane*

functioning and a specific application involving a sandwich membrane separator were described above in dealing with dialysis.[48, 49] It is schematically depicted in Figure 3.43B; as can be seen, it shares features of dialysis (*e.g.* two aqueous phase streams are involved and a single separation unit is used) and liquid–liquid extraction (*e.g.* mass transfer results from the solubility in the immiscible liquid phase supported by the membrane).

Liquid–liquid extraction with none of the typical elements of a liquid–liquid extractor has also been implemented on the FIA manifold schematically depicted in Figure 3.44A,[171] which has two salient features: (*a*) the location of the detector flow-cell in the loop of the injection valve; and (*b*) the programmed iterative change of the flow direction from injection by use of a peristaltic pump controlled by a timer or microprocessor. In the load position, the organic phase fills the loop of the injection valve, thereby establishing the signal that is used as the baseline, and the sample stream is continuously introduced and merged with a reagent stream, after which it is driven to waste. In the inject position, two liquid–liquid interfaces are established within the valve. The flow circulates left to right for a preset time (*e.g.* 10 s), as shown in Figure 3.44B. Only one of the halves of the organic plug is monitored, and the interface is never allowed to pass through the flow-cell in order to avoid parasitic signals. Between points 1 and 3 in Figure 3.44B takes place a whole cycle that yields an FIA peak: the top (I) and bottom (CP) correspond to the instants in which the interface is closest to the detection point (Figures 3.44B.2 and 3.44B.4) and at the centre of the organic plug (Figures 3.44B.1 and 3), respectively. The multipeak thus obtained provides a wealth of information that can be exploited both for theoretical studies and for the implementation of analytical methods based on a variety of parameters such as: (*a*) the signal at peak *n*; (*b*) signal increment between two peaks (rate measurements); (*c*) slope of the envelope of the minima; (*d*) derivative of the transient signal after one cycle; (*e*) signal of the first minimum (dilution method); and (*f*) summation of the signals of *n* maxima (concentration method). A manifold similar to that shown in Figure 3.44 was used for the automatic determination of surfactants in water over the concentration range 0.08–10.0 μg ml^{-1} on the basis of the formation of an ion-pair with Methylene Blue and subsequent extraction into chloroform.[101] The methanol stream in Figure 3.44A was used to flush the system between successively inserted samples. This singular methodology was successfully used to study the kinetics of extraction into chloroform of 23 anions with ferroin as counter-ion; not only the mechanism, but also a correlation between the anion size and its extraction rate was established. This in turn allowed the development of differential kinetic methods for the determination of mixtures of anions, which is unaffordable by other technical variants of liquid–liquid extraction.[172] A manifold similar to that in Figure 3.44 but with the detector positioned between the injection valve and the waste and mass transfer taking place from the organic to the aqueous phase was used to determine total

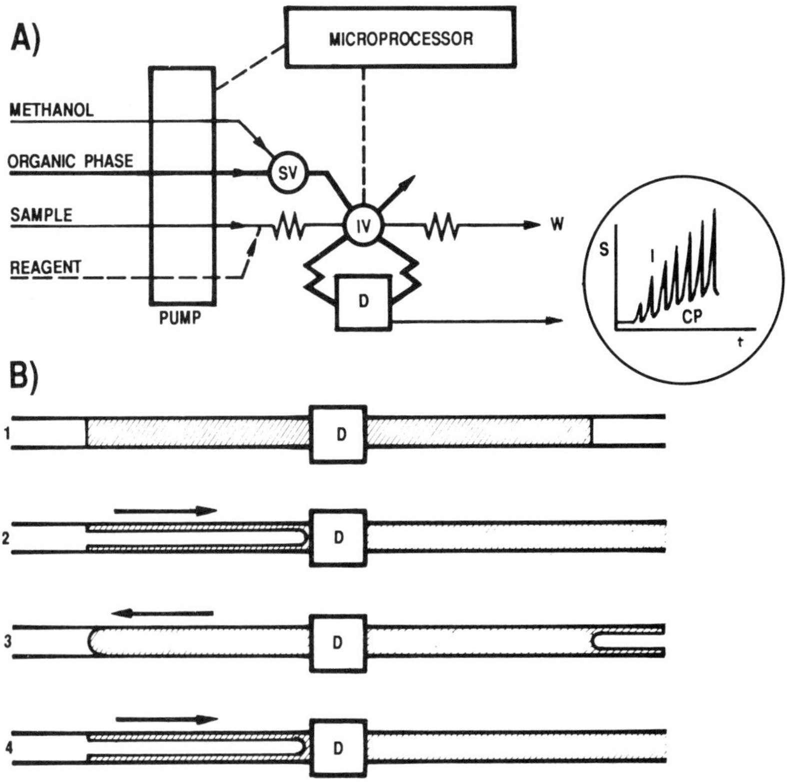

Figure 3.44 *Continuous liquid–liquid extraction without phase separation based on the iterative reversal of the flow direction.* (A) *Scheme of the FIA manifold.* (B) *Functioning of the liquid–liquid extraction system:* (1) *injection,* (2–4) *reversal of flow direction. SV, selecting valve; IV, injection valve; D, detector; W, waste; S, signal; t, time; I, interface; CP, centre of the organic loop*

polyphenols in olive oil.[173] The sample was directly aspirated and diluted with a hexane stream, after which it filled the loop of the injection valve. The aqueous phase stream contained Folin–Ciocalteau as reagent in a $NaHCO_3$ medium. Continuous photometric monitoring of the aqueous phase provided a multi-peak recording due to the extraction of the polyphenols and their reaction in the aqueous medium. In order to improve the functioning of the system, the flow-rate of the outgoing stream was also controlled via the peristaltic pump, which was programmed for the iterative reversal of the flow direction. This automatic methodology offers major advantages over its manual counterpart, namely: (*a*) it avoids time-consuming sample manipulation, which is introduced directly into the system; (*b*) it features significantly increased sample throughout and precision and substantially decreased sample and reagent consumption; and (*c*) it allows polyphenols to be determined over a broad concentration range with no special modifications.

Occasionally, authors use incorrect terms to refer to continuous liquid-liquid extraction systems, which can led to misinterpretation (*e.g.*, when extraction is carried out manually and an aliquot of the extract is subsequently injected into an FIA manifold[134, 174]).

3.3.6 On-line Coupling of a Continuous Extractor to a Chromatograph

The continuous liquid-liquid extractor is one of the modules most frequently coupled on-line to other analytical separation instruments. Chapter 2 dealt with the association of liquid–liquid extraction and hydride generation (Figure 2.11B) and vaporization (Figure 2.18). The function of the extractor in these combined systems usually has to do with interference removal or solvent changeover.

An extractor can be coupled to a gas or liquid chromatograph in three different ways, namely: (*a*) indirectly, by collecting the flow of the phase of interest in vials after the sample has been aspirated or injected into the FIA manifold and subsequently injecting their contents manually into the chromatograph; (*b*) on-line, via a sample turntable acting as an interface between the two continuous separation systems; and (*c*) also on-line, but directly, *i.e.* with no interface or discontinuity between the extractor and the chromatograph. Simplicity and the degree of automation of these three alternatives increase in the same order as they have been mentioned. With a few exceptions[90, 116] direct on-line coupling (*c*) has given rise to the most interesting applications in this context, some of which are commented on below.

Figure 3.45 illustrates the ways in which a manifold including a liquid-liquid extractor can be coupled to a chromatograph. Of the four potential alternatives (pre-and post-column coupling to a gas or liquid chromatograph) one (post-column coupling to a gas chromatograph) is unfeasible in practice. The chief aims of pre-column arrangements are automatic concentration, sample clean-up and derivatization, while post-column arrangements are intended to facilitate derivatization and reagent removal. In turn, derivatization of the analytes is aimed at facilitating chromatographic separation and detection. The examples described below have been classified according to the nature of the mobile chromatographic phase used.

3.3.6.1 Continuous Extractor/Gas Chromatograph

Despite the abundance of gas chromatographic methodologies involving a liquid–liquid extraction as a preliminary step to the introduction of a sample aliquot into the chromatograph injector, few attempts have so far been aimed at automation. This is probably a result of the technical difficulties posed by this hyphenated technique.

A rotary valve is usually employed as interface in these systems. The

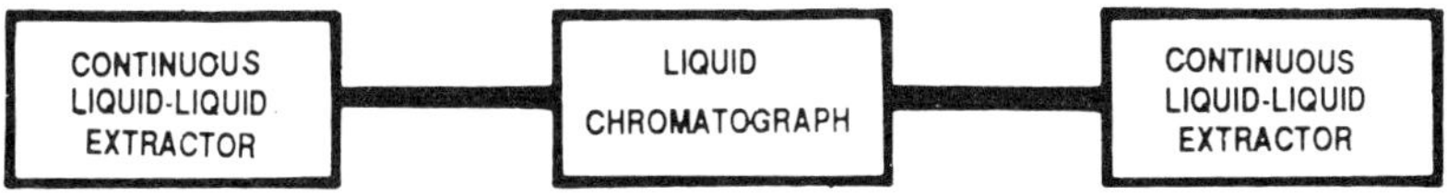

Figure 3.45 *Different alternatives to the direct on-line coupling of a liquid–liquid extractor to a chromatograph*

organic phase stream from the extractor fills the loop of the valve, which is also crossed by a carrier gas stream that is driven to the chromatograph. In the inject position, the gas sweeps the liquid microvolume to the chromatograph. The most serious problem encountered arises from the vaporization of this microvolume.

In 1985, Roeraade[175] reported the first direct coupling of an extractor to a gas chromatograph for the determination of traces of aliphatic and aromatic hydrocarbons in waste water using a complex set-up involving two high-pressure pumps and a large injected volume but providing high-quality results in terms of sensitivity (p.p.t. level) and precision. Later, Folgelquist[176] exploited Roeraade's achievements to develop a simpler extractor–GC system for the determination of halocarbons (chloroform, bromodichloroethylene, dibromochloromethane, bromoform, and 1,1,1-trichloroethane) in sea water. The main shortcomings of this system are as follows: (*a*) the chromatograph must be altered by piercing the furnace wall, connecting two columns serially or changing the carrier gas inlet, which renders it unusable for routine injections with a syringe; and (*b*) the addition of an internal standard or other reagent required for pre-column derivatization cannot be performed in an automatic fashion.

Recently, Ballesteros *et al.*[177] reported a new approach aimed at overcoming most of the pitfalls of the above-described systems. It is illustrated in Figure 3.46. An HPLC injection valve with a small sample volume (4 μl) was connected to the chromatograph via a stainless steel tube (25 cm long, 0.5 mm ID) with a soldered needle of the same material that was directly inserted into the injection port of the chromatograph. The tube was heated from 25 to 175°C by means of an electrical resistance. The carrier gas from the container was split into two lines connected to one of the ports of the injection valve and to the injection port of the chromatograph, respectively. A stop-clock was used to shut the flow of the

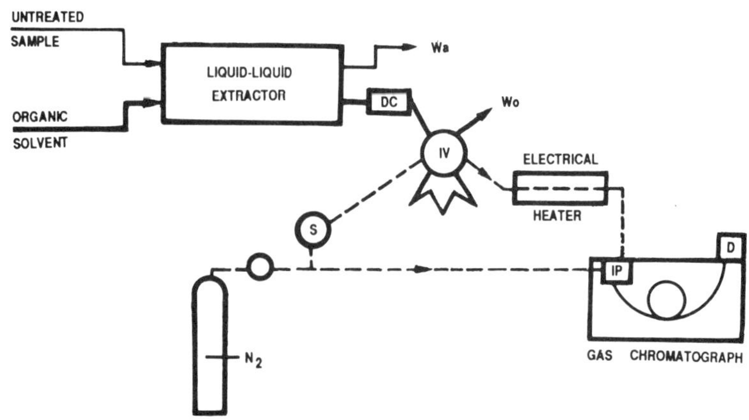

Figure 3.46 *On-line column coupling of a liquid–liquid extractor to a gas chromatograph. DC, desiccating minicolumn; IV, injection valve; IP, injection port; D, chromatographic detector; W_a and W_o, aqueous and organic phase waste, respectively; S, stopcock*

carrier gas to the valve, thereby allowing the chromatograph to be used for manual injections. Water traces were prevented from entering the chromatograph by using a desiccating column filled with pellets of sodium aluminium silicate. The organic phase contained naphthalene as internal standard. By using ethyl acetate as extractant, four phenols (phenol, 3,4-dimethylphenol, 2-t-butylphenol, and 2-t-butyl-4-methylphenol) were determined in water with a detection limit of 0.5 μg ml^{-1} and an r.s.d of 0.8–3.0%. One of the most salient assets of this novel approach is the possibility of implementing derivatization simultaneously with the liquid-liquid extraction. This requires using n-hexane as extractant, which contains dissolved acetic anhydride as derivatizing reagent for the determination of mixtures of phenols, cresols and chlorophenols with a detection limit of 0.2 μg ml^{-1} and an r.s.d. similar to that achieved with no derivatization. These automatic methods allow the extraction and derivatization of phenol compounds at a sampling rate of 30 samples h^{-1}; however, the actual sample throughput is determined by the furnace temperature program used.

3.3.6.2 Continuous Extractor/Liquid Chromatograph

As noted earlier, these two systems can be coupled in a pre- or post-column arrangement for the same or a different purpose. This association has been much more widely used than the previous one.[178] A comprehensive description is beyond the scope of this book, so only a few representative examples of the on-line coupling of a liquid–liquid extractor to an HPLC instrument (Figure 3.47) are discussed.

In *pre-column arrangements* (Figure 3.47A), the outgoing organic stream containing the analyte fills the loop of the chromatographic injection column in an assembly similar to that depicted in Figure 3.46. The sample injection, not shown in Figure 3.47A, requires the valves to be operated in perfect synchronicity as the flow plug that contains the sample is reduced and the chromatographic injection must take place while this resides in the loop. Dolan *et al.*[179] reported a continuous automatic air-segmented assembly for the determination of fat-soluble vitamins (A_1, D_2 and E) in tablets. The manifold used for this purpose consists of three modules: (*a*) an automatic device (Solid-Prep, from Technicon) for dissolution of the tablets in ethanol/water; (*b*) a continuous liquid–liquid extractor transferring the analytes to a hexane stream; and (*c*) a liquid chromatograph, the injection valve of which is filled with the immiscible solvent. Recently, Farran *et al.*[180] developed an automatic continuous methodology on the basis of a set-up similar to that in Figure 3.47A for the determination of organophosphorous pesticides (fenthion, azinphos methyl, and drazion) with a detection limit of 0.4–0.09 mg l^{-1} and an r.s.d. of 3.4–5.3%, using n-heptane as organic phase. The sample throughput thus achieved was 15 samples h^{-1}.

Most of the methods based on this coupling were of post-column type. Figure 3.47B shows two schemes representative of most, though not all the possibilities in this respect. The essential difference between them lies in the location of the confluence point for the flow from the

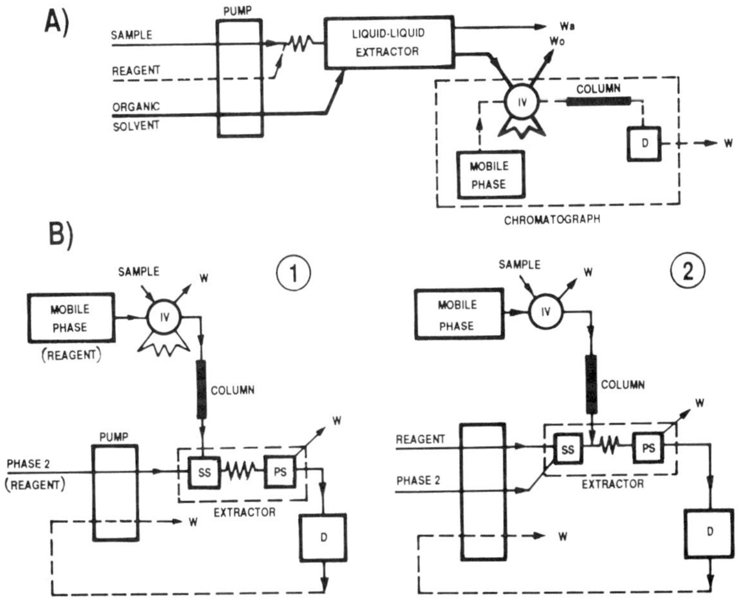

Figure 3.47 *On-line pre-* (A) *and post-column coupling* (B) *of a continuous liquid–liquid extractor to a liquid chromatograph. IV, injection valve; D, detector; SS, solvent segmenter; PS, phase separator; W, waste*

chromatographic column, which can be at the solvent segmenter (Figure 3.47B.1), or somewhere between this and the extraction coil (Figure 3.47B.2), although it can also be placed before the elements of the continuous extractor, particularly in continuous air-segmented configurations.[181, 182] As a rule, the post-column extractors used in HPLC have two main purposes: (*a*) to concentrate the analyte in the eluate, thereby reducing band broadening; and (*b*) to enable otherwise impossible analytical detection when the excess reagent interferes by giving the same signal as the reaction product (*e.g.* with ion-pairs) or the solvent must be changed (*e.g.* with special detectors such as a mass spectrometer).[183, 184]

In the configuration shown in Figure 3.47B.1, the chromatographic eluate is introduced into the derivatizing zone simultaneously with the liquid–liquid segmentation. The nature of the carrier to be used (phase 2) will depend on the type of chromatography to be implemented: an immiscible organic solvent and an aqueous phase for reversed-phase and normal-phase chromatography, respectively. The counter-ionic reagent (coloured or fluorescent) can be included in either stream used. Thus, 9,10-dimethoxyanthracene-2-sulphonate (DAS) was used dissolved in the mobile phase for the determination of hydroxyanthrazine in urine[185] and of secoverine in biological samples;[186] the organic solvent extracted the fluorescent ion-pair to be monitored, while the excess fluorescent counter-ion remained in the aqueous phase, which was driven to waste. By using the reagent (Methyl Orange with photometric detection, or DAS with fluorimetric detection) dissolved in aqueous carrier merged at the solvent segmenter with the organic chromatographic mobile phase, cationic surfactants were determined in manually pretreated environmental samples that were eventually dissolved into chloroform.[187]

The chromatographic eluate is introduced into the liquid–liquid extractor at some point between the segmenter and the extraction coil (Figure 3.47B.2) when the reagent is included in the carrier of the continuous post-column system. In this way tertiary amines, drugs, and chloro- and bromo-phenylamine with DAS,[188] secondary amines of pharmaceutical interest such as histidine, chloroxamine, and fluoxamine with dansyl chloride,[189] and erythromycin and its ethylsuccinate in serum with naphthotriazoledisulphate,[190] have been determined using fluorimetric detection in all instances.

References

1. W.B. Furman, 'Continuous Flow Analysis. Theory and Practice', M. Dekker, New York, 1976.
2. J.K. Foreman and P.B. Stockwell, 'Automatic Chemical Analysis', Ellis Horwood, Chichester, 1975.
3. D.O. Cooney, E.J. Davis, and S.S. Kim, *Chem. Eng. J.*, 1974, **8**, 213.
4. D.O. Cooney, S.S. Kim, and E.J. Davis, *Chem. Eng. Sci.*, 1974, **29**, 1731.

5. B. Bernhardsson, E. Martins, and G. Johansson, *Anal. Chim. Acta*, 1985, **167**, 111.
6. L.C. Craig and K. Stewart, *Biochemistry*, 1965, **4(12)**, 2712.
7. C.F. Mandenius, B. Danielsson, and B. Mattiasson, *Anal. Chim. Acta*, 1984, **163**, 135.
8. T.D. Gibson and J.R. Woodward, *Anal. Chim. Acta*, 1988, **213**, 61.
9. W.E. Neeley, *Clin. Chem.*, 1972, **18**, 509.
10. W.R. Wright, J.C. Rainwater, and L.D. Tolle, *Clin. Chem.*, 1971, **17**, 1010.
11. M.A. Kenny and M.H. Cheng, *Clin. Chem.*, 1972, **18**, 352.
12. W.A. Osborne, *J. Physiol.*, 1906, **34**, 84.
13. R.M. Rosenberg and I.M. Klotz, in 'A Laboratory Manual of Analytical Methods of Protein Chemistry' Vol.2, ed. P. Alexander and R.J. Block, Pergamon Press, New York, 1960, p. 131.
14. I.M. Klotz, F.M. Walker, and R.B. Pivan, *J. Am. Chem. Soc.*, 1946, **68**, 1486.
15. G. Scarchard, *Ann. N.Y. Acad. Sci.*, 1949, **51**, 660.
16. F. Karush, *J. Am. Chem. Soc.*, 1050, **72**, 2705.
17. H.H. Stein, *Anal. Biochem.*, 1965, **13**, 305.
18. A.H. Kadish and D.A. Hall, *Clin. Chem.*, 1965, **11**, 869.
19. E. Martins and G. Johansson, *Anal. Chim. Acta*, 1982, **140**, 29.
20. E. Martins, M. Bengtsson, and G. Johansson, *Anal. Chim. Acta*, 1985, **169**, 31.
21. M.M. Joe, D. Sakai, and R.A. Moffit, 'Simultaneous Determination of Nitrogen, Calcium and Phosphorus in Fluid Milk Products', Technicon Symposia 1966, Vol. 1, Mediad. Incl, New York, 1967, pp. 595-597.
22. W.D. Basson and J.F. van Staden, *Analyst (London)*, 1978, **103**, 296.
23. W.D. Basson and J.F. van Staden, *Analyst (London)*, 1979, **104**, 419.
24. J.F. van Staden and W.D. Basson, *Lab. Pract.*, 1980, **29**, 1279.
25. E.H. Hansen and J. Ruzicka, *Anal. Chim. Acta*, 1976, **87**, 353.
26. R.Y. Xie and G.D. Christian, *Anal. Chem.*, 1986, **58**, 1808.
27. L. Gorton and L. Ogren, *Anal. Chim. Acta*, 1981, **130**, 45.
28. M. Masoom and A. Townshend, *Anal. Chim. Acta*, 1984, **166**, 111.
29. Q. Chang and M.E. Meyerhoff, *Anal. Chim. Acta*, 1986, **186**, 81.
30. R.A. Durst, *Anal. Letts*, 1977, **10**, 961.
31. K.P. Ang, *J. Chem. Soc.*, 1959, 3822.
32. T.D.B. Morgan, G. Stedman, and P.A.E. Whincup, *J. Chem. Soc.*, 1965, 4813.
33. A. Tanaka, K. Mashiba, and T. Deguchi, *Anal. Chim. Acta*, 1988, **214**, 259.
34. M.A. Koupparis and P. Anagnostopoulou, *J. Autom. Chem.*, 1984, **6**, 186.
35. P.E. Macheras and M.A. Koupparis, *Anal. Chim. Acta*, 1986, **185**, 65.
36. S.J. Gluck, *Anal. Chim. Acta*, 1988, **214**, 315.
37. B. Olsson, H. Lundbäck, and G. Johansson, *Anal. Chim. Acta*, 1985, **167**, 123.
38. H. Lundbäck and B. Olsson, *Anal. Lett.*, 1985, **18(B7)**, 871.
39. J. Lankelma and H. Hoppe, *J. Chromatogr.*, 1978, **149**, 587.
40. UK Patent Application 2 124 370A.
41. T.C. Butler, in 'Antiepileptic Drugs: Quantitative Analysis and Interpretation', ed. C.E. Pippenger, J.K. Penry, and H. Kutt, Raven Press, New York, 1978, pp. 268.
42. P.M. Kabra, B.E. Stafford, and L.J. Marton, *Clin. Chem.*, 1977, **145**, 1284.
43. D.C. Turnell and J.D.H. Cooper, *J. Autom. Chem.*, 1985, **7(4)**, 177.
44. F.R. Nordmeyer and L.D. Hansen, *Anal. Chem.*, 1982, **54**, 2605.
45. D.C. Turnell and J.D.H. Cooper, *J. Chromatogr.*, 1987, **395**, 613.

46. D.C. Turnell, J.D.H. Cooper, B. Green, G. Hughes, and D.J. Wright, *Clin. Chem.*, 1988, **34**, 1816.
47. J.D.H. Cooper, D.C. Turnell, and B. Green, *J. Chromatogr.*, 1988, **456**, 53.
48. R.G. Melcher, *Anal. Chim. Acta*, 1988, **214**, 299.
49. G. Audunsson, *Anal. Chem.*, 1986, **58**, 2714.
50. K.H. Kroner and M.R. Kula, *Anal. Chim. Acta*, 1984, **163**, 3.
51. U. Nalback, H. Schiemenz, W.W. Stamm, W. Hummel, and M.R. Kula, *Anal. Chim. Acta*, 1988, **213**, 55.
52. N. Ahlmann, A. Niehoff, U. Rinas, Th. Scheper, and K. Schügerl, *Anal. Chim. Acta*, 1986, **190**, 221.
53. P.K. Dasgupta, *Anal. Chem.*, 1984, **56**, 96.
54. H. Hwang and P.K. Dasgupta, *Anal. Chem.*, 1986, **58**, 1521.
55. P.K. Dasgupta and H.C. Yang, *Anal. Chem.*, 1986, **58**, 2839.
56. P.K. Dasgupta and H. Hwang, *Anal. Chem.*, 1985, **57**, 1009.
57. A. Ríos, M.D. Luque de Castro, and M. Valcárcel, *Anal. Chem.*, 1986, **58**, 663.
58. P.K. Dasgupta, R. Vithanage, and K. Petersen, *Anal. Chim. Acta*, 1988, **215**, 277.
59. V.J. Nau and T.A. Nieman, *Anal. Chem.*, 1979, **51**, 424.
60. D. Pilosof and T.A. Nieman, *Anal. Chem.*, 1980, **52**, 662.
61. D. Pilosof and T.A. Nieman, *Anal. Chem.*, 1982, **54**, 1698.
62. U. Ungerstedt, *Current Separations*, 1986, **7(2)**, 43.
63. U. Ungerstedt, 'Measurement of Neurotransmitter Release *in Vivo*', ed. C.A. Marsden, John Wiley , 1984, pp. 81–105.
64. U. Tossman and U. Ungerstedt, *Acta Physiol. Scand.*, 1986, **128**, 9.
65. J.B. Justice, Jr., and D.B. Neill, *Ann. N.Y. Acad. Sci.*, 1987, 170.
66. S.A. Wages, W.H. Church, and J.B. Justice Jr., *Anal. Chem.*, 1986, **58**, 1649.
67. W.H. Church and J.B. Justice Jr., *Anal. Chem.*, 1987, **59**, 712.
68. D.O. Scott, M.A. Bell, and C.E. Lunte, *J. Pharm. Biomed. Anal.*, 1989, **7(11)**, 1249.
69. W.G. Kuhr and J. Korf, *Anal. Chim. Acta*, 1988, **205**, 53.
70. S.D. Menacherry and J.B. Justice Jr., *Anal. Chem.*, 1990, **62**, 597.
71. M. Valcárcel and M. Silva, 'Teoría y Práctica de la Extracción Líquido-Líquido', Alhambra, Madrid, 1984.
72. L.C. Craig, *Anal. Chem.*, 1950, **22**, 1346.
73. M.D. Luque de Castro, *J. Autom. Chem.*, 1986, **8**, 56.
74. M. Valcárcel in Chapter 7 of 'Developments in Solvent Extraction', ed. S. Alegret, Ellis Horwood, Chichester, 1989.
75. M. Valcárcel and A. Gómez-Hens 'Técnicas Analíticas de Separación', Reverté, Barcelona, 1988.
76. Y. Ito and N.D. Conway, *Anal. Chem.*, 1984, **56**, 534A.
77. Y. Ito, *Trends Anal. Chem.*, 1986, **5**, 142.
78. W.B. Furman 'Continuous Flow Analysis. Theory and Practice', M. Dekker, New York, 1982.
79. J. Ruzicka and E.H. Hansen, 'Flow Injection Analysis', 2nd Edn., Wiley, New York, 1988.
80. M. Valcárcel and M.D. Luque de Castro 'Flow Injection Analysis. Principles and Applications', Ellis Horwood, Chichester, 1987.
81. IUPAC 'Compendium of Analytical Nomenclature', Blackwells, Oxford, 1987.
82. S. Motomizu and M. Oshima, *Analyst (London)*, 1987, **112**, 295.

83. J. Kawase, *Anal. Chem.*, 1980, **52**, 2124.
84. F.F. Cantwell and J.A. Sweileh, *Anal. Chem.*, 1985, **57**, 329.
85. D.T. Burns, N. Chimpalee, and M. Harriot, *Anal. Chim. Acta*, 1989, **225**, 123.
86. T.M. Rossi, D.C. Shelly, and I.M. Warner, *Anal. Chem.*, 1982, **52**, 2124.
87. L. Nord and B. Karlberg, *Anal. Chim. Acta*, 1984, **164**, 233.
88. B. Karlberg, *Anal. Chim. Acta*, 1986, **180**, 16.
89. L. Nord, K. Backstrom, L.G. Danielsson, P. Ingman, and B. Karlberg, *Anal. Chim. Acta*, 1987, **194**, 221.
90. L. Nord, S. Johansson, and H. Brotell, *Anal. Chim. Acta*, 1985, **175**, 281.
91. L. Nord, M. Sundgren, and A. Torstensson, *J. Pharm. Biomed. Anal.*, 1989, **4**, 421.
92. J.M. Carter and G. Nickless, *Analyst (London)*, 1970, **95**, 148.
93. V. Wallace, *Anal. Biochem.*, 1967, **20**, 411.
94. J.L. Burguera and M. Burguera, *Anal. Chim. Acta*, 1983, **153**, 207.
95. F.H. Bergamin, J.X. Medeiros, B.F. Reis, and E.A.G. Zagatto, *Anal. Chim. Acta*, 1978, **101**, 9.
96. B. Karlberg and S. Thelander, *Anal. Chim. Acta*, 1978, **98**, 1.
97. T.M. Rossi, D.C. Shelly, and I.M. Warner, *Anal. Chem.*, 1982, **54**, 2056.
98. L. Nord and B. Karlberg, *Anal. Chim. Acta*, 1980, **118**, 285.
99. T. Imasaka, T. Harada, and N. Ishibashi, *Anal. Chim. Acta*, 1981, **129**, 195.
100. J. Toei, *Analyst (London)*, 1988, **113**, 1861.
101. K. Backstrom, L.G. Danielsson, and L. Nord, *Anal. Chim. Acta*, 1986, **187**, 255.
102. L. Fossey and F.F. Cantwell, *Anal. Chem.*, 1982, **54**, 1693.
103. K. Backstrom, L.G. Danielsson, and L. Nord, *Anal. Chim. Acta*, 1985, **169**, 43.
104. S. Motomizu and K. Korechika, *Anal. Chim. Acta*, 1989, **220**, 275.
105. L. Fossey and F.F. Cantwell, *Anal. Chem.*, 1983, **55**, 1882.
106. K. Ogata, K. Taguchi, and T. Imanad, *Anal. Chem.*, 1982, **54**, 2127.
107. G. de Ruiter, J.H. Wolf, U.A.Th. Brinkman, and R.W. Frei, *Anal. Chim. Acta*, 1987, **192**, 267.
108. Y. Sahlestrom and B. Karlberg, *Anal. Chim. Acta*, 1986, **179**, 315.
109. H.L. Nekimken, B.F. Smith, G.D. Jarvinen, E.J. Peterson, and M.M. Jones, *Anal. Chem.*, 1988, **60**, 1390.
110. S. Motomizu, M. Onoda, M. Oshima, and T. Iwachdo, *Analyst (London)*, 1988, **113**, 743.
111. S. Motomizu, M. Oshima, and T. Kuroda, *Analyst (London)*, 1988, **113**, 747.
112. R.H. Atallah, J. Ruzicka, and G.D. Christian, *Anal. Chem.*, 1987, **59**, 2909.
113. R.H. Atallah, G.D. Christian, and S.D. Hartenstein, *Analyst (London)*, 1988, **113**, 463.
114. M. Valcárcel and M. Gallego, Chapter 5 of 'Flow Injection Atomic Spectroscopy', ed. J.L. Burguera, M. Dekker, New York, 1989.
115. K. Fujiwara, *Anal. Chim. Acta*, 1988, **212**, 245.
116. M. Bengtsson and G. Johansson, *Anal. Chim. Acta*, 1984, **158(2)**, 147.
117. K. Backstrom, L.G. Danielsson, and L. Nord, *Analyst (London)*, 1984, **109**, 323.
118. B. Karlberg and S. Thelander, *Anal. Chim. Acta*, 1980, **114**, 129.
119. A. Sodergren, *Analyst (London)*, 1965, **91**, 113.
120. P.D. Taegye, *Tenside Deterg.*, 1975, **12**, 41.
121. T. Tsuda and M. Yamanaka, *Yukagaku*, 1979, **28**, 398.
122. R. Sawyer, P.B. Stockwell, and K.B.E. Tucker, *Analyst (London)*, 1970, **95**, 879.

123. K.B.E. Tucker, R. Sawyer, and P.B. Stockwell, *Analyst (London)*, 1970, **95**, 730.
124. L. Nyberg, *J. Pharm. Pharmacol.*, 1970, **22**, 500.
125. S. Ahuja, C. Spitzer, and F.R. Brofazi, *J. Pharm. Sci.*, 1970, **12**, 1833.
126. E.R. Rutter, *Clin. Chem.*, 1972, **18**, 616.
127. D.J. Blackmore, A.S. Curry, T.S. Hayes, and E.R. Rutter, *Clin. Chem.*, 1971, **17**, 896.
128. B. Karlberg, P.A. Johansson, and S. Thelander, *Anal. Chim. Acta*, 1979, **104**, 21.
129. T. Kato, *Anal. Chim. Acta*, 1985, **175**, 339.
130. D.T. Burns, N. Chimpalee, and M. Harriot, *Anal. Chim. Acta*, 1989, **225**, 449.
131. R. Montero. Unpublished results.
132. R. Santelli, M. Gallego, and M. Valcárcel. Unpublished results.
133. K. Ogata, S. Tanabe, and T. Imanari, *Chem. Pharm. Bull.*, 1983, **31(4)**, 1419.
134. J.L. Burguera, M. Burguera, L. Cruz, and O.R. Naranjo, *Anal. Chim. Acta*, 1986, **186**, 273.
135. M. Gallego and M. Valcárcel, *Anal. Chim. Acta*, 1985, **169**, 161.
136. J.L. Burguera and M. Burguera, *Anal. Chim. Acta*, 1983, **153**, 207.
137. K. Ogata, K. Taguchi, and T. Imanari, *Bunseki Kagaku*, 1982, **31**, 641.
138. G.E. Pacey, Y.P. Wu, and K. Sasaki, *Anal. Biochem.*, 1987, **160**, 243.
139. K. Kimura, S. Iketani, H. Sakamoto, and T. Shono, *Anal. Sci.*, 1988, **4**, 221.
140. American Public Health Association. Washington DC, 1980, pp. 530.
141. J. Kawase, A. Nakae, and M. Yamanaka, *Anal. Chem.*, 1979, **51**, 1640.
142. M. del Valle, J. Alonso, J. Bartrolí, and I. Martí, *Analyst (London)*, 1988, **113**, 1677.
143. S. Motomizu, Y. Hazaki, M. Oshima, and K. Toei, *Anal. Sci.*, 1987, **3**, 265.
144. M. Gallego, M. Silva, and M. Valcárcel, *Anal. Chem.*, 1986, **58**, 2265.
145. P. Martínez-Jiménez, M. Gallego, and M. Valcárcel, *Anal. Chim. Acta*, 1988, **215**, 233.
146. T. Iwachido, M. Onoda, and S. Motomizu, *Anal. Sci.*, 1986, **2**, 493.
147. O. Klinghoffer, J. Ruzicka, and E.H. Hansen, *Talanta*, 1980, **27**, 169.
148. H. Koizumi and Y. Suzuki, *Anal. Sci.*, 1988, **4**, 537.
149. Y. Sahlestrom, S. Twengstrom, and B. Karlberg, *Anal. Chim. Acta*, 1986, **187**, 339.
150. J.S. Canham and G.E. Pacey, *J. Am. Oil Chem. Soc.*, 1987, **64**, 1004.
151. T. Mise, *J. Flow Injec. Anal.*, 1988, **5**, 87.
152. M. Harriot and D.T. Burns, *Anal. Proc.*, 1989, **26**, 315.
153. D.T. Burns, N. Chimpalee, M. Harriot, and G.M. McKillen, *Anal. Chim. Acta*, 1989, **217**, 183.
154. D.T. Burns, N. Chimpalee, and M. Harriot, *Anal. Chim. Acta*, 1989, **217**, 177.
155. D.T. Burns, N. Chimpalee, and M. Harriot, *Anal. Proc.*, 1989, **26**, 4.
156. J. Coello, L.G. Danielsson, and S. Hernández-Cassou, *Anal. Chim. Acta*, 1987, **201**, 325.
157. L. Sun, L. Li, and Z. Fang, *Fenxi Huaxue*, 1985, **13**, 447.
158. L. Fossey and F.F. Cantwell, *Anal. Chem.*, 1985, **57**, 922.
159. P.A. Johansson, B. Karlberg, and S. Thelander, *Anal. Chim. Acta*, 1980, **114**, 215.
160. J.A. Sweileh and F.F. Cantwell, *Anal. Chem.*, 1985, **57**, 420.
161. M. Valcárcel and M. Gallego, *Quim. Anal.*, 1987, **6**, 321.
162. M. Gallego, M. Silva, and M. Valcárcel, *Fresenius' Z. Anal. Chem.*, 1986, **323**, 50.

163. M. Gallego, M. Silva, and M. Valcárcel, *Anal. Chim. Acta*, 1986, **179**, 341.
164. K. Kina, K. Shiraishi, and N. Ishibashi, *Talanta*, 1978, **25**, 295.
165. E. Pelizzeti and E. Pramuro, *Anal. Chim. Acta*, 1985, **169**, 1.
166. M.H. Memon and P.J. Worsfold, *Anal. Chim. Acta*, 1986, **183**, 179.
167. M.H. Memon and P.J. Worsfold, *Anal. Chim. Acta*, 1987, **201**, 345.
168. J. Toei, *Talanta*, 1989, **36**, 691.
169. J. Ruzicka and A. Arndal, *Anal. Chim. Acta*, 1989, **216**, 243.
170. P.R. Danesi in Chapter 9 of 'Developments in Solvent Extraction', ed. S. Alegret, Ellis Horwood, Chichester, 1988.
171. F. Cañete, A. Ríos, M.D. Luque de Castro, and M. Valcárcel, *Anal. Chem.*, 1988, **60**, 2354.
172. F. Cañete, A. Ríos, M.D. Luque de Castro, and M. Valcárcel, *Anal. Chim. Acta*, 1989, **224**, 169.
173. J.A.G. Mesa, P. Linares, M.D. Luque de Castro, and M. Valcárcel, *Anal. Chim. Acta*, 1990, **235**, 441.
174. Y. Hirai and K. Tomokuni, *Anal. Chim. Acta*, 1985, **167**, 409.
175. J. Roeraade, *J. Chromatogr.*, 1985, **330**, 263.
176. E. Fogelqvist, M. Krysell, and L.G. Danielsson, *Anal. Chem.*, 1986, **58**, 15.
177. E. Ballesteros, M. Gallego, and M. Valcárcel, *Anal. Chem.*, 1990, **62**, 1587.
178. 'Reaction Detection in Liquid Chromatography', ed. I.S. Krull, M. Dekker, New York, 1986.
179. J.W. Dolan, J.R. Grant, N. Tanaka, R.W. Giese, and B.L. Karger, *J. Chromatogr. Sci.*, 1978, **16**, 616.
180. A. Farrán, J. de Pablo, and S. Hernández, *Anal. Chim. Acta*, 1988, **212**, 123.
181. J.C. Gfeller, R. Frei, J.M. Huen, and J.P. Thevenin, *J. High Res. Chromatogr.*, 1978, **41**, 213.
182. R. Frei, J.F. Lawrence, U.A.Th. Brinckman, and I.L. Honingberg, *J. High Res. Chromatogr.*, 1979, **42**, 11.
183. B.L. Karger, D.P. Kirby, P. Vouros, R.L. Foltz, and B. Hidy, *Anal. Chem.*, 1979, **51**, 2324.
184. D.P. Kirby, P. Vouros, B.L. Karger, B. Hidy, and C. Petersen, *J. Chromatogr.*, 1981, **203**, 139.
185. C. van Buuren, J.L Lawrence, U.A.Th. Brinckman, I.L. Honingberg, and R.W. Frei, *Anal. Chem.*, 1980, **52**, 700.
186. R.J. Reddingius, G.H. de Song, U.A.Th. Brinckman, and R.W. Frei, *J. Chromatogr.*, 1981, **205**, 77.
187. C. de Ruiter, J.C.H.F. Hefkens, U.A.Th. Brinckman, R.W. Frei, M. Evers, E. Matthijs, and J.A. Meijer, *Intern. J. Environ. Anal. Chem.*, 1987, **31**, 325.
188. J.F. Lawrence, U.A.Th. Brinckman, and R.W. Frei, *J. Chromatogr.*, 1979, **171**, 73.
189. C.E. Werkhoven-Goewie, U.A.Th. Brinckman, and R.W. Frei, *Anal. Chim. Acta*, 1980, **114**, 147.
190. K. Tsuji, *J. Chromatogr.*, 1978, **158**, 337.

CHAPTER 4

Liquid–Solid Systems

4.1 Introduction

The non-chromatographic continuous separation systems described in this chapter are based on mass transfer between a liquid and a solid phase. As a rule, it is the analytes or their derivatives which take part in the separation process which, however, also occasionally involve interfering species. These systems rely on a variety of phenomena (sorption, electrodeposition-stripping, precipitation, leaching) that call for quite varied configurations. Nevertheless, they share some basic aspects in addition to the type of interface involved, aspects which are commented on below.

The processes involved in liquid–solid separations can be purely chemical (*e.g.* precipitation/filtration/dissolution), physico-chemical (*e.g.* sorption and ion-exchange) or electrochemical (*e.g.* stripping techniques). Mass transfer can take place in a reactor (*e.g.* sorption, ion-exchange) or in the detector itself (*e.g.* in stripping techniques implemented in a continuous fashion or in integrated separation–detection systems).

The continuous variants of the above techniques are formally resemblant of chromatography—particularly ion exchange and sorption—with the basic difference of the degree of discrimination accomplished: between analytes in chromatography and overall here. A continuous liquid–solid separation always takes place in two steps that involve retention (ion-exchange, sorption, electrodeposition) or precipitation/dissolution, and elution, electrode stripping or precipitate dissolution, respectively. One exception to this rule is late on-line leaching systems which involve a single-step process.

The active solid phase is a permanent part of the continuous systems in most alternatives except precipitation, where it is generated *in situ*. As a rule, the sample to be treated is liquid and is introduced into the analytical system by injection or continuous aspiration. Only in one technique, *viz*. on-line leaching, is the sample solid and constitutes the second phase, which is somewhat resemblant of the introduction of gas samples dealt with in Chapter 2.

Most of the methodologies developed for continuous analysers based on a liquid–solid interface are implemented at room temperature or with mild heating to accelerate the kinetics of the physico-chemical phenomena involved in mass transfer. Occasionally, an external source of energy is required; such is the case with on-line leaching, which is aided by application of an electric discharge or an ultrasonic field, and of stripping techniques, the performance of which is boosted by applying a potential difference.

Most of the continuous systems described in this chapter can be considered to derive from batch processes involving intensive human participation. A critical comparison of the results accomplished by the two versions of each separation technique on application to the same analyte and type of sample is of great interest in evaluating their respective advantages and disadvantages. Wherever applicable, comments in this respect are made below.

The following sections deal with non-chromatographic separation systems techniques implemented in automatic continuous analytical systems. The first two sections are concerned with the two types of sorption systems most widely used. There follows a description of continuous configurations based on electrochemical stripping. Finally, two recently developed solid–liquid separation techniques that have aroused much interest from researchers and analysts alike, namely precipitation/filtration/dissolution and on-line leaching, are commented on.

4.2 Ion Exchange

Ion exchange is one of the oldest separation techniques known. Unlike other separation techniques, it allows the ionic composition of the analytical solution to be changed without introducing undesirable interfering species. Ion exchangers are most frequently used as packing materials for chromatographic columns,[1] though they are also employed as active materials in sample pretreatment processes (trace enrichment, clean-up, storage, guard function) applied prior to introduction into the chromatographic system.* In addition to this basic function, ion exchange is widely used to preconcentrate the species of interest prior to its analytical determination and for a variety of other purposes such as facilitating multi-determinations and/or speciation, accomplishing indirect determinations, and integrating reaction(retention) and detection, all of which are discussed in this section.

The solid–liquid interface arising from the presence of the ion exchange material allows passage of the species to be retained by the solid via a physico-chemical mechanism in which electrostatic attraction phenomena are prevalent, but are also accompanied by adsorption, chelate-formation and other processes.

* Insofar as these functions can also be performed by substances other than ion exchangers, all of them are dealt with jointly in Section 4.3, devoted to sorption systems coupled on-line to liquid chromatographs.

The easy experimental implementation of ion-exchange techniques and the simplicity of the basic units required make them quite affordable in contrast to other separation alternatives. These assets are further enhanced by implementation in continuous systems, which also result in minimal consumption of sample, reagents, and exchange material.

4.2.1 Ion exchangers

The exchange material is the indispensable element in developing an on-line ion-exchange separation procedure; it is usually placed into a column whose geometric features are suited to the continuous system to be used. Some recently developed methodologies use no columns, but place the ion exchanger within the detector flow-cell, so the separation and/or concentration process takes place simultaneously with detection.

4.2.1.1 Types of Ion Exchanger

Cationic exchangers are the most widely used in continuous systems insofar as their chief goal is the preconcentration and separation of metal species in general and heavy metals in particular. Among cationic exchangers (strong and weak acids, chelating agents with an active group), chelating species are the most selective and thus also the most frequently used. The chelating group can be supported on an organic polymer (Chelex-100, Dowex A-1, Muromac A-1, Resin 122) or on an inorganic material such as controlled pore glass (CPG), silica gel, or alumina.

Anionic exchangers (bases of small, medium, or great strength) are less commonly used with continuous systems and are normally intended to remove anionic interferences with the determination of cationic species. The anionic exchangers most widely used are of organic nature (Amberlite and De-Acidite FF, among others). Some inorganic exchangers (*e.g.* alumina) can act as cationic or anionic, depending on the pH of the medium.

The amount of exchange material required to construct an ion-exchange set-up is a function of the type of exchange material and the dimensions (length and diameter) of the column to be used. By way of example, a 35 mm long column with an ID of 2.8 mm takes 36 mg of Chelex-100 in ammonium form, 10 mg of 8-quinolinol-based resin, or 110 mg of Resin 122. When the exchange material is placed in the detector flow-cell, the amount to be used will depend on whether it adheres to the cell walls (*ca.* 0.5 mg) or fills part of the cell (between 0.3 and 1.0 mg, depending on the cell volume).

4.2.1.2 Features of Ion Exchangers

The influence of the exchanger features on the performance of a continuous separation system is rather different to that it exerts on a batch separation unit.

Particle size plays a major role in ion-exchange separations: the smaller the particle size, the more efficient the separation will be. However, as particle size decreases, pressure within the system increases, which requires making a compromise between the two extremes. The mesh size of the chelating resins used in this type of separation ranges between 50–100 for Chelex-100, Dowex A-1, and Resin 122. On the other hand, the particle size of inorganic exchangers usually varies between 100 and 200 μm for CPG, 150–250 μm for silica gel, and 75–125 μm for activated alumina.

The *swelling capacity* of ion exchangers is of great significance to continuous systems since changes in the resin volume result in concomitant changes in the pressure within the system that can in turn give rise to problems such as erratic blockage, leakage and flow-rate oscillations. The elastic gel structure of ion-exchange resins makes them particularly vulnerable to swelling. However, of such resins some (*e.g.* Chelex-100) are affected to a greater extent than others (*e.g.* Muromac A-1) with the same exchange group (iminodiacetate); thus, the resin Chelex-100 undergoes volume changes of up to 100% in passing from its H^+ to its NH_4^+ form.[2]

The *breakthrough capacity* of an exchanger is defined as the amount of exchangeable species that can be extracted per unit mass of the solid under the prevailing working conditions before it becomes detectable in the eluate. Breakthrough is directly related both to the features of the separation system (column dimensions, particle size) and to the working conditions (flow-rate, temperature, nature, and concentration of the counter-ion). Insofar as those factors inherent in the separation unit that increase breakthrough capacity also increase pressure within the system, a compromise solution must be adopted in every instance.

The preconcentration factor, also known as 'enrichment factor', is directly related to the breakthrough capacity of an exchanger. The 'concentration efficiency' is an alternative to the expression of the preconcentration factor which is obtained by multiplying it by the sampling frequency (in reciprocal minutes).

The kinetics of the exchange step are a key to choosing a suitable exchange material for use in continuous systems, in contrast to conventional preconcentration systems, in which the whole retained solute is eluted prior to measurement; thus, provided the breakthrough capacity is not reached, the amount of eluted analyte will be identical with that originally present in the sample. Conversely, continuous systems involve on-line elution; this gives rise to a transient signal, the features of which depend on the elution rate—the lower the rate is, the more difficult quantification of the analyte from the elution profile will be. On the other hand, the kinetics of analyte retention influence both the selectivity of the separation process and sample throughput. In this respect, the flow-rate is a key variable in suiting the system to the kinetics of the process concerned.

Comparative studies[2] made on various exchangers (Chelex-100, 8-

quinolinol immobilized on CPG, and Resin 122) have shown the efficiency of the separation process to vary widely with the element involved (Cu, Zn, Pb, Cd), as well as between exchangers used with the same element. In fact, 8-quinolinol on CPG yielded the highest concentration factors, probably as a result of the faster exchange of its surface chelating groups, which expedite elution of the analyte that is ultimately contained in a smaller solvent volume. Element recovery was found to decrease drastically with decreasing stability constant of the complexes formed—hence the null or little interference of alkaline and alkaline-earth metals with the separation of heavy metals—and with the presence of species acting as potential ligands for the cations to be retained. On the other hand, and for a given chelating group, the selectivity achieved depended on the support onto which the exchanger was attached. In this respect, 8-quinolinol was found to be more selective supported on silica than on CPG.[3]

4.2.2 Continuous Configurations

The development of an analytical determinative methodology involving an ion-exchange separation step entails using a series of basic units which are described below. Some, such as columns, are typical of separation techniques; others, such as valves and detection systems, are adapted for use with these types of system. These units, in addition to those typical of continuous methodologies, are assembled to construct the different types of configurations described below.

4.2.2.1 Columns

The central unit of a conventional on-line ion-exchange system is a mini-column packed with exchange material and in which the exchange proper takes place on passage of the solution. The column, which can be tubular of conical in shape, is usually made of glass, Teflon, Tygon or PVC and measures between 10 and 50 mm in length and 1.5–3.2 mm in ID.

The location of the column in a continuous system is a function of the analytical goal. For interference removal, the column is preferentially placed at the beginning of the configuration, close to the zone aspiration sample—if the sample is not injected—and before the confluence point with any streams of derivatizing reagents, or prior to the injection system in an FIA manifold. If the chief goal is to isolate the analyte from the sample matrix, the column should be placed as close as possible to the detection system, particularly when no derivatization reaction is required. For analyte preconcentration, the column should be located within the injection system of an FIA manifold and time-based injection should be done in order to ensure appropriate performance. Finally, for other specific purposes, the column can be placed in less common locations (*e.g.* between two detectors, making up a conductivity cell to obtain differential measurements proportional to the concentration of acid or base in the unknown sample[4]).

Construction of the column varies with its shape (Figure 4.1), but is never especially laborious. Tubular columns (Figure 4.1A) are made by placing a small plug of glass wool, dimethylchlorosilane-treated glass wool, nylon netting, gauze, or Teflon wool in one of the ends and aspirating a usually aqueous suspension of the exchange material with the aid of a syringe until the tube is filled completely or partly, depending on the swelling capacity of the exchanger. Another plug of the filter material is placed in the other end. The difference in diameter between the column and the rest of the continuous system calls for the use of customized small connecting tubes of an intermediate size or more elaborate—and also more efficient—commercially available connectors (UNF plastic or stainless steel swagelok connectors), on the wider end of which is placed a porous disc (PVC, glass), which is fixed to the connector together with the column by means of epoxy glue (Figure 4.1B). The narrower end is fitted to the rest of the system. This type of connector is more resistant to pressure changes and easier to handle.

Continuous circulation of a liquid in the same direction tends to compact the exchange material held in the column and hence to increase pressure within the system. This problem can be overcome by placing the column upright and passing the solution upstream or by carrying out elution in the opposite direction of retention.

Conical columns (Figure 4.1C) can be readily constructed from pipette tips fitted with any of the above-mentioned materials, which are connected to the manifold via tubes of appropriate diameter. The direction of

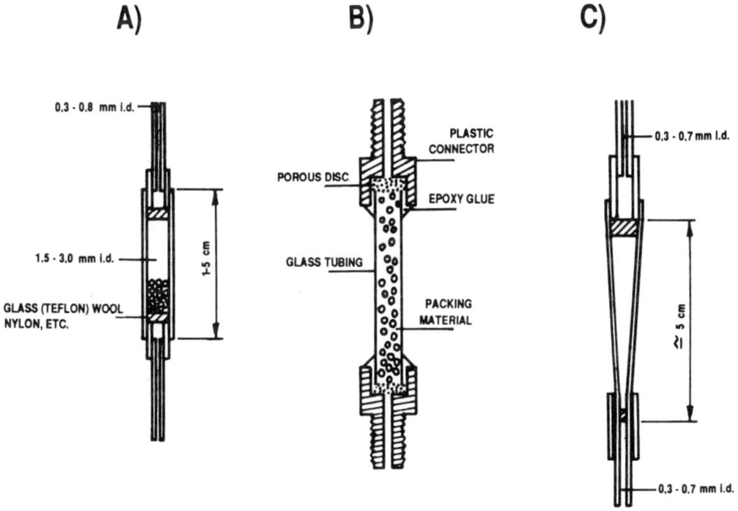

Figure 4.1 *Types of ion-exchange columns.* (A, B) *Tubular.* (C) *Conical*

the column in the manifold is arranged in such a way that the adsorbed analytes are eluted from the wider to the narrower end of the column in order to minimize dispersion of the analyte in the column. For similar reasons, sample loading is effected countercurrent in order that the bulk adsorbed analyte may be retained at the narrower end of the column, thereby also avoiding compaction of the exchange material.

4.2.2.2 Valves

Valves are not indispensable for the construction of continuous manifolds including an on-line ion-exchange separation system. However, they overcome some of the problems encountered in this respect and contribute to improved performance. Valves typically used with other techniques (injection, selecting, and diverting) are suitably modified for adaptation to specific purposes.

Injection valves. The injection units used in this type of system range from single valves or switches accommodating the exchange column in their loops (Figures 4.2A and B) to two valves arranged in series or parallel and intended to effect sequential injection of the sample and eluent (serial arrangement) or increase sample throughput by using an exchange column in each loop (parallel arrangement). The functions of two coupled valves can also be served by a single eight-port valve or a dual switching system.

Simple injection valves (rotary or switching) can be readily modified to accommodate the exchange column in their loop. This requires controlling the sample volume to be passed through the column, which can be accomplished by using a timer or a computer to program accurately the interval over which the sample is passed through the loop and hence the sample volume (time-based mode) or by using a reverse operational mode and having the sample circulate continuously through the system in such a way that the sample volume passing through the column is determined by the time the valve is kept in its inject position. This type of valve can be coupled serially to a selecting valve so that, once the analyte has been eluted and before a fresh sample is introduced into the loop, a washing cycle is effected by passing appropriate washing and rinsing solutions through the column to regenerate the exchanger.

Two serially arranged injection valves located prior to a column have been used to inject the sample and then the eluent, consumption of which is minimized, thereby avoiding potential hazards (*e.g.* when using strong acids or bases as eluents) arising from its continuous presence in the system (Figure 4.2C). The two valves can be identical (loop-based injection) or different (*e.g.* for preconcentration, which requires one of them to be of the time-based type). This kind of assembly can be replaced with an eight-way rotary valve, which acts as illustrated in Figure 4.3. In position A, the sample is inserted into a carrier stream that drives it to the column while the eluent fills the corresponding loop. In position B, the eluent

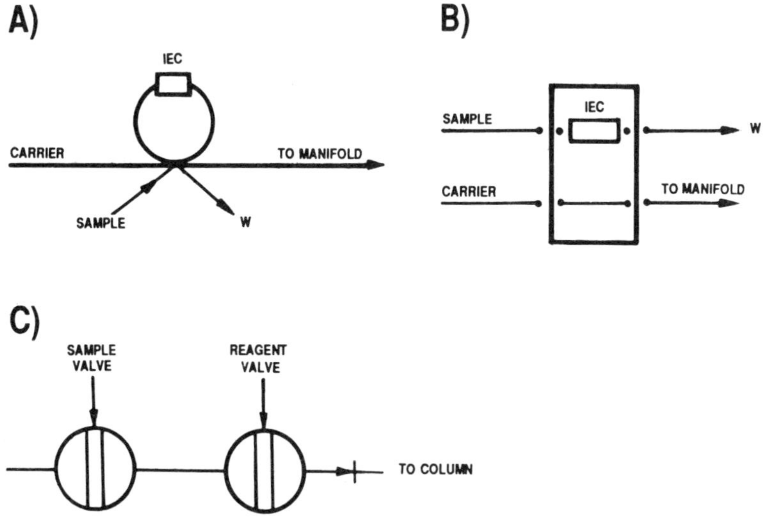

Figure 4.2 (A) *Rotary valve, and* (B) *injector-switch accommodating an ion-exchange column in their loops.* (C) *Serially arranged valves for the sequential insertion of sample and eluent (the sample valve can perform time-based injections)*

held in loop L_E is led to the column by the carrier as the sample fills loop L_S. This operational scheme results in a simplified manifold.

Two injection valves arranged in parallel and accommodating a column in their loop have been used to increase sample throughput by synchronizing their load and unload operations. This can also be accomplished by using a simpler configuration including an eight-way rotary valve or a three-layer sandwich valve. Figure 4.4 shows the two positions the valve can take: one for retention and the other for elution. The gain in terms of sample throughput is not too significant taking into account that the two steps usually have a different duration—elution is normally much shorter—so the system has to be suited to the lengthier. This shortcoming can be circumvented by using two peristaltic pumps, the functioning and operational sequences of which are illustrated in Figure 4.5. The operation starting at position A coincides with the start of sampling, which is effected with the aid of peristaltic pump P_1, at the end of which the pump is stopped and pump P_2 is started to propel the eluent solution through column 1, thereby eluting the analyte retained there, the solution is then driven to the manifold (Figure 4.5B). Then, pump P_2 is stopped and P_1 is restarted and kept functioning for the same time as P_2, which results in the analyte being eluted and driven to column 2. After that time, the valve is switched back to position A and a new cycle is started. The time over which P_2 is functioning is used for sample changeover. The water circulated in the preconcentration step is employed to flush the system. The analyte is diluted in the opposite direction to preconcentration, thereby minimizing column compaction arising both from build-up of exchange

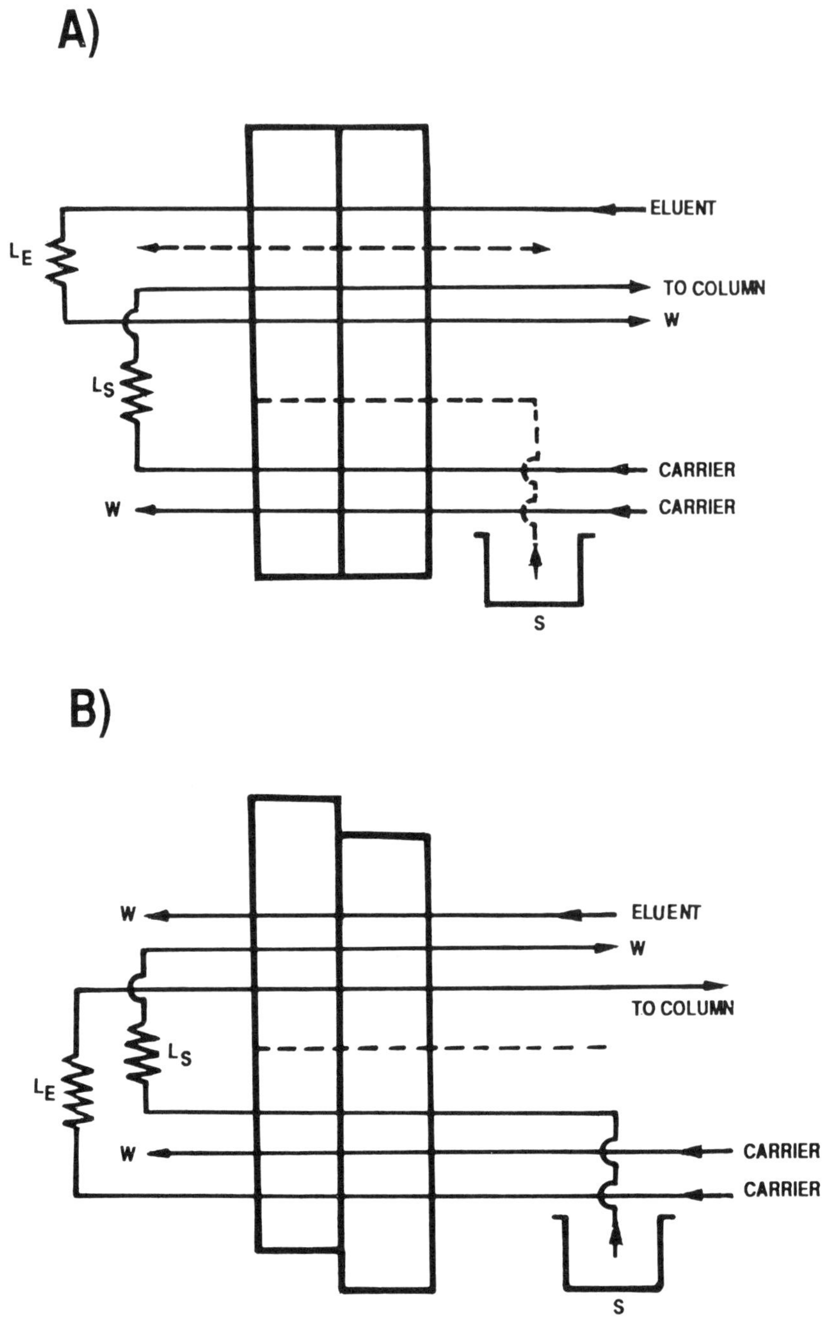

Figure 4.3 *Eight-port injection valve for the sequential injection of sample* (A) *and eluent* (B)

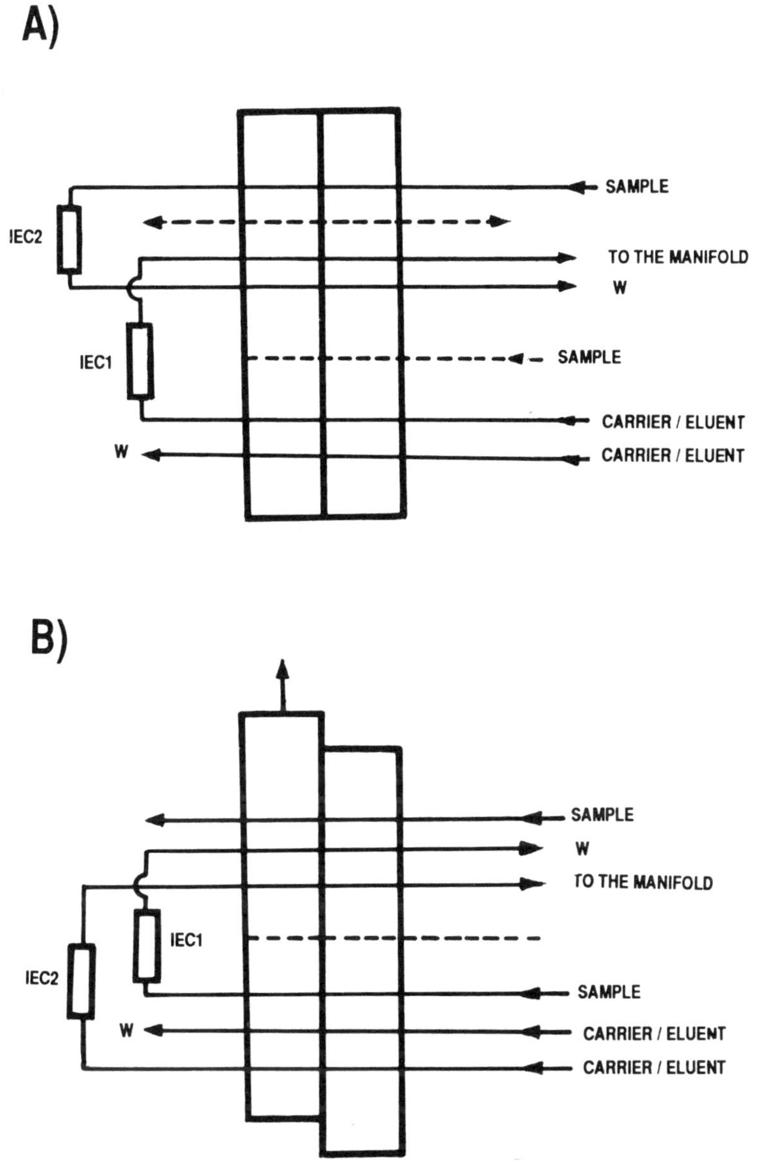

Figure 4.4 *Use of an eight-way valve to increase sample throughput. The time intervals for the two positions (*A *and* B*) are identical*

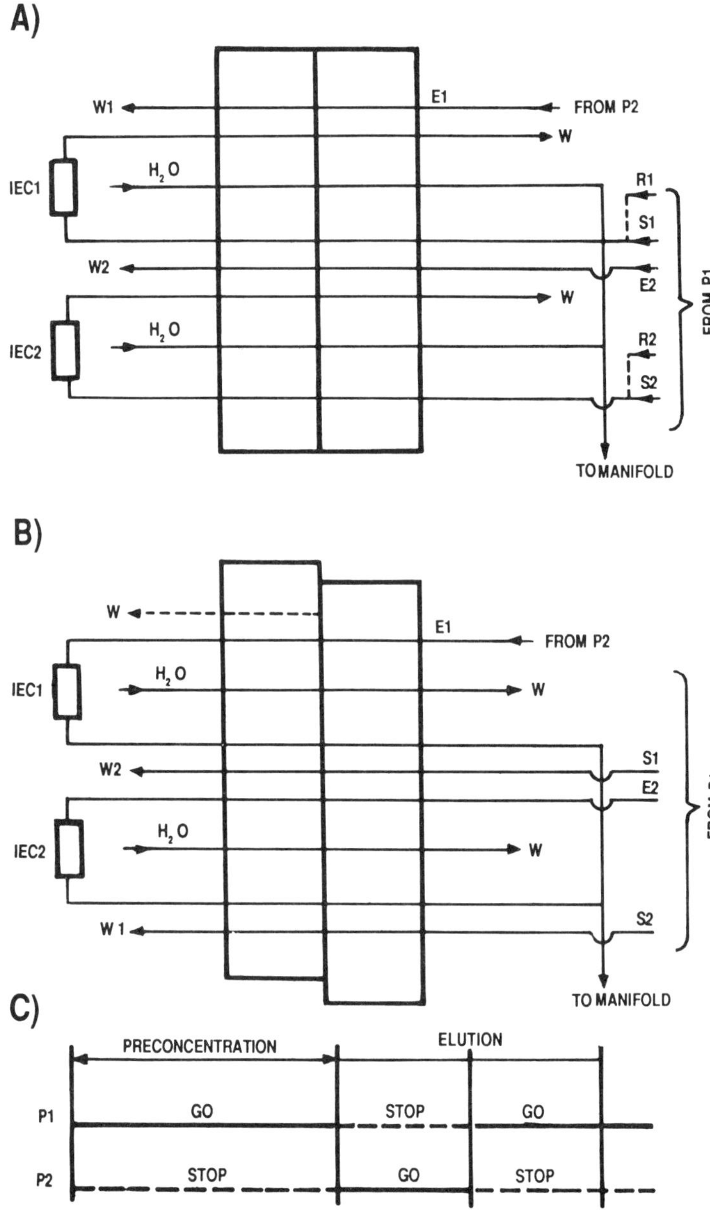

Figure 4.5 *Valve for dual-column ion-exchange preconcentration.* (A) *Sampling and preconcentration mode.* (B) *Elution mode. (C) Stop-and-go sequences of the pumps*

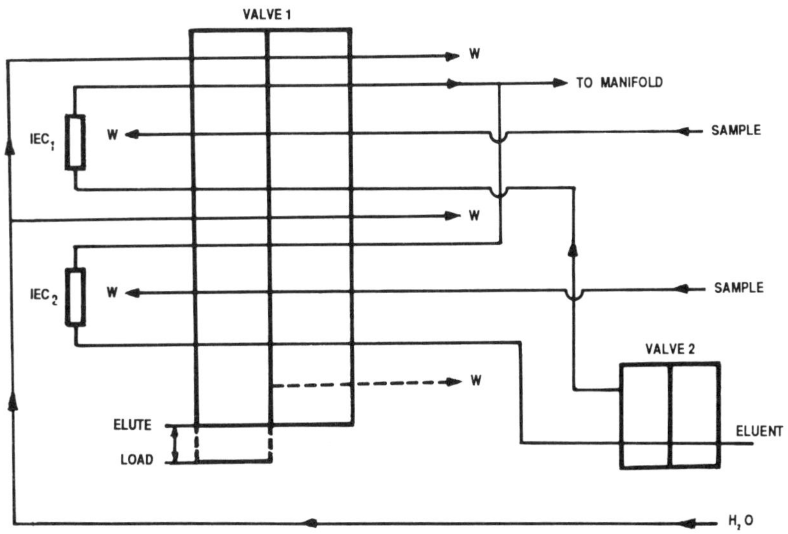

Figure 4.6 *Serially arranged valves for preconcentration in a single stage in the two columns of valve* 1, *and the sequential supply of both with the aid of valve* 2

exchange material in the exit zone and from swelling. By including a second valve (an ordinary valve such as that shown in Figure 4.6) to supply a given volume of eluent to each column sequentially, the above system can be simplified and eluent consumption substantially decreased.

A somewhat more sophisticated injection system was recently used by Lei *et al.*[5] for sorbent isolation and elution with an immiscible solvent. Hypodermic steel needles are inserted into the loading and waste ports of the valve, and the conductivity between the two metal probes thus created is monitored in such a way that the loop is filled with a conductive aqueous phase rather than an immiscible phase, electrical resistance drops automatically switching another ordinary injection valve included in the system.

Selecting valves. These valves can be used for a variety of purposes, namely: (*a*) Altering the nature of liquid circulated through the system (sample/eluent, carrier/eluent, carrier/sample), which is particularly attractive for configurations without an injection system. (*b*) Selecting sample/standard or sample/blank solutions prior to introduction into the injection system. The location and function of this type of valve, which endows configurations with increased flexibility, is determined by the type of manifold used and hence by the specific purpose.

Diverting valves. These are less versatile than selecting valves and are generally used to waste the sample matrix once this has been passed through the column and the species of interest has been isolated, thereby preventing the matrix from reaching the detector and yielding parasitic signals or even poisoning or fouling electrochemical detectors.

4.2.2.3 Detection Systems

Atomic spectroscopic techniques are the most commonly used for detection purposes in connection with on-line ion-exchange separation systems.[6, 7] Taking into account that most of the species to be separated or preconcentrated are metals, atomic detectors make the most convenient, expedient, and economic alternative to quantitation. Of these techniques atomic absorption spectrometry is the most commonly used, followed by inductively coupled plasma emission spectrometry and, at some distance, by flame emission spectrometry. Photometry has been virtually the sole molecular optical technique used with this type of system, while electroanalytical techniques have only occasionally been used and for very specific purposes.

4.2.2.4 Types of Continuous Configurations

The continuous configurations most frequently coupled on-line to ion-exchange separation systems include a unit for injection of sample and/or eluent. The wide variety of manifolds designed for this purpose testifies to the versatility and flexibility of these configurations for adaptation to specific problems. Drawbacks arising from the use of a given manifold can usually be overcome by introducing diverting or selecting valves or by increasing the number of injection valves or their complexity. Air segmentation occasionally boosts the performance of these systems in some aspect, to the detriment of simplicity.

Those manifolds in which samples are continuously aspirated into the system are simpler, but also less versatile. As a rule, configurations used with no derivatization reaction prior to detection (*i.e.* those based on atomic or electrochemical detectors) are quite straightforward.

Configurations with continuous aspiration of the sample. Those configurations in which the sample is continuously introduced into the system call for an aspiration unit to obtain a constant flow-rate in order to control accurately the sample volume that is passed through the column over a given interval. Usually, the sample is merged with a buffer stream that adjusts it to the optimal pH for the subsequent exchange process according to the features of the column packing (Figure 4.7A). Elution is effected as the sample and buffer aspiration probes are immersed in an eluent reservoir (Figure 4.7A). When the detection system calls for a previous derivatizing reaction, the set-up must include a reagent stream merging with the main line immediately after the column. The length of the reaction tube to be used between this merging point and the detector will be dictated by the features of the derivatization reaction concerned (Figure 4.7B). The alternate functioning of the two propulsion systems avoids probe changeover between the preconcentration and elution steps (Figure 4.7C). Pump P_1 (Figure 4.7C) works during the preconcentration step by propelling the sample and conditioner along the system. After this

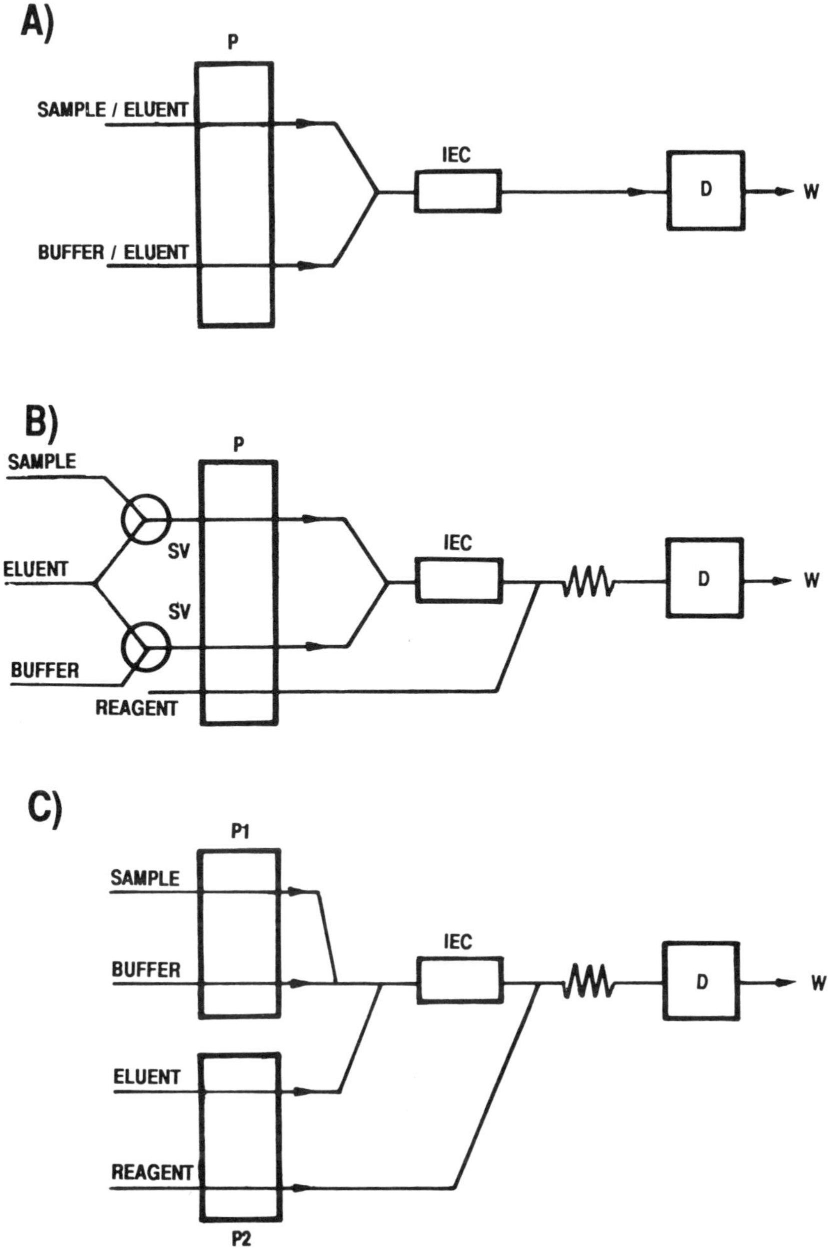

Figure 4.7 *Configurations with continuous aspiration of sample including on-line ion-exchange separation.* (A) *With sequential aspiration of sample + buffer and eluent and no derivatization reaction.* (B) *With a derivatization reaction and continuous aspiration of the reagent and sequential aspiration of the sample + buffer and eluent.* (C) *With a derivatization reaction and alternate functioning of the two pumps for the preconcentration* (P1) *and elution steps* (P2). SV = *selecting valves*

stage, this pump is stopped and P_2 is started to propel the eluent during the second stage. If a derivatization reaction need be implemented, then the reagent carrier is merged with the main stream immediately after the column in order to avoid dispersion of the analyte plug as far as possible.

Configurations with injection of the sample and/or eluent. The large variety of configurations of this type designed so far are dealt with below in increasing order of complexity (or performance).

The simplest design in this context is represented by those configurations that consist of a single unit of each basic type (propulsion and injection system, and column). One such configuration is depicted schematically in Figure 4.8A. The eluent itself acts as carrier, so no selecting valve is required for operation in the normal FIA mode. When the carrier is different from the eluent or the reversed mode is used, a selecting valve (SV) allows the nature of the carrier/eluent or carrier/sample stream to be changed according to whether the normal or reversed FIA mode is applied. Eluent consumption can be reduced and the selecting valve dispensed with by intercalating an additional valve for the eluent between the sample injection valve and the column (Figure 4.8B).

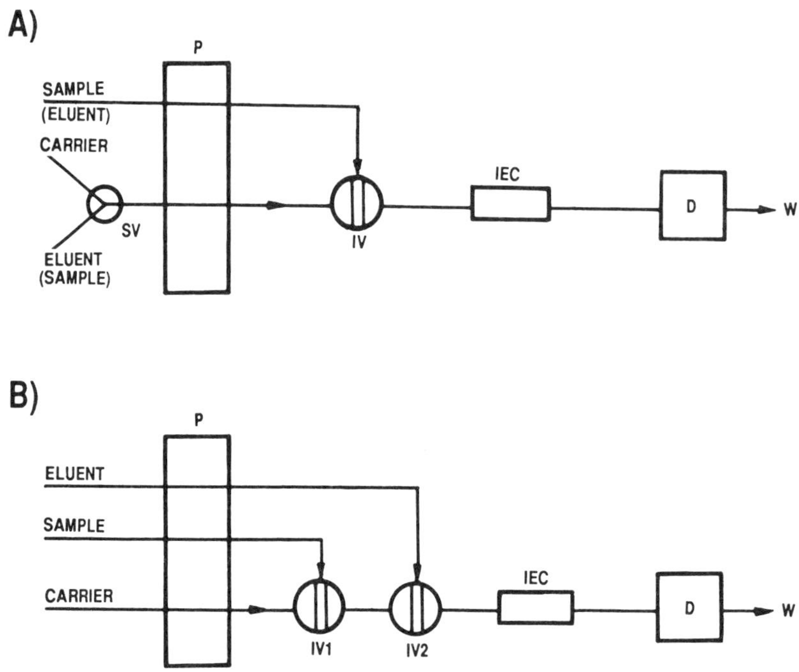

Figure 4.8 *Simple configurations with sample injection including an on-line ion-exchange system.* (A) *With selecting* (SV) *and injection valves operating in the normal or reversed mode (in brackets).* (B) *With two serially arranged injection valves for sample and reagent*

The most serious drawbacks posed by these configurations are as follows:

(*a*) Uneven mixing of sample and carrier (generally a buffer), which results in pH changes along the sample plug and in partial separation of the analyte in the column.

(*b*) A gradual increase in the column compactness at the exit by effect of the continuous flow in the same direction and the sequential swelling and contraction of the exchange material.

(c) The obtainment of parasitic signals and the risk of clogging the burner outlet (atomic detectors) and poisoning or fouling the sensor (electrochemical detectors).

The first of the above-mentioned problems can be circumvented by using an additional channel to inject the sample into a carrier (*e.g.* water) that is later merged with the buffer stream so as to ensure even mixing in normal FIA configurations. On the other hand, the buffer stream is merged with the sample in reversed FIA manifolds.

Problems (*b*) and (*c*) can be avoided in three basic ways, as illustrated in Figure 4.9:

(1) By using selecting valves to reverse the direction of the flow and drive it to different points as required (Figure 4.9A). In the preconcentration step (1), the sample matrix is directly led to waste on leaving the column and the eluent gives rise to the baseline on continuous passage through the detector. In the elution step (2), the eluent is passed through the column in the opposite direction as before and drives the eluate to the detector.

(2) The use of two propulsion systems acting sequentially is a viable alternative to that of selecting valves. As P_2 is started, the preconcentration step is initiated by injecting the sample plug into the system while water is circulated through the detector—the main stream is prevented from taking other directions in passing by the different merging points by aspirating the waste W_2 stream. Finally, P_2 is stopped and P_1 is started on switching the injection valve to its load position while the eluent drives the analyte to the detector.

(3) The simplest solution involves placing the column in the loop of the injection valve and loading it by aspiration. Time-based injection is indispensable in this case for an accurate control of the sample volume that passes through the column.

Flow-injection configurations including two columns are normally required for two basic purposes, namely: (*a*) To implement preconcentration and interference removal. In this case, the columns are arranged serially in the manifolds and in the more appropriate order as a function of the features of the sample—and the derivatization reaction if needed. One of the simplest configurations of this type is constructed by inserting the column for interference removal (SC) in the line of the injection valve and before it, placing the preconcentration column (PC) in the loop of the injection valve. Elution is effected in the opposite direction to

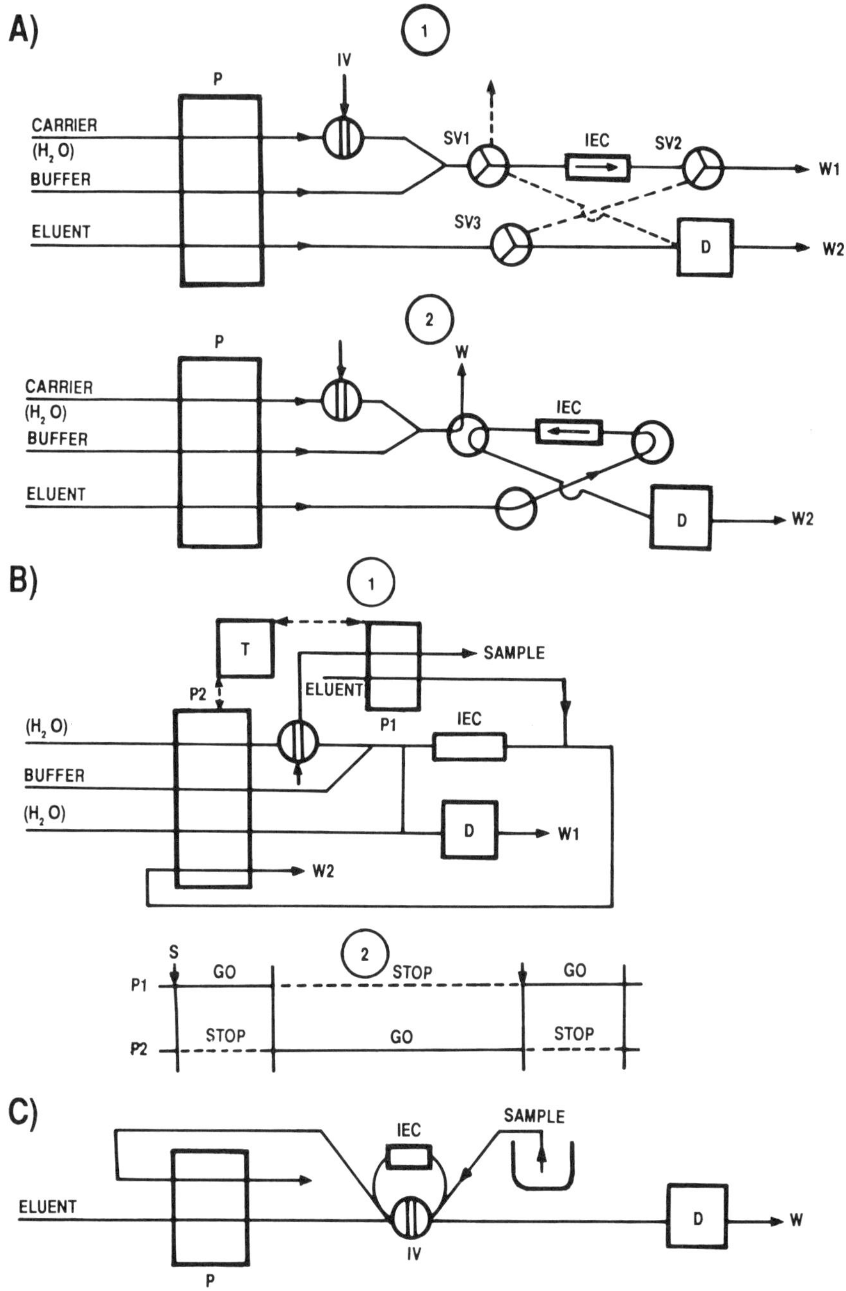

Figure 4.9 *Different ways of carrying out elution in the opposite direction to preconcentration in order to prevent the sample matrix from reaching the detector.* (A) *With different selecting valves:* (1) *preconcentration step;* (2) *elution step.* (B) *With alternate functioning of the two pumps:* (1) *manifold;* (2) *programmed functioning of the two pumps.* (C) *With the column placed in the loop of the injection valve*

preconcentration (Figure 4.10). Changes in the compactness of SC have no effect on performance. The SP unit can be regenerated during the elution step by using a selecting valve to switch between the sample and the regenerating solution. (*b*) To increase the sampling frequency. For this purpose, the columns—logically identical—are placed in two loops of one of the eight-port injection valves depicted in Figure 4.6, which results in the advantages and disadvantages commented on above.

Air-segmented configurations. This type of configuration can be used indifferently with sample aspiration or injection. Any of the above-described manifolds can be modified for this purpose by placing the aspiration unit in a zone where the air segments can act to separate successive samples and the air-removal system (debubbler) as close to the detector as possible. In this way, carry-over between samples is avoided and a pseudo-washing step is implemented. However, air-segmentation is a major source of irreproducibility and introduces additional technical complications, which usually do not compensate for the advantages it offers and leads users to opt for including a washing step between successively inserted samples.

4.2.3 Applications

The on-line association of an ion-exchange system to a continuous analyser has so far been virtually exclusively used in the area of inorganic analysis, especially in connection with metal species in general and heavy metals in particular.

The chief goal of this association is the preconcentration and/or separation of the analyte. Other punctual, but also interesting objectives include multi-determination (or speciation in some cases), indirect determinations, and interference removal. Fulfillment of these objectives requires designing configurations suited to the type of sample and detection system involved, some of which are described below.

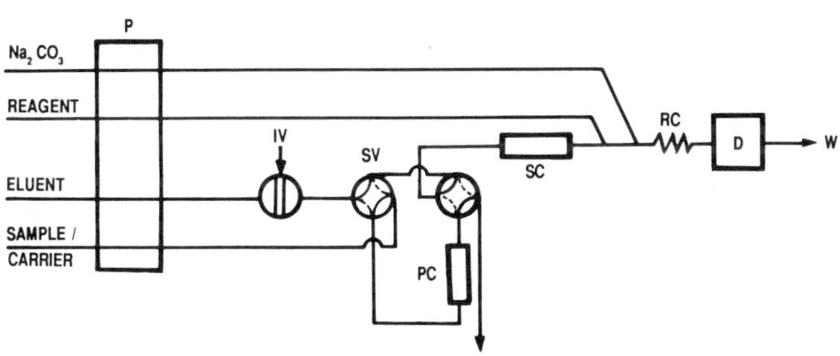

Figure 4.10 *Reversed-FIA systems for the photometric determination of cobalt with protocatechuic acid after concentration* (PC) *and separation* (SC) *of the analyte.* IV = *injection valve.* SV = *dual four-way selecting valve*

4.2.3.1 Preconcentration and/or Separation

The determination of heavy metals by atomic detection usually calls for a preconcentration step. The need for no derivatization reaction allows considerably simpler manifolds to be used for this purpose.

The continuous configuration with sample aspiration depicted in Figure 4.7A was used for the ICP/AES determination of cadmium. An analyte preconcentration factor of 25 was achieved by using 5 ml of sample and an iminodiacetate chelating column (1 cm long, 1.6 mm ID). The sample frequency was 25 h^{-1} and the detection limit 0.05 ng Cd/ml.[8] Manifolds with two serially coupled injection valves for insertion of sample and eluent have been frequently used, both in straightforward versions (two identical loop-based injection valves)[9-11] and in time-based injection modes[12,13] for the preconcentration/determination of heavy metals in sea water,[9,10] aluminium in haemodialysis fluids,[10] and Cu[12] and Cd[13] in natural water. Also, rather simple FIA systems (Figure 4.8A) have been used for the determination of lead in water by AAS[14] and of other heavy metals (Al, Cr, Fe, Ti and V) by ICP/AES.[15]

The use of two pumps working alternately to dispense with selecting valves allowed the development of methods for the preconcentration/determination of copper and zinc by ICP/AES[16] and of various metal ions by AAS.[17]

Columns have been placed in the loops of ordinary injection valves for AAS and FAAS determinations by effecting elution in the same[18] or in the opposite direction to preconcentration.[19]

Systems with eight-port valves and two identical columns located in their loops (Figure 4.4) have been used for the preconcentration/determination of some heavy metals by AAS[2,20] and ICP/AES[21]—in this last case with simultaneous multi-determination. Elution was effected in the same[20] or the opposite direction[2] as preconcentration, or with an additional injector.[21]

The ICP/AES technique has been used to develop some preconcentration/determination methods for anions such as sulphate, which was quantified at levels as low as 2.8 ng ml^{-1} in 2 ml of sample of natural and boiler feed water[22] by using a reversed FIA configuration. Several oxo-anions have also been determined by similar procedures.[23]

Continuous preconcentration/separation and photometric determination have also been used to develop determinative methods for cations. A configuration similar to that depicted in Figure 4.7C was used for preconcentration on chelating Dowex A-1 resin and elution with HCl in the photometric determination of calcium and magnesium in brine based on a complex-formation reaction with an azo ligand. The total concentration of calcium and magnesium was determined in one sample; then, in a second sample, calcium was masked with a ligand buffer containing excess Ba^{II} and EGTA, and magnesium was determined, calcium being thus quantified by difference. The two alkaline earths were thus determined in 2.5 M sodium chloride over the ranges 1–30 ng ml^{-1} (Mg) and 8–120 ng ml^{-1} (Ca). A time-based injection unit including a column packed with Amberlite IR-120 was used to construct a two-channel manifold for the determination of ammonium ion in rain water with contents below 200 ng ml^{-1} at a rate of 40 samples h^{-1}.[25] Also, a system including serially arranged valves similar to that depicted in Figure 4.6 was used for separation/preconcentration of uranium from geological samples on Levextrel CL-5209 resin (preconcentration factor = 10), with subsequent elution and formation of a complex with Arsenazo III. Uranium concentrations above 3 ng ml^{-1} were determined at a rate of 30 samples h^{-1}.[26]

Cobalt was determined in sea water by two interesting methods. In one, Yamane *et al.*[27] used the configuration shown in Figure 4.10 to carry out the overall process, in which 50 ml of sample previously conditioned with sodium acetate and ammonia was pumped through the preconcentration column (silica-immobilized 8-quinolinol) and drained to waste by means of a dual four-way valve in its load position (solid line in SV). After this step, SV was switched to its elution position and 200 µl of 0.5 M HCl was inserted through IV to elute cobalt and other metal ions retained in the separation column (SC), which was packed with a strongly acidic cation-exchange resin. Eluted cobalt catalysed the reaction between the ingredients of the indicator mixture (protocatechuic acid/hydrogen peroxide), the resulting absorbance change being monitored at 480 nm. One washing cycle per injection of 200 µl of 0.5 M HCl was included with SV in the elution position in order to flush any traces of metals such as Mn and Mg. The detection limit achieved was 5×10^{-1} ng ml^{-1}.

In the second method for cobalt, an injection valve similar to that depicted in Figure 4.9C, which accommodates 8-quinolinol immobilized on Fractogel in its loop and through which elution is performed in the opposite direction to preconcentration, was used as part of a manifold in which a 0.5 M HCl solution was used as eluent/carrier to drive the eluate to a confluence point converging with a stream of gallic acid and hydrogen peroxide in an alkaline medium that catalysed the oxidation of the organic compound. By measuring the luminescence generated at 640 nm, the catalyst was determined at the picomole level. A selecting valve located before the injection valve allowed inclusion of a washing cycle with 1 M HCl + 0.1 M HNO_3 followed by rinsing with high-purity water between successively inserted samples.[28]

A complex system including an ISE and two chelating columns inserted in the carrier and conditioner channels and a third packed with the same exchanger and used for analyte preconcentration was employed for the potentiometric determination of copper. The three columns were connected serially with the potentiometric cell and a diverting valve placed between the two to prevent the sample matrix from reaching the detection cell and damaging the detector. Detection limits of 10^{-7} and 10^{-8} M copper were achieved for sample volumes of 5 and 25 ml, respectively, at sampling rates of 12 and 5 samples h^{-1}.[29]

4.2.3.2 *Interference Removal*

Removal of potential interferences from a given sample matrix by using an appropriate exchange material has been included in the determination of different analytes in various types of sample.

An air-segmented configuration with sample aspiration was used for the colorimetric determination of free acids after removal of hydrolysable species. A column packed with a mixture of cationic (Bio-Rad AG 50W-X8) and anionic (Bio-Rad AG 1-X4) exchangers was used to separate the different hydrolysable species before the sample was merged and thoroughly mixed with the indicator solution (Thymol Blue), debubbled and eventually monitored at 534 nm.[30]

The UV-absorption determination (λ_{max} = 271 nm) of hydrochlorothioazide pharmaceutical preparations containing hydralazine hydrochloride and reserpine requires these last two active principles to be previously removed since they absorb in the same spectral zone as the analyte. By placing a 15 cm long, 1 mm ID glass column packed with Amberlyst 15 (40–60 mesh) immediately after the aspiration point, the two interferents can be conveniently removed. The configuration includes a mixing device that ensures thorough mixing and can be used with and without air segmentation.[31]

The AAS determination of calcium in water is seriously interfered with by the presence of anionic species such as sulphate, phosphate, and silicate. This problem has been addressed by using an extremely simple method that can be implemented on a configuration with a single channel through which a carrier (water) stream is circulated and into which the sample (25 µl) is inserted. Short lengths of tubing connect the injection system serially with the column (Amberlite IR-400 or De-Acidite FF, both in chloride form) and the atomic absorption spectrophotometer used in order to minimize dispersion of the analyte. The interference from silicate is more effectively overcome by masking with hydrofluoric acid. The column is regenerated at appropriate intervals by injection of 0.1 M HC1.[32]

The most appropriate location of the column for interference removal purposes is in front of the sample injection unit, which avoids dispersion of analyte in the carrier when it is the sample plug rather than the continuous flow circulating in that position prior to injection, which is passed through the column. This is the arrangement used, among others, for the photometric determination of selenium based on its catalytic effect on the reduction of picrate ion by sodium sulphide, which is seriously interfered with by heavy metals. A column packed with cation exchanger (Dowex 50W-X8, 100–200 mesh, in acidic form) successfully separates these species, thereby allowing the catalyst to be determined over the range 0.1–1.0 μg ml^{-1}, with a detection limit of 5 ng ml^{-1}.[33]

4.2.3.3 Multi-determinations

The use of an ion-exchange column in a continuous analyser provides a means for developing determinative methods for several substances or different chemical forms of a given substance (speciation).

One simple way of accomplishing the sequential determination of zinc and cadmium relies on the inhibitory effect of these metals on the Co^{II} catalysis of the luminol/hydrogen peroxide system. The analytes, injected into an acid stream through the injection valve (Figure 4.11A), are retained as chloro complexes in the Amberlite IRA-400 column (2 cm long, 3 mm ID) used. The sequential injection of 500 μl of NaOH and 1 M HNO_3 elutes zinc and cadmium, respectively. The high luminescence baseline arising from continuous passage of the Co^{II}/luminol/H_2O_2 stream through the detector is diminished in proportion to the amount of

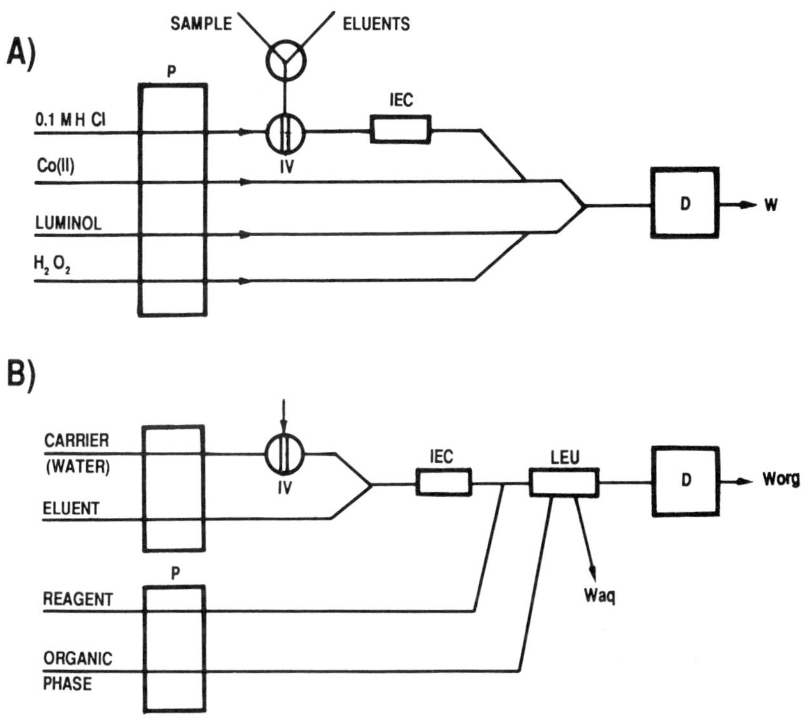

Figure 4.11 (A) *FIA configuration for the sequential determination of zinc and cadmium based on their inhibitory effect on the Co^{II}-catalysed reaction between luminol and hydrogen peroxide. The analytes, retained on the column* (IEC) *are sequentially separated by changing the eluent.* (B) *Configuration with ion-exchange* (IEC) *and liquid extraction (LEU) units for the sequential determination of sodium and potassium*

Zn or Cd present in the eluate, the negative peaks obtained in the overall process being related to the concentration of the analytes in the sample.[34]

A sequential determination of sodium and potassium in tap and river water based on column separation, liquid–liquid extraction and photometric detection was developed by Motomizu and Onoda[35] by using the configuration depicted schematically in Figure 4.11B. Analyte separation was based on the differences in retention times resulting from elution from a silica gel C-200 column by means of a benzo-18-crown-6 stream. The complex formed an ion-pair with tetrabromophenolphthalein ethyl ester anion that was extracted into 1:3 chlorobenzene:benzene, the absorbance of the organic phase being monitored at 620 nm. The sampling rate achieved was 15 samples h^{-1}.

Chromium speciation into Cr^{III} and Cr^{VI} can be readily accomplished in a sequential manner by retaining either chemical form in an appropriate exchanger. The speciation methods proposed for this purpose use atomic detection, either in association with an activated alumina column in its acidic form—with a high affinity for oxoanions—and sequential injection of sample and eluent (0.1 M HNO_3) plus ICP/AES detection,[36] or by using a chelating exchanger (8-quinolinol immobilized on porous glass) and an eight-port valve with sample and eluent loops, and flame atomic spectroscopic detection.[37] Two peaks per injected sample are obtained in both cases; the first corresponds to the sample plug containing the unretained species and the second to the eluate. Insofar as not only speciation, but also preconcentration of the retained species are achieved, one can choose between the two depending on which species is present at the lower concentration.

Speciation of free and complexed copper by AAS can be accomplished in a single-channel system including two simple injection valves arranged serially with each other and with a column packed with 60–100 mesh chelating Resin 122 containing salicylic acid functional groups. Each injected sample provides two peaks, the first corresponding to complexed copper and the second to free copper eluted from the resin with nitric acid. A sampling rate of 90 samples h^{-1} and excellent precision are noteworthy assets of this method.

Obviously, eliminating the effects from the sample matrix on the detector is impossible in these cases.

4.2.3.4 Indirect Determinations

Conversion techniques allow the development of interesting continuous methods by on-line coupling to ion-exchange systems. One such method for the determination of sulphide is based on its precipitation with excess cadmium and retention of the excess on a column, so that only CdS reaches the atomic absorption spectrophotometer used—where it yields the analytical signal. The manifold used for this purpose includes an eight-port rotary valve for the simultaneous injection of sample and reagent Cu^{II} solution into two channels which are merged prior to reaching a column packed with 8-quinolinol azo-immobilized on controlled pore glass that retains excess reagent. The signal provided by cadmium from CdS is proportional to the amount of sulphide ion present in the sample. An injection valve located between the merging point and the column allows insertion of an acid solution intended to regenerate the column.[39]

One flow injection method for the determination of EDTA by flame absorption spectrophotometry was developed by using two similar manifolds into which sample and regenerating reagent were sequentially injected by means of two serially coupled valves (Figure 4.2C) or a single laboratory-built eight-port valve (Figure 4.3). In the first step of the process, a Cu^{II} regenerating solution was injected onto an on-line column packed with Chelex-100 resin. Excess copper not retained in the column was washed by means of an ammonia carrier. In the second step, the EDTA sample was injected and the analyte displaced an equivalent amount of Cu^{II} from the chelating column. Eluted copper was then determined by using a FAA spectrometric detector located downstream. The signal obtained was proportional to the EDTA concentration in the sample. The detection limit and sample throughput thus achieved were 0.1 μg ml^{-1} and 45 samples h^{-1}, respectively, with a precision better than a relative standard deviation of 1.5% for 0.5 μg ml^{-1} of EDTA.[40]

4.2.3.5 Special Applications

A few very interesting specific applications warrant separate discussion in order to exemplify further the versatility with which ion exchange endows continuous systems.

The total ionic concentration of water can be determined by using a two-channel reversed FIA system through which the sample is continuously circulated and merged with an acid–base indicator stream. The loop of the injection valve accommodates an exchange column packed with Dowex 50-X cationic resin in its proton form, so that, while the valve is in the injection position, the sample passes through the column, where cationic species are retained and an equivalent amount of protons is released; these protons bring about a colour change in the indicator which is detected on passage of the fluid segment through the flow-cell of a photometric detector.[41]

Control of the total salt concentration in some industrial processes is a key to production control. The system used for the on-line monitoring of this parameter is the starting point to determine its value and act on the process. The above-described configuration including two conductivity cells between which a column packed with Ag 50Wx16 (proton form) or AGIx10 (OH^- form) was placed and used for this purpose. The conductivity difference between the two cells was a measure of the amount of analyte contained in the sample plug and thus retained on passage through the column.[4]

One uncommon use of exchange materials is as redox reagents for the generation of arsine for subsequent atomic spectrometric determination. A column packed with a strong macroporous anion-exchange resin (Amberlyst A-26) in tetrahydroborate form and placed immediately after the sample injection point acts as a polymer-supported reductant. The analysis cycle comprises regeneration, washing, and injection of the acidified sample, and can be repeated every four minutes. The detection limit of As^{III} is 60 pg for a 40 μl sample and a linear working range up to 4 ng is achieved. The column can be used for at least three months with no significant loss in performance.[42]

One very interesting approach involves sensing an immiscible plug, isolating it and driving it to the detector by using a miscible carrier and performing further chemistry on route if desired. A *sui generis* FIA manifold (Figure 4.12) was designed for the determination of mercaptans in gasoline; the most outstanding feature of the system is valve V_2, the loop loading and waste ports of which accommodate stainless steel needles between which conductivity is monitored. The sample circulates through the system and passes through a 16–50 mesh Amberlite URA-401-S column in its OH^- form, where the analytes are concentrated in the sorbent; meanwhile, an alkaline solution fills the loop of valve V_1. The overall sample volume, preconcentrated before elution takes place is a function of the inject/load cycle of V_1. As V_1 is switched to the inject position, the alkaline plug elutes the mercaptan captured on the microcolumn and regenerates the column. Then, as this aqueous extract passes through V_2, the aqueous phase is isolated by automatically actuating the valve and inserted into an alkaline carrier stream. An $NaHCO_3$ buffer solution is merged with the main stream and then with one of 5,5′-dithiobis(2-nitrobenzoic acid) (DTNB). The resulting reaction product is monitored at 412 nm. Since V_2 is activated by the injected immiscible liquid from V_1, the cycling period of the two valves is identical, *i.e.* a fresh extract is injected into the aqueous system every two minutes. A similar manifold was used for the determination of aniline in benzene.[5]

4.2.4 Integration of Retention and Detection

This novel application of exchange materials outside their ordinary scope is arousing growing interest as the foundation for a revolutionary type of continuous regenerable sensors.[43]

Ion-exchange absorptiometry is the immediate precedent of integrated retention/detection.[44] In this technique, the exchange material is used for preconcentration/determination purposes in a variant of the classical bath technique. Essentially, it involves mixing the exchanger with the sample-reagent solution in order to retain the coloured product formed in the reaction. Once retained, the exchanger is rinsed and the eluate is filtered and manually introduced into a conventional photometric cell for measurement. Then, the cell is cleaned and the exchanger is sent to waste

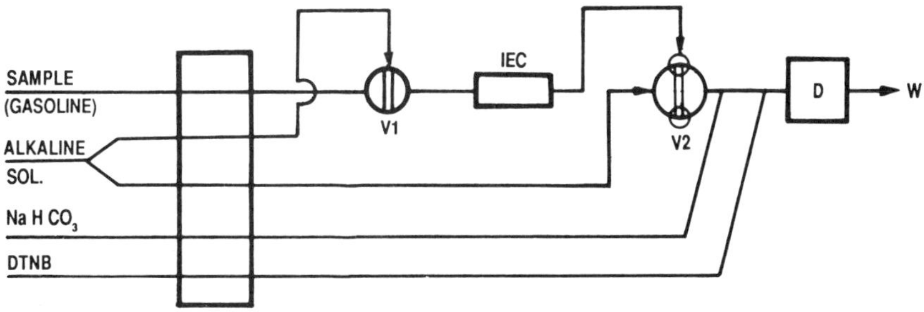

Figure 4.12 *Manifold for the determination of mercaptans in gasoline by sorbent isolation and elution with an immiscible solvent. (For details, see text)*

or, if possible, regenerated. Though ingenious, this technique is slow and tedious as each analysis takes between one and two hours. On the other hand, the continuous variant is extremely simple, convenient and rapid. It only requires introducing the exchange material in an appropriate flow-cell and passing the sample and any reagents required through it in such a way that the analyte (or its reaction product) is retained simultaneously with measurement of the signal it yields. This is equivalent to simultaneous concentration of the species of interest and detection, *i.e. in situ* concentration as opposed to conventional preconcentration processes. On changing the carrier or its composition, the retained species is eluted and the exchanger is rendered ready for a new analysis.

4.2.4.1 Technical Features

The optical techniques used so far with this type of system demand special technical features from the exchanger and the measuring cell as well.

The exchange material should be transparent in the measurement region, so it should be of neither too large nor too small a size to avoid compaction problems arising from increasing pressure in the system. Polymer cross-linkage determines the absorbing capacity of resins, while the nature of the polymer conditions its emission power. Easy swelling is a desirable feature here as it decreases the material density while increasing its transparency. The lack of general rules for choosing the most appropriate material for a given purpose call for individual assessment, as shown by the examples described below.

Flow-cells to be used in this context should have some special features—however, some commercially available cells are more than adequate in performance. Thus, their lightpath should be quite short (0.2–1.5 mm) in order that the amount of exchanger placed in the cell may not saturate the detector capacity while leaving a wide range for use by the analyte. The cell width should be quite small so that virtually the whole exchange material lies in the optical path—the lateral portions not covered by the lightpath will not contribute to the signal and will therefore detract from the sensitivity. Large inner volumes have an adverse effect on the sensitivity and sample throughput as they call for larger sample volumes to obtain a given signal level and result in longer elution times for the retained species and hence in long-tailed recordings.

In order to pack the cell, one of its ends is blocked with nylon-fibre, glass wool, or a similar material and a suspension of exchanger is then introduced, usually by means of a syringe. Insofar as the substance to be retained is first concentrated at the top of the exchanger (stream inlet), the lightpath must also be placed in that zone. Sensitivity will decrease as the distance between the resin top and the lightpath increases. Low exchanger levels (below the lightpath) provide measurements for the solution.

4.2.4.2 Types of Systems

Chemical sensors bases on integrated reaction and detection can be classified into two broad groups according to whether or not they involve a derivatization reaction prior to the retention/detection proper. Figure 4.13 depicts the three basic types of configuration schematically.

Without a prior derivatization reaction. This is the simpler alternative and involves retaining the analyte itself in the exchanger or immobilizing a reagent (usually a chelating ligand) to form the measurable product. In order to ensure adequate performance in this last case, the ligand should remain bound to the exchanger on elution, so that only the analyte is eluted, which can be accomplished by using a displacement reaction.

Yoshimura, the proponent of ion-exchange absorptiometry,[44] developed a photometric flow-through sensor for copper to show the feasibility of incorporating the technique into a flow analysis system. The sensor, based on the direct retention of Cu^{II} on a 100–200 mesh Bio-rad AG 50W-X12 resin packed in the flow-cell of a photometric detector, was tested by inclusion in a manifold with two serially coupled injection valves, one of which was used to inject the sample (4.5 ml) into a weakly acidic stream (1.4×10^{-3} M HNO_3), while the other was employed after the analyte was retained and measured to effect elution by injection of 5 ml of 2 M HNO_3. An interesting comparative study of the sensitivity afforded by a sensor compared to the conventional absorptiometric method was carried out on the basis of the expression

$$SR = (\varepsilon_{res} V)/(\varepsilon_{sol} S l)$$

where ε_{res} and ε_{sol} are the molar absorptivities of the resin and the solution, respectively, V the injected sample volume, S the area of flow-cell cross-section perpendicular to the lightpath (in cm^2) and l the lightpath (in cm) for solution absorptiometry. Cells with a diameter of 1.5 cm provide sensitivity ratios (SR) of up to 250.[45]

Another sensor of this type developed for the determination of the

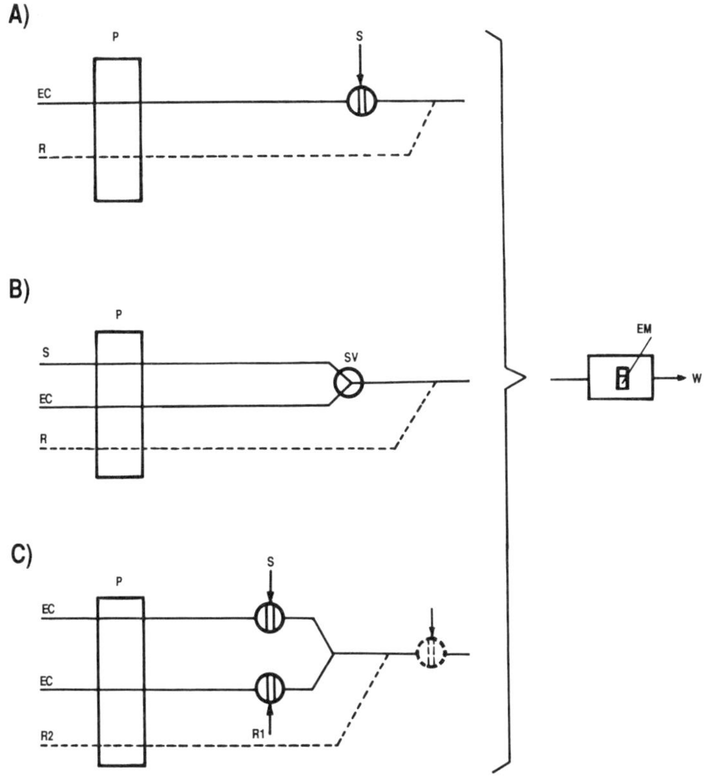

Figure 4.13 *General configurations for integrated reaction (and/or retention) and detection with the exchange material* (EM) *located in the flow-cell of a non-destructive detector.* (A) *Single-channel FIA configuration.* (B) *Completely continuous configuration with selecting* (SV), *sample* (S) *and reagent valves* (R). (C) *Merging-zones FIA configuration for the simultaneous injection of sample and reagent. The dashed line represents an additional reagent or injection valve*

same analyte relies on a rather different chemical principle and is used in one of the configurations depicted in Figures 4.13A and B. Measurements are made on the reaction between Cu^{II} in the sample and 1-(2-pyridylazo)-2-naphthol (PAN) immobilized in Dowex 50W X4-400 resin packed in a photometric flow-cell of 1.0 mm lightpath. The ligand is displaced by the thioglycolic acid contained in the carrier, which effects rapid elution (sample throughput, 110 samples h^{-1})

$$(Cu^{2+})_{sol} + (PAN)_{res} \rightleftharpoons [Cu(PAN)_2]_{res} \; (\lambda_{max} = 546 \text{ nm})$$

$$\nwarrow \text{thioglycolic acid} \nearrow$$

Two different methods are applied depending on the concentration of the analyte in the sample. The manifold in Figure 4.13A is used for samples with copper concentrations between 0.1 and 26.01 g ml^{-1}; a sample volume of 300–900 μl is injected and peak-height measurements are made. The manifold in Figure 4.13B is used for more dilute samples (with copper concentrations of up to 1 ng ml^{-1}); the sample is continuously circulated through the system and the absorbance decrease resulting from retention of the analyte by complex formation (kinetic concentration curve or *in situ* concentration) is measured. The tangent to the initial portion of the curve obtained allows one to determine the analyte concentration while avoiding the two most serious shortcomings of preconcentration processes: high sample and time consumption. Sampling rates of 50 samples h^{-1} can readily be achieved, with a reproducibility level of 3% as the r.s.d.[46]

A similar fluorimetric sensor was used for the determination of beryllium using Morin supported on Dowex-1 X4-200 resin as chelating ligand and 0.02 M HNO_3 as eluent.[47]

With a prior derivatization reaction. The use of a derivatization reaction prior to retention offers advantages such as the need for less resin manipulation—no prior immobilization of the reagent is required—and greater versatility of the chemical step as the ligand need not be permanently retained on the exchanger, which overcomes a major drawback; on the other hand, it has also some shortcomings such as increased reagent consumption resulting from its continuous circulation through the system or simultaneous injection with each sample (Figure 4.13C).

Undisputable evidence of the gains involved in using integrated retention and detection with a given chemical system is provided by the determination of iron based on the formation of a complex with thiocyanate; on removal of the complex from the solution by retention on a Dowex 1-X2-200 anion exchanger, the complex-formation equilibrium is dramatically displaced to the right. The configurations depicted in Figures 4.13A and B (dashed line included) were used for this purpose. In the former case, a 2 ml sample was injected into an eluent/masking stream (0.2 M NaF/0.5 M NaOH/0.02 M NaAcO/0.002 M Na_2EDTA) that was subsequently merged with a reagent stream (0.4 M NH_4SCN/0.5 M HCl) to form the iron complex along a reactor located prior to the cell, a complex which was retained on the resin. The rapid elution effected after measurement allowed a sample throughput of 30 samples h^{-1} to be achieved. The method was applied to the determination of iron in water and wines of different type (white, rosé). The manifold in Figure 4.13C was resorted to when the analyte concentration was very low or the sample matrix contributed to the total absorbance of the peak; the *in situ* concentration and kinetic measurements afforded allowed both pitfalls to be overcome. The use of these methods allows the analyte to be determined over the range 10–400 ng ml^{-1} with an r.s.d = 3% for 20 μg ml^{-1}. The two were contrasted with a reference method[48] and the recoveries obtained on several standard additions to the assayed samples testified to their usefulness.[49]

A similar photometric method was developed for the determination of ammonia in soils based on using the phenol/hypochlorite system to form Indophenol Blue.[50]

The configuration depicted in Figure 4.13C was also used to test two flow-through fluorimetric sensors designed for the determination of fluoride and cyanide. The former anion was determined by formation of the ternary complex Zr^{IV}/Calcein Blue/F^-, which was accomplished by simultaneously injecting 500 μl of the binary complex Zr^{IV}/Calcein Blue and the sample containing the fluoride ion into two streams of 5×10^{-3} M HCl and water, respectively. The product, formed along the reactor located after the merging point, was retained on 40–120 mesh DEAE Sephadex anion exchanger resin and was subsequently eluted after measurement by injecting 0.4 M HCl via a valve situated prior to the detector. The analyte determination range (1–40 ng ml^{-1}), precision (r.s.d. = 1%) and sample throughput (30 samples h^{-1}) of the method compare favourably with those afforded by a probe-type sensor based on the same chemical system,[51] which it surpassed amply.[52]

The determination of cyanide was based on its reaction with pyridoxal-5-phosphate (P5P) and retention on QAE Sephadex resin of the pyridoxic acid formed. The reagent, also fluorescent, was not retained by the exchanger. The reaction took place on simultaneous injection of the analyte and reagent in an acetate medium into two water streams subsequently merging with each other and further down the system with a 0.5 M HCl stream. Because of the buffering effect of the sodium acetate in which the reagent was dissolved, retained pyridoxic acid exhibited no fluorescence until the sample + reagent matrix in the sodium acetate medium left the cell, thereby eliminating any interference from other fluorescent species. The fluorescent protonated species was formed by the action of the hydrochloric medium once the buffer had been removed from the flow-cell; then, the measured product was eluted by the acid. The determination range was linear between 50 ng ml^{-1} and 3.0 μg ml^{-1} (r.s.d = 1.39%) and the sample throughput was 10 samples h^{-1}. In addition, the method proved to be rather selective.[53]

When the species to be retained on the material with which the cell is packed is a catalyst (usually an enzyme), the overall process involves no exchange step as the reaction develops in the solution in contact with the catalyst, which is permanently bound to the support.[54]

4.3 Sorption Systems Coupled On-line with Liquid Chromatographs

Sorption is a broad concept used to describe any type of mechanism, whether physical, physico-chemical, or chemical, whereby a compound or a group of related compounds are retained on an active solid support (solid-phase extraction). Columns packed with sorption materials retain a given group of compounds as a whole as the sample is passed through them. This has facilitated the development of sample handling procedures of a wider potential than liquid–liquid extraction, evaporation or steam distillation. Retention of the species concerned, whether analytes or interferents, on this type of column should be a reversible process in order to allow the retained species to be readily swept and the column to be re-used.

There are two general alternatives to the use of sorption columns. In the batch mode, the component of interest is concentrated on a suitable column, usually quite long, and then eluted with a small volume of solvent. After appropriate treatment (*e.g.* partial evaporation of the solvent or derivatization of the solute), an aliquot is injected into the chromatographic system. These procedures have the inherent disadvantages of decreased sensitivity—if only an aliquot is injected, losses arising from thermal decomposition (evaporation stage) and risk of contamination. These shortcomings are normally overcome by using internal standards. In addition, their automation is rather complex and calls for the use of robots[55] with a few exceptions.

Many of the drawbacks of off-line processes are overcome on converting them to on-line processes, particularly those related to the ease of automation of sample handling and trace enrichment, which enable large-scale screening and monitoring of organic species of interest. In the continuous mode, the sorption column is an integral part of the analytical system and is referred to as 'pre-column'.

The most common function of solid-phase extraction columns are as follows:

(1) *Trace preconcentration*. One of the most common goals in this context is the enrichment with compounds of interest by using appropriate sorbents to isolate and preconcentrate them prior to chromatographic detection. The functioning of these columns is formally resemblant of column chromatography; however, no discrimination of the retained analytes is possible in sorption systems. Compounds with a strong affinity for the sorbent are adsorbed on a small segment of the sorption bed, and breakthrough rarely occurs, even if large volumes, high sampling rates, and short columns are involved. One typical application in this context is the enrichment of traces of relatively non-polar organic compounds in an aqueous solution by using a hydrophobic sorbent such as porous polymer, alkylsilane-modified silica, or activated carbon. These systems allow the use of sample volumes as large as 1 l and 2–4 mm long pre-columns. Taking into account the typical injection of an ordinary liquid chromatograph, this represents preconcentration factors of up to 10 000. Preconcentration of more polar compounds calls for the use of smaller sample volumes in order to avoid breakthrough and hence the loss of trace components. Alternatively, a derivatizing reagent can be used to convert the polar component into a less polar derivative.

(2) *Sample clean-up*. Differences in the interaction between the compo-

nents of a given sample and the sorbent are the basis for their separation, even during the loading step. Thus, weakly retained phenol compounds or acid substances can be readily separated from phthalate esters strongly adsorbed on a non-polar sorbent.

(3) *Sample storage.* Thanks to the relatively inert character of many sorbent materials, particularly porous polymers, some compounds can remain adsorbed for long periods of time with no significant alteration. This is of special interest when samples have to be collected in remote places and therefore require long-term storage. In these cases, the dramatically decreased sample size (*e.g.* from 1 l of solution to 1 g of loaded sorbent in a cartridge) is a major asset and reduces the risk of contamination and losses arising from decomposition.

(4) *Protection of the analytical system.* On-line solid-phase extraction makes an excellent means of protecting the analytical column (usually quite expensive) as the pre-column can act as a protective filter lengthening the useful lifetime of the separation unit.

(5) *Pre-column derivatization.* Pre-columns used to enact a derivatization reaction intended to facilitate subsequent discrimination of the species involved on the analytical column can be packed with a variety of materials suited to the reaction concerned (*e.g.* a sorbent impregnated with the reagent, a solid redox agent, or a support for retention of a catalyst).

4.3.1 The Sorbent Material

Of the different types of material used as supports in sorption systems (divinylbenzene polystyrene resins, foamed plastics, silicates, controlled pore glass, cross-linked cellulose, and activated carbon), silicates offer the widest range of properties of interest to the solid–liquid extraction of organic and inorganic species. Their growing interest arises from their low cost, their suitability to organic and inorganic species contained in liquid or gas samples, and the possibility of using them as such or modified by physical or chemical adsorption of chelating agents. In addition, the surface of these sorbents can be altered to obtain cross-linked alkyl groups[56–58] and other functional groups.[59, 60]

Selection of the most appropriate support for each application requires considering some physico-chemical aspects of the material in question in relation to the features of the analyte and the sample matrix, namely:

(*a*) The nature of the solvated bonded phase (polarity, ion-exchange capacity in relation to the liquid phase).

(*b*) The nature of the functional groups of the compound(s) of interest, which may result in a wider range of effects from a weak hydrophobic interaction (dispersion) to dipole induction, hydrogen or ionic bonding, or even covalent bonding.

(*c*) Secondary interactions between the compound(s) of interest and the bonded phase, which may play a major role in the overall interaction.

(*d*) Interactions between the sample matrix components and the bonded phase, which may result from a scarcely active solid phase and occasionally detract from selectivity.

(*e*) Interactions between the compound(s) of interest and the sample matrix, which determine the efficiency of the separation and must be checked by performing recovery experiments on samples of known concentration.

The influence of other features such as the breakthrough capacity, which is determined by the nature of the sorbent, and the analyte retention capacity, is similar to that discussed in Section 4.2.1.

Band broadening is a concept closely related to pre-columns and results from their on-line connection to a chromatographic system; the extent to which this phenomenon takes place depends on the column dimensions, the particle size of the packing material and its nature—whether it is a single substrate or a mixture of sorbents. In practice, the sorbent mixture used and the pre-column dimensions are the two factors most markedly influencing band broadening in on-line solid phase extraction systems. Major band broadening resulting from selection of inappropriate sorbents or inadequate column dimensions can be reduced or eliminated altogether by implementing a gradient elution step to obtain a peak compression effect at the top of the analytical column.

The purity of the support is a key to the performance of pre-concentration processes. Although silicates can be obtained with a high degree of purity, some impurities may be incorporated by the reagents used in their chemical modification; this calls for a pretreatment step to remove any traces of the reagents, either by Soxhlet extraction or by pre-washing with a suitable organic solvent or mineral acid.

The rigidity of the support is a physical feature of great significance insofar as elution is usually performed at pressures that may result in deformation of the extracting material and hence in altered physico-chemical properties. As a rule, the more rigid the support is, the higher the flow-rate at which elution can be effected and hence the shorter the analysis time. Thus, divinylbenzene–polystyrene resins (*e.g.* Chelex 100) tend to swell, contract, or even lose their shape on contact with the mobile phase. These problems are minimized by mixing with a silica gel of about the same particle size. On the other hand, foamed plastics are quite rigid and preserve their rigidity even at high elution flow-rates, while silicates, controlled pore glass and cross-linked cellulose that contain chelating substituents must inevitably be used at lower elution flow-rates.

The kinetics of the retention–elution process plays a major role in the selection of the support as noted in Section 4.2.1.

The column dimensions to be used for optimal performance are related to its own volume (V_p), the volume (V_c) and number of plates (N_c) of the analytical column, and the capacity factor (k') for the analyte through the equation

$$V_p/V_c \leqslant (1 + k')/2N_c^{1/2}$$

Additional band broadening arising from the pre-column should also be duly taken into account when using short fast LC, microbore, or highly efficient separation columns.

The particle size of the stationary phase with which the column is packed is usually the same as that used with the analytical column (5–10 μm) if the two phases are identical. However, other considerations such as low back-pressures—and the resultant high sampling rates, manual packing, and field sampling, and the prevention of clogging the pre-column with dirty or viscous samples may compel the analytical chemist to accept some additional band broadening for the benefit of using larger particles (20–50 μm).

Other aspects to be taken into account in selecting the sorbent material are the phenomena involved in liquid–liquid extraction, according to which, sorbents can be classified into six groups, namely: hydrophobic sorbents, ion-exchangers with functional groups, ion-pairing reagents, ligand-exchange sorbents, size-exclusion sorbents, and others.

The hydrophobic sorbents most commonly employed were packing materials in sorption columns such as C_{18}, C_8, and C_2 bonded silicas and styrene–divinylbenzene copolymers (*e.g.* XAD-2 and PRP1), which retain non-polar and medium-polarity aromatic compounds from aqueous samples. Partial clean-up can be accomplished by flushing the loaded sample with an aqueous solution of an acid, base, or buffer, while elution can be efficiently effected, either directly or via gradient steps involving the organic solvent used as mobile phase for the analytical column.

Ion exchangers are typically used in pre-columns for the extraction of strong and weak cationic species (sulphonic and carboxylic acids, respectively) by bonding to their silica or polymer-bound functional groups, and for the extraction of strong (quaternary ammonium salts) or weak anionic species (secondary amines) on a suitable support. Aqueous samples of low ionic strength and a pH suited to the type of exchanger used are introduced into a stream of water or the organic solvent also of low ionic strength and the appropriate pH. Elution is effected with water or with a water/organic eluent mixture of high ionic strength and/or a pH ensuring neutralization of the analytes or the exchanger.

Ion-pairing reagents (sulphonates or ammonium salts with non-polar carbon chains) are widely used in this context on account of their ability to extract ionic compounds with the opposite charge via their ionic sites. These reagents can be added to the sorbent or the sample solution, after which elution is effected as described above.

Ligand-exchange interaction between the stationary phase and the solute of interest takes place via the formation of co-ordination bonds in the co-ordination sphere of a metal ion, which is thus immobilized on a support (*e.g.* a silica- or polymer-based material) containing functional groups at ion-exchange or ligand-exchange sites. The sample can be aqueous or organic and contain small amounts of other competing species (*e.g.* complexing or precipitating agents for the metal), and the pH is a key to the quantitative retention of the analyte. The sample can be transported by using an appropriate aqueous or organic carrier, and elution can be effected by introducing a pH change and/or a stronger ligand or an excess ligand, or by using another metal to displace the analyte.

Size-exclusion pre-columns (*e.g.* Sephadex) can be used to separate compounds of different molecular sizes—only those with the smaller sizes penetrate through the gel pores and are retained. Both transport of the sample and elution are effected with aqueous solutions.

Other types of device such as wide-pore columns are used for the extraction of proteins and other compounds of high molecular weights. Thus, phenylboronic acid-modified silica retains *cis*-diol functional groups selectively. Elution is accomplished by simply changing the pH.

4.3.2 Equipment

The equipment involved in using a sorption column coupled on-line to a chromatograph is much more complex than that typically used in off-line assemblies, which only require the pre-column, whether laboratory-made or commercially available (Sep-PAK cartridges, Baker-10 Spe, Bond-Elut, or Chrom-Prep systems), and, at the most, a unit allowing a low pressure to be applied onto the top of the column or some vacuum on the outlet side. On-line assemblies also include one or more high-pressure switching valves, solvent selecting valves, and a small additional pump.

4.3.2.1 Columns

The columns ordinarily used for on-line solid-phase extraction are of small dimensions in order to avoid additional band broadening during the elution of the analytical column. The shorter the pre-column is, the smaller will be the distortion to the baseline caused by negative peaks arising from the liquid contained in the void volumes of the pre-column and reaching the analytical column. Small pre-column volumes also result in easier packing and smaller amounts of packing material. Some of the pre-columns used so far are as long as 300 mm; however, most frequently they range between 5 and 40 mm in length and are commercially available as cartridges or in ready-packed forms. The active material should be retained on screens (sieves) rather than frits because the latter are prone to clogging, particularly with biological samples.[61-63] These shortcomings arise not only from the use of packing materials of small particle size, but also from the small pores of sintered metal frits used for closing pre-columns. Because of their small pore size and relatively large thickness, sintered frits clog very easily; this problem can be overcome by sealing the pre-column with fine-mesh wire gauzes with mesh sizes just small enough to retain the packing.[64-67] This type of column is marketed by such firms as Bischoff, Knauer, and Hewlett-Packard. They should be easy to pack or be available as readily replaceable cartridges.

Pre-columns can be readily constructed from small LC columns according to the schemes in Figures 4.14A and B. The design in Figure 4.14A is older and allows the length of the packed bed to be adjusted as required by means of the PTFE tube included (4). This type of pre-column can be readily packed manually by using a microspatula or a syringe filled with a slurry of the packing material. Portions of the material are successively introduced into the column, which is closed at one end, and the packing is finally compacted by tapping the column on a hard surface. The pre-column depicted in Figure 4.14B is an improved design that enables use of small cartridges that can be manually assembled and replaced. Figure 4.14C shows a scheme of the Trace Enrichment Cartridge (TEC) manufactured by Gilson. It consists of a 316L stainless steel tube packed, as standard, with 20 mg of Hypersil C_{18} bonded 10 μm silica. The inner chamber, 1.6 × 5 mm, is bound by woven 316L stainless steel mesh and PCTFE (Kel-F) sealing plugs. The TEC is positioned in its holder as shown in the Figure and mounted on the injection valve of the chromatograph.

Automatic sample treatment on a pre-column followed by direct insertion into the analytical column can only be accomplished in routine analyses if the two columns feature a sufficiently long service life. This requisite usually affects analytical columns, which are particularly expensive (the manual packing of pre-columns scarcely influences the costs). The service life of the pre-column is of special interest when used for automated night-time or week-end operation because depleted columns have to be replaced manually. This parameter is determined by the total cumulative injected sample volume, also called 'total volume load', or by the overall number of individual injections of the samples after which the column must be replaced. The need for replacement is marked by a large pressure drop (greater than 30 bar), broadened and occasionally split peaks, and declining recoveries—provided these are not due to faulty performance from the analytical column. The maximum overall volume that can be processed by a pre-column depends on such factors as (*a*) the nature of the sample matrix; (*b*) the packing material (the nature of the chemically bound phase and its particle size and shape—spherical or fragmented); (*c*) the pre-column length; (*d*) the nature of the flushing liquid; (*e*) the duration of the flushing stage; (*f*) whether straight or back-flushing is effected; (*g*) the type of column closure used; and (*h*) the switching speed of the high-pressure valve.

For routine analyses, the column should be ideally replaced at regular intervals as preventive maintenance of the HPLC after a given amount of sample has been passed, particularly if the sample is of biological origin. Pre-column changeover takes little time (usually a few minutes) and the new pre-column can be equilibrated for high recovery by using two or three injections of spiked standards. By way of example, a standard pre-

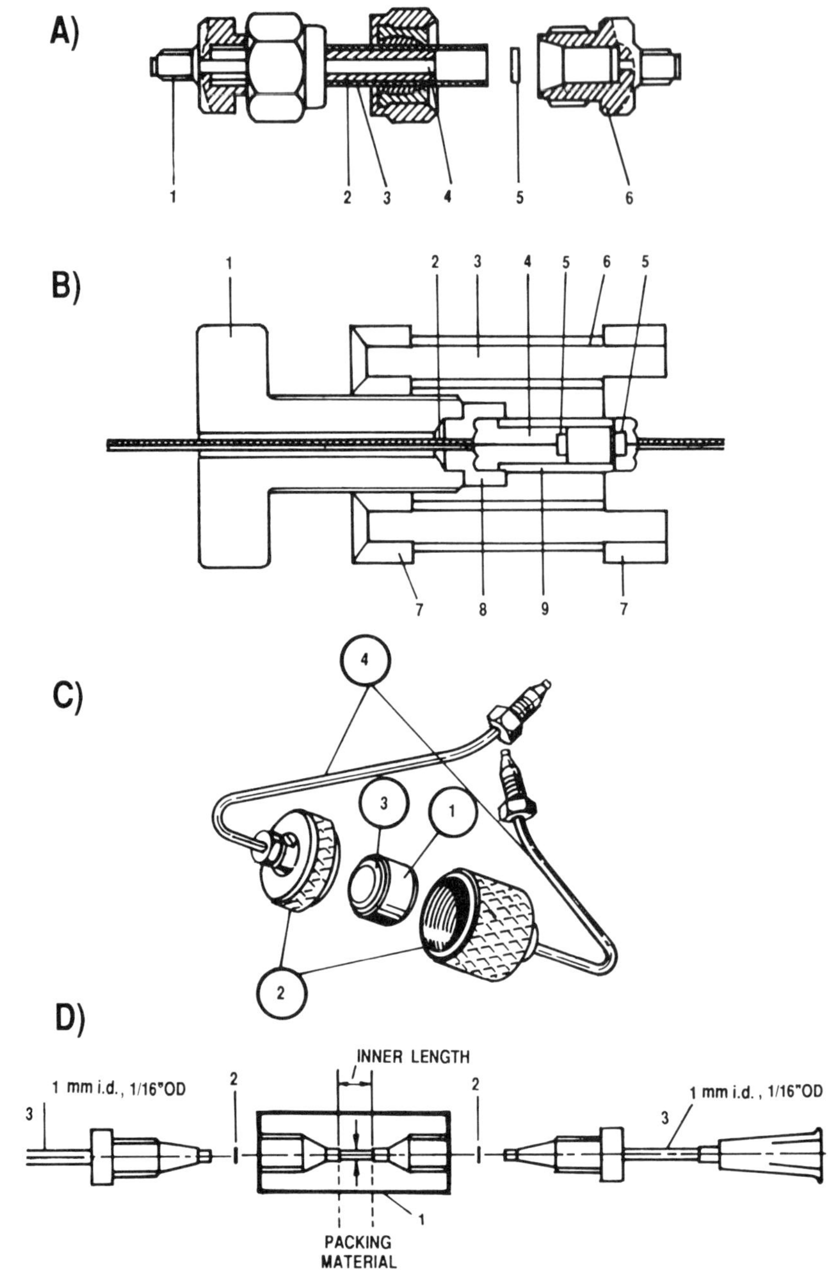

Figure 4.14 *Types of pre-columns used in on-line solid–liquid extraction.* (A) *Laboratory-built design:* (1) *1/4 × 1/16 in swagelok;* (2) *variable-length, 4.6 mm OD PTFE rod;* (3) *stainless steel tube 1/4in × 45mm;* (4) *stainless steel capillary 1/6in OD, 0.25mm ID* (5) *20μm PTFE or stainless steel frit;* (6) *1/4 × 1/16 in swagelok.* (B) *Improved laboratory-built pre-column design:* (1) *hand-tightened screw;* (2) *1/16 in OD, 0.25 mm ID stainless steel capillary;* (3) *bolt of column holder;* (4) *4.6 mm ID PVDF rod;* (5) *stainless steel*

column (10 × 4.6 mm, 25–40 μm RP-2) can be used to process between 60 and 80 200 μl serum samples; this is equivalent to a total column loading of 12–16 ml, *i.e.* much larger than the volume afforded by the single use of off-line extraction cartridges recommended by their manufacturers.

There are a variety of commercially available pre-columns. Upchurch Scientific, Inc. and Brownlee Labs, Inc. market columns furnished with stainless steel frits. The former firm manufactures 20 × 2 mm ID stainless steel units that can be readily dry- or slurry-packed by means of an adapter. The latter firm markets 16 types of pre-packed cartridges that can be readily mounted on 30 × 4.6 mm pre-columns. Recently, this firm launched a new type of guard cartridge (15 × 3.2 mm) pre-packed with 7 μm materials, in addition to eleven types of bonded silicas and porous polymers for on-line preconcentration purposes. On the other hand, the pre-columns manufactured by Bischoff GmBH, Waters Inc., and Chrompak International are furnished with stainless steel screens and consist of a cartridge holder for readily replaceable pre-packed cartridges of dimensions ranging between 0.5 and 2 cm (length) and 4.6 mm ID for the first firm, 4 × 6.0 mm ID for the second and 10 × 2 mm ID (40 μm particle size) for the third. While the Chrompak cartridges are basically intended for preconcentration, those from Waters are conceived for protection of the analytical column.

The miniaturization trend detected in chromatographic systems has also reached pre-column designs, where it is intended to eliminate or minimize band broadening by decreasing the pre-column diameter rather than its length.[68–71] Inner diameters between 0.5 and 1.0 mm have been used by some authors in designing their own customized columns. One such micro pre-column is depicted in Figure 4.14D. The body is made of stainless steel and resembles an ordinary connector. The packing material is placed between two screens which are fitted between the connector body and the 1/16 in OD connecting capillaries. Its dimensions range between 2 and 5 mm (length) and 0.5–1.0 mm (OD), and they can be readily packed manually by means of a syringe filled with a thin slurry of the active material.

4.3.2.2 Pre-column Pumps

An inexpensive high-pressure pump with a single piston is usually adequate for introducing samples into the pre-column as it need not be pulse-free—the perturbations from a highly pulsating flow can be overcome by

frit of variable diameter depending on the column diameter; (6) *plug of stationary phase;* (7) *column holder;* (8) *stainless steel cap;* (9) *pre-column proper (1 cm long, 1/4 in OD, 2.2–4.6 mm ID stainless steel tube).* (C) *Exploded view of TEC and its holder:* (1) *TEC;* (2) *stainless steel holder;* (3) *PCTFE seals;* (4) *75 × 1.6 × 0.5 mm stainless steel tube.* (D) *Micro pre-column* (1)*; with syringe adapter for hand-packing* (2)*; and screen* (3).

using a damping capillary or piece. Peristaltic pumps are inadequate here because they are not sufficiently pressure-resistant, so long analytical series may result in rather high pressures.

4.3.2.3 Valves

High-pressure switching valves are indispensable to the automatic functioning of a sorption column connected on-line to a chromatograph. Any six-way valve provided with six connectors and featuring an upper pressure limit of about 400 bar can in principle be used in these configurations, Latek, Kontron, and Gynotek currently market models suitable for this purpose. Valve switching is usually accomplished pneumatically or, even better, by introducing helium with the aid of a solenoid valve.

The solvent-selecting and injection valves occasionally used in on-line sorption/chromatographic configurations are of the same type as those typically employed with other systems; the former usually include between two and eight channels.

4.3.2.4 Column-switching Assemblies

The large number of manufacturers that market equipment for on-line solid–liquid extraction is clear proof of the widespread use of pre-column systems. Such firms as Waters Inc., Kontron Ltd., Spark Holland, and Hewlett-Packard commercialize equipment covering a wide range of possibilities. Each type of analysis is addressed by using the appropriate number of solvent-selecting valves, switching valves and columns. Because of their variety, these configurations are difficult to discuss systematically, so only a few examples illustrating different degrees of complexity and performance levels are discussed below by way of state-of-the-art overview of this type of equipment.

Column switching devices are generally intended to allow on-line combination of all the stages that are normally carried out off-line with solid–liquid cartridges. For this purpose, the required amount of sample is transferred from the sample loop onto the pre-column by means of a metering pump that serves this column and effects the flushing immediately afterwards. The pre-column containing the purified sample is thus 'switched to' (*i.e.* coupled with) the main column or analytical column with the aid of a high-pressure pump where the sample mixture is presented for separation and analysis proper. When sample transfer is over, the pre-column is disconnected from the analytical column and equilibrated with the washing liquid, which renders it ready for dosing the next sample. Time-programmed column switching by the high-pressure valve is effected by the switching contacts of an HPLC module or a computer, or else by external contacts combined into a compact device. When interference removal rather than analyte preconcentration and/or matrix removal is required, the operation of the column set-up is somewhat different: it is switched to the analytical system while the sample is flushed and collected in the loop of the injection valve of the chromatograph. Then, the pre-column is disconnected from the main column to effect elution of the interferents and equilibration with a suitable solution.

The design of switching systems must always be suited to the requirements of the chromatographic process involved, which thus determines its complexity. Figure 4.15 shows the scheme of a straightforward system for this purpose with elution in the same (A) and in the opposite (B) direction to retention. This latter alternative is also the more appropriate as it minimizes band broadening while avoiding excessive compaction of the packing material,

Tandem assemblies are particularly useful to boost sample throughput insofar as they allow a sample to be prepared while the previous one is being analysed, which results in the obtainment of uninterrupted successive chromatograms. Figure 4.16 shows an assembly of this type in which the two pre-columns work alternately via high-pressure valves in such a way that while pre-column 1 is coupled to the analytical column, pre-column 2 is loaded with the next sample after equilibration. The most serious problem posed by this type of configuration is that it only allows working in an isocratic regime—gradient elution calls for a more complex configuration.[72]

Serially arranged pre-columns are used whenever potential interferents with the analyte preconcentration are to be removed prior to this process. The configuration depicted in Figure 4.17 allows the two stages to be carried out on-line with the analytical system in quite a short time.

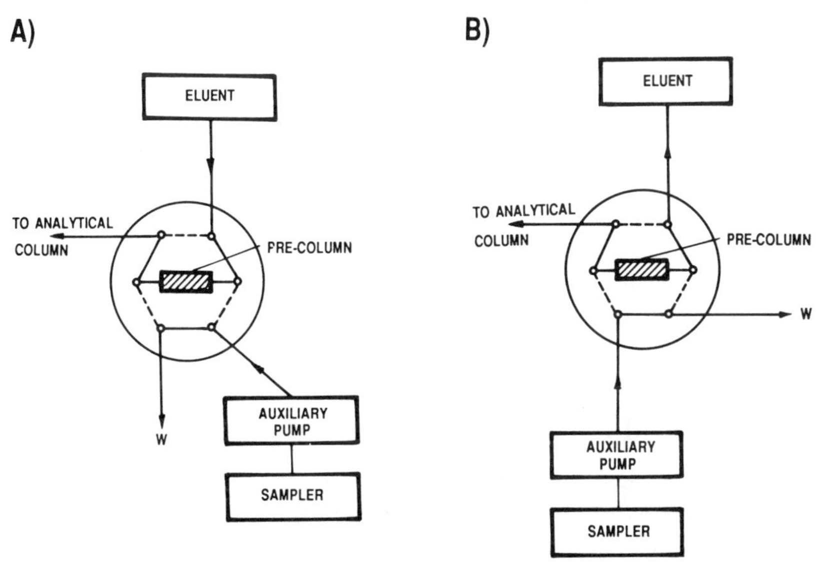

Figure 4.15 *Six-port switching valves for automatic sample processing with a pre-column.* (—) *sampling mode;* (- - -) *flush mode. Elution in the direct* (A) *or reversed mode* (B)

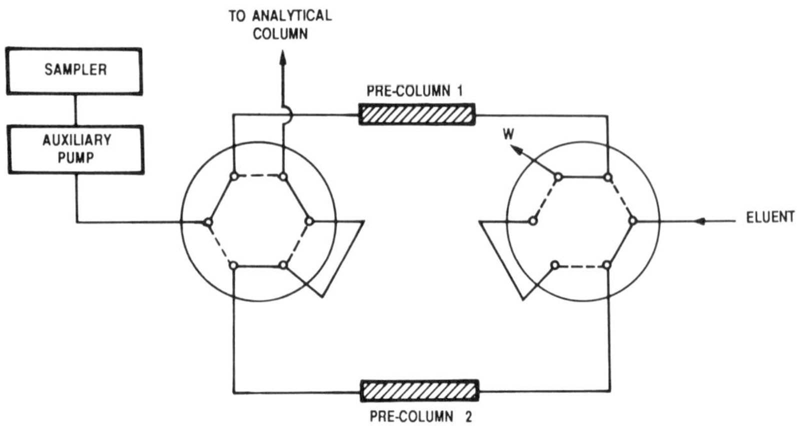

Figure 4.16 *Scheme of switching valves for the automatic alternate processing of two samples*

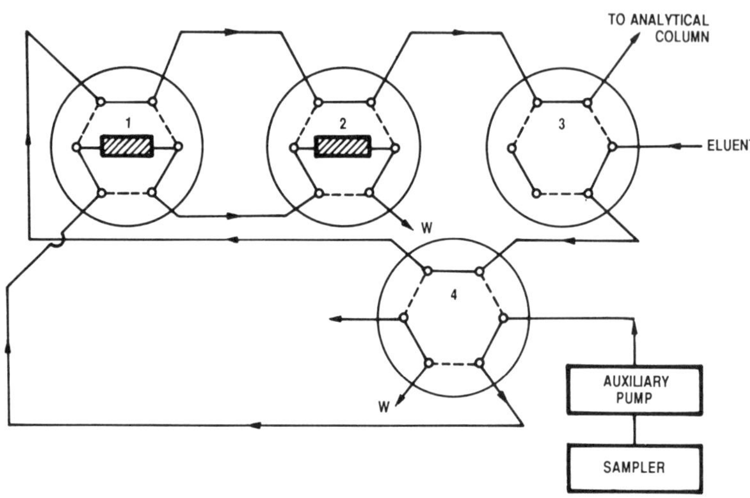

Figure 4.17 *Column-switching system for on-line group separation and trace enrichment*

4.3.3 Typical Examples

The broad scope of application of sorption systems coupled on-line to chromatographs[73-75] defies any global description, so only a few representative examples of interest to various areas of great social concern such as environmental and clinical chemistry are commented on below in order to provide a general picture of the potential of these systems.

As a rule, the sample volumes typically used range between 1 ml and 1 l and samples are of a varied nature, from aqueous (waters and biological fluids included), to extracts from solid, liquid, or even gas matrices.

Many environmental pollutants are toxic at trace levels, so their determination often calls for the use of on-line preconcentration or enrichment systems. Phenylurea herbicides were successfully preconcentrated from river water samples on a pre-column packed with Lichrosorb RP-18 (11 × 2 mm ID, 10 μm particle size) included in an assembly similar to that shown in Figure 4.15A. Sample volumes of 10 ml containing analyte concentrations between 1 and 17 p.p.b. were analysed in this way. The reproducibility of the method was not good (9–15%) as a result of the electrochemical system used for detection, which provided some spurious results arising from considerable fouling of the working electrode.[76] The same set-up was used for the analysis of airborne herbicides (phenoxyacetic acids), which were sampled in glass scrubber flasks containing distilled water and subsequently preconcentrated in quartz columns (2.2 × 3.6 mm ID, 37–50 μm, C_{18} bonded phase silica) that were preconditioned with acetic acid and into which the samples were introduced by means of a syringe thanks to the prevailing low back-pressure. The detection limit thus achieved for 2 ml of sample was 5–10 $\mu g\ l^{-1}$.

Ion-exchange columns are quite efficient in isolating very polar and ionic compounds; in addition, they provide enhanced selectivity. In this respect, polymer-based ion exchangers have been shown to surpass silica-based materials for this purpose.[73, 74] However, some caution should be exercised in using these exchange materials for the analysis of real samples because the hydrophobicity of the polymer matrix may give rise to secondary interactions resulting in non-polar compounds also being loaded onto the material surface sample, thereby severely reducing the capacity of the active material. One extra complication may arise from the presence of interfering cations blocking sites of the material intended to be occupied by the analytes. The serial column arrangement depicted in Figure 4.17 effectively circumvents this problem. Up to 29 pollutants present at trace levels in waste water can be preconcentrated by using three serially arranged pre-columns (Figure 4.18). Also, up to ten non-polar compounds (anthraquinone and other benzene derivatives) were trapped on a C_{18} pre-column, and up to eight compounds of medium polarity such as substituted aromatics were retained on a PRP_1 pre-column while the other eleven—all of ionic character—were retained on the ion exchanger Aminex A5. The sample was pumped serially through the three pre-columns in the following order: $C_{18} \rightarrow PRP_1 \rightarrow$ Aminex A-5; however, some of the compounds destined for the last two columns (PRP_1 and Aminex A-5) were retained by the previous ones (C_{18} and PRP_1, respectively). Each desired group was separated by flushing the pre-columns serially with dilute perchloric acid after introduction of the sample. In this way, only two compounds were present in the fractions of more than one column. The overall process consisted of nine steps in which, after complexation or precipitation of alkaline-earth cations and Fe^{III} in the sample with oxalate and EDTA and filtering out if required, the sample was passed through the three serially arranged pre-columns, which were subsequently washed with perchloric acid and methanol. The retentate of each column was then eluted separately by applying a continuous gradient of acetate and methanol, and the analytes were isolated on passage through a C_{18} chromatographic column. The diode array system used for detection provided the recording shown in Figure 4.18, which was yielded by 5 ml of a standard solution containing 200 p.p.b. of each analyte. The first 28 minutes of the chromatogram correspond to the ion-exchange column (polar aniline fraction); the interval between 28 and 58 minutes corresponds to the PRP_1 column (other aromatics of medium polarity), and the final interval, 58–90 minutes, corresponds to the C_{18} pre-column (non-polar compounds). Some of these compounds were found to occur in the waste waters assayed and were quantified by standard-addition procedures.[77]

Any type of biological sample (*e.g.* serum, plasma, urine, hemodialysates,[78] synovial fluid,[79] milk, saliva, faeces,[88] supernatants from tissue homogenates and tissue homogenates liquefied with subtilisin or another protease,[80] gastric juices, cerebrospinal fluid, fermentation broths and liquors,[81] and incubation solutions from *in vitro* systems[82]) that is free from suspended particles can in principle be processed on pre-columns.

Ethynyl steroids used as contraceptives are usually ingested at very low concentrations (10–50 μg) and are excreted very slowly via urine, in which they occur at concentrations of only a few ng ml^{-1}. In addition, HPLC methods for the determination of these analytes in urine are often interfered by the poor resolution between norethindrone, norgestrel, and ethynylestradiol. The method reported by Nielen *et al.*[83] circumvents these shortcomings by using two 10 × 2 mm ID serial pre-columns [PRP_1 and Ag^I-oxine] (Figure 4.17). The experimental procedure consists of the following steps: (1) trapping of the ethynyl steroid and various contaminants on the PRP_1 pre-column while inorganic and water-soluble constituents are driven to waste; (2) flushing of the PRP_1 pre-column with 50:50 methanol:water to ensure that any contaminants that might interfere with the sorption process are flushed to waste; (3) conditioning of the second pre-column with 70:30 methanol:water and elution of the retained steroids from the PRP_1 pre-column in the forward-flush mode to the Ag^I-oxine pre-column, where they are trapped by complexation in a narrow zone while organic interferents are flushed to waste; (4) washing of the second pre-column with methanol/water first and then water alone in order to flush completely any traces of organic solvent; (5) back-flush desorption from the Ag^I-oxine pre-column to the C_{18} analytical column with the aid of 15:85 acetonitrile:water; and (6) application of a step gradient with 50:50 acetonitrile:water. In order to increase sample throughput, the next run was started while the current separation was still in progress. In this way, analyte concentrations over the range 5–10 ng ml^{-1} were determined in 20 ml urine samples with satisfactory results.

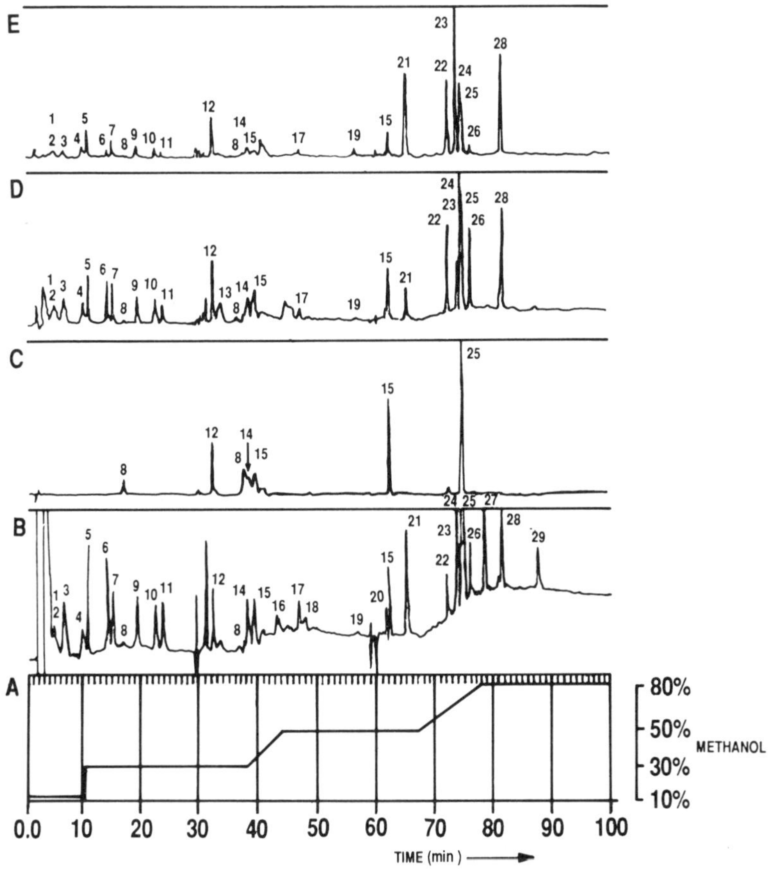

Figure 4.18 (A) *Chromatogram yielded by a 5 ml standard solution containing 200 p.p.b. of each of various pollutants. Gradient elution with 0.1 M potassium acetate (pH 8.0) and methanol (10–80%) as indicated. Detection at 222* (B), *390* (C), *244* (D) *and 295 nm* (E). *Compounds:* (1) p-*aminophenol;* (2) p-*phenylenediamine;* (3) m-*phenylenediamine;* (4) *4-methyl-*m-*phenylenediamine;* (5) o-*phenylenediamine;* (6) *aniline;* (7) p-*anisidine;* (8) p-*nitroaniline;* (9) *3-amino-4-ethoxyacetanilide;* (10) *anisidine;* (11) o-*tolidine;* (12) *picramic acid;* (13) o-*chloroaniline;* (14) p-*nitrophenol;* (15) *3,5-dinitro-*o-*cresol;* (16) m-*cresol;* (17) *nitrobenzene;* (18) p-*chlorophenol;* (19) p-*chloronitrobenzene;* (20) *pentachlorophenol;* (21) o-*dianisidine;* (22) *2- aminoanthraquinone;* (23) *3,3′-dichlorobenzidine;* (24) *3-amino-9-ethylcarbazole;* (25) p-*aminoanthraquinone;* (26) *1-aminoazobenzene;* (27) p-*dichlorobenzene;* (28) *2-phenylaminonaphthalene; and* (29) *1,2,5-trichlorobenzene*

The problem arising from the above-mentioned interferents was solved by using a diode array detector for simultaneous monitoring at various wavelengths—the absorbance maxima of the interferents and the analytes lay no less than 40 nm apart.

A configuration similar to the previous one was used for the determination of barbiturates in the same type of sample by using a 2 × 4.6 mm ID PRP_1 and a 5 × 3.0 mm ID Aminex A-28 pre-column. Similarly, the former pre-column was used to isolate the analytes from inorganic and organic water-soluble components of the sample, while the latter was employed to preconcentrate the barbiturates and separate them from neutral organic compounds. Quantitative discrimination was subsequently achieved on the analytical column, and the analytes were detected by UV spectrophotometry at 220 nm.[84]

The development of narrow-bore (< 2 mm ID) liquid chromatographic separation columns has resulted in the concomitant development of micro pre-columns intended to overcome the main shortcoming of liquid chromatography, namely the loss of sensitivity resulting from the decreased sample volumes (*ca.* 1 µl) and lightpaths of the UV flow-cells used.

Nielen *et al.*[90] designed a switching valve accommodating an internal micro pre-column for automated sample handling and trace enrichment in narrow-bore liquid chromatography. The system combines low dispersion with the capability of injecting plasma directly with the need for no off-line clean-up. The proponents compared the results provided by trace enrichment and direct loop injection of 100 and 0.5 µl, respectively, in terms of band broadening and enrichment factors. By way of example, they applied their system to the automated analysis of plasma samples containing clobazam and demethylclobazam at the low-p.p.b. level.[90]

4.4 Stripping Techniques

Electrolytic preconcentration is one of the major alternatives to the determination of metal traces. The analyte concentration at an electrode can be increased by a factor of up to 10^6 with respect to its concentration in the bulk solution while keeping the sample matrix in it. This process can be coupled to the stripping of the deposited species, which normally consists of two steps: preconcentration or separation proper, and stripping or determinative step. Thus, stripping techniques allow the separation, preconcentration and determination of one or several analytes and therefore match the performance of chromatographic techniques, even though their scope of application is somewhat narrower.

Species can be separated at an electrode by reduction (metals) or oxidation (anions forming insoluble salts with the oxidized form of a metal electrode, and metal oxides and hydroxides). The features of this first step determine the stripping process proper; this will be anodic when the deposited species is stripped on oxidation and cathodic in the opposite case.

Stripping can be accomplished in a variety of ways that result in as many technical variants of this technique, namely voltammetric, potentiometric, chronopotentiometric, coulometric, *etc*., all of which are normally implemented in a batch fashion.

Continuous stripping offers unquestionable advantages over manual manipulation in terms of sample throughput and sample and reagent consumption, in addition to those derived from its medium-exchange capability (deposition occurs while the sample is flowing through the detector and the stripping measurement step is effected after the sample volume has passed through the detector). These advantages are even more outstanding when complex samples requiring special treatment are dealt with. One of the most common pitfalls of stripping methods is peak overlap arising from the presence of species that are stripped at very close potentials. In most cases, good chemistry is the best solution and is applied in one of the three following ways:

(*a*) By masking (complexing) one of the species and thus shifting its electrode potential. Masking reagents can be readily incorporated into continuous systems via merging points.

(*b*) By carrying out the plating operation in one base electrolyte and replacing it with a secondary electrolyte allowing better selectivity during the stripping process. This alternative offers multi-element capabilities and reasonably high selectivity, but provides low reproducibility in the manual mode as a result of the species of interest undergoing the redox process on which measurements are based to some extent during the transfer. This problem is fully overcome by continuous systems, which eliminate one of the electrolytes by flushing with the next one.

(*c*) By using silica-immobilized algae cells for on-line column separation prior to the preconcentration step.[91]

4.4.1 Equipment

Continuous stripping configurations are assembled from a series of basic units such as valves, cells, electrodes, potential (or intensity) application systems and measurement devices, which give rise to two basic types of set-up according to whether samples are introduced by injection or aspiration.

4.4.1.1 Valves

Stripping configurations include two types of valves, namely:

Injection valves. These are normally four-way rotary valves that function to insert the sample into the continuous system and, occasionally, to insert a solution for conditioning of the working electrode.

Selecting valves. These are indispensable for sample/electrolyte change-over in configurations where the sample is introduced by aspiration. On the other hand, their purpose in configurations with sample injection is usually to switch between different electrolytes (plating and stripping), samples or standards, or to implement a washing cycle. They are normally placed prior to the injection valve or in its loading line. As a rule, they are not indispensable to configurations with sample injection. While six-port valves are by far the most commonly used for this purpose, eight-port valves provide added versatility and allow switching between the two sample introduction modes as required while preserving their channel-switching capabilities.

4.4.1.2 Cells

The cell is the most important unit in a stripping configuration and should accordingly meet a series of basic requirements, namely:

(*a*) It should be readily disassemblable and re-assemblable in order that the electrodes—particularly the working electrode—can be easily cleaned as frequently as required.

(*b*) It should not leak, not even in working with rather viscous solutions.

(*c*) Its internal potential drop should be minimal. This is of utmost importance insofar as the IR drop affects both the electrolytic and the stripping step of all stripping techniques. Only in potentiometric stripping is the second step not affected by the IR drop because no current passes through the cell during the stripping. However, the sample electrolysis is done in non-deoxygenated solutions. Since the electrolysis current is usually of the order of 10 μA, the internal resistance of the cell should be lower than 1 kΩ so that the electrolysis potential may be controlled to within ±10 mV.

(*d*) Because the electrolyte-containing tubes of the flow system act as antennae, the flow-cell should be isolated from noise, mostly on-line frequency, picked up by these electrolytes.

Wall-jet and thin-layer cells meet all of the above requirements and are thus the most commonly used with stripping systems.

Wall-jet cells, a denomination coined by Glauert,[92] are arranged in such a way that the solution is introduced as an impinging jet perpendicular to the electrode surface. By using a working electrode radius much larger than that of the solution inlet,[93] the sensibility is substantially increased thanks to the high linear flow-rates across the narrow inlet; however, some authors suggest that the mere geometric relationship of the solution impinging perpendicularly onto the electrode surface will not result in the classical wall-jet behaviour. In fact, less than 100% of the electrode exhibits wall-jet behaviour unless its surface is 2.3 inlet diameters away from the inlet.

Figure 4.19 shows the scheme of a typical wall-jet cell. The relative position of the reference and auxiliary electrodes is not unique; in fact, they can be placed in the cell itself or at the flow inlet or outlet.

Cells of this type including a liquid electrode pose a special difficulty arising from the need to collect used drops without introducing undue

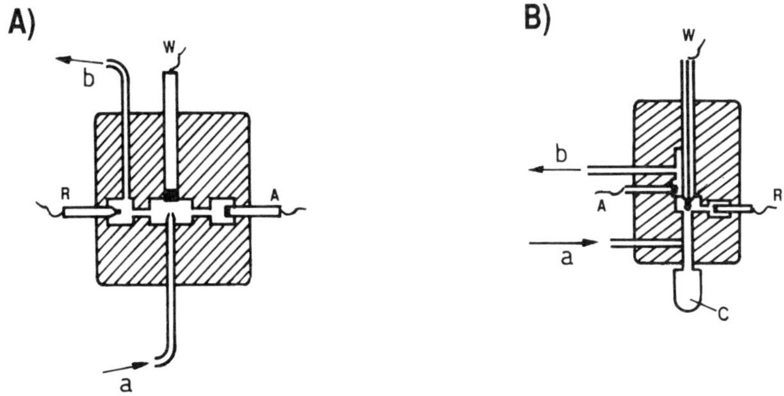

Figure 4.19 *Wall-jet cells used in continuous stripping methods.* (A) *With a solid working electrode.* (B) *With a mercury gold electrode.* W, R, *and* A = *working, reference and auxiliary electrode.* a = *inlet,* b = *outlet,* c = *mercury reservoir*

dead volumes. One of the ways of solving this problem is illustrated in Figure 4.19. Drops are separated in a wider section of the cell and are collected in a mercury reservoir at the bottom. This type of cell is more frequently used with liquid electrodes because any other type of flow impingement (in the same direction as, or perpendicular to, the drop fall) favours separation from the capillary and thus calls for smaller drops, which in turn results in decreased sensitivity. The optimal drop size/flow velocity ratio is normally a compromise between two extremes.

Wall-jet cells of large volumes (up to 100 ml) have so far proved to be inadequate for use in continuous systems.[94]

The thin-layer cells commonly used in continuous stripping systems are usually between 0.1 and 0.5 mm thick, *i.e.* much thicker that conventional cells—normally 100 μm thick—and hence prone to markedly higher dispersion. Unlike in wall-jet cells, in thin-layer cells the sample containing the electrochemically active substance is passed over the working electrode.

Figure 4.20 shows general types of finite-diffusion flow-cells with their reference and auxiliary electrodes arranged in different ways and their working electrode in the typical position of thin-layer cells. Figure 4.20A depicts the most usual geometry, with both the reference (R) and the auxiliary electrode (A) mounted downstream. This design is relatively inexpensive, but may suffer from *IR* drop problems. While its three-electrode system does not compensate for any significant solution resistance, it does have the advantage of not drawing current through the reference electrode. Figure 4.20B.1 shows a modified version of the previous design in which a planar auxiliary electrode faces the working electrode. The thin-layer channel serves as a salt bridge to the reference electrode, but since no current passes along the thin layer, there is no resulting *IR* drop

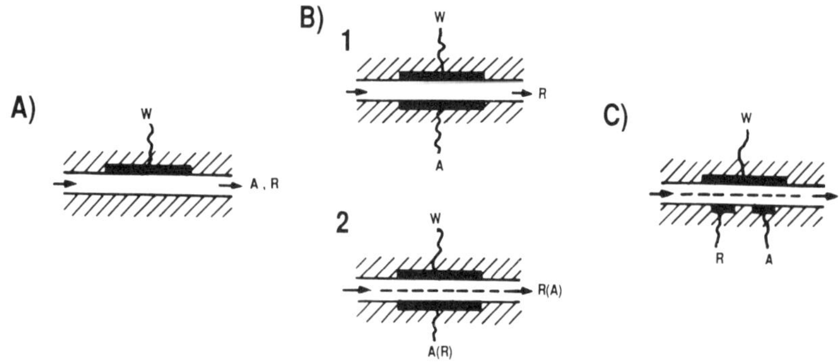

Figure 4.20 *Configurations of thin-layer cell differing in the relative position of their electrodes.* (A) *With a single electrode in the cell.* (B) *With two electrodes in the cell and the third located downstream* (1) *or in a cell with a porous diffusion membrane allowing ionic migration while blocking the flow of the electrolyte solution* (2). (C) *With the three electrodes on the cell walls and a porous membrane.* W, R *and* A = *working, reference, and auxiliary electrode*

in the solution. Some attention should be paid to problems arising from short transit times compared to the diffusion across the thin layer solution with large electrode areas, low flow-rates and very thin gaskets, for which the geometry depicted in Figure 4.20C can be of some advantage. Placing the reference electrode on the wall of the thin channel can best be accomplished by using an ionic junction constructed from a porous material permitting ions to migrate while blocking the flow of the electrolyte solution [Figure 4.20B.2, bracketed data]. The configurations in Figure 4.20B are superior on fundamental grounds because the current intensity across the electrode is not significantly influenced by the *IR* drop and the potential of the auxiliary electrode remains low as a result of the low impedance between A and W. The configuration depicted in Figure 4.20C is perhaps most satisfactory from a fundamental point of view. However, construction difficulties currently make this type of cell less desirable for routine work.

Oxygen removal is usually an essential requirement of stripping techniques in general and voltammetric stripping in particular. This is because the reduction of oxygen results in a severely sloping baseline that obscures the metal stripping peaks. The method most widely used for oxygen removal in stripping techniques involves purging the solutions with an inert gas prior to introduction into the flow system. However, oxygen has been found to diffuse back into the flow system through tubing and joints, thereby adversely affecting detection limits.[95] A number of methods have so far been tested to remove oxygen effectively from the flow system; some involve the use of continuous de-aeration devices based on nitrogen purging;[95, 97] others rely on electrochemical scrubbers for this purpose.[98, 99] Some methods discriminate against the effect of oxygen (*e.g.* subtractive stripping with dual electrodes[100] or computer-controlled instrumentation[101] that requires special equipment to be used). An effective method for oxygen removal uses a flow design with nitrogen purging and involves the deposition of mercury to form a renewable preplated mercury film electrode, and that of metals in the presence of oxygen, followed by medium exchange with nitrogen purging prior to stripping analysis.[102]

4.4.1.3 Electrodes

The working electrode is a key to cell performance and its shape and nature should be suited to the stripping mode to be used and the features of the analyte to be determined.

This electrode can be placed in three basic positions with respect to the flow stream, namely: *flow-by*, where the electrode surface is parallel to the direction of the flow; *flow-onto*, where the electrode surface is perpendicular to the flow direction; and *flow-through*, with two variants according to whether the electrode is a hollow cylinder or a porous matrix of some sort through which the solution flows.

Glassy carbon electrodes have so far been the most commonly used as working electrodes in stripping techniques. However, workers are showing increasing interest in another two types of electrode for this purpose:

(*a*) Reticulated vitreous carbon (RVC) electrodes, the porosity of which results in a high surface electrode/solution volume ratio. Reticulated vitreous carbon is an open-pore material with a honeycomb (foam) structure that competes advantageously with other materials typically used with porous electrodes (*e.g.* platinum chips, sintered silver discs, graphite powder or granules, amalgamated nickel chips or parallel screens of gold, platinum,and carbon).

(*b*) Carbon fibre electrodes, which are widely used at present for various purposes, one of which is to exploit the minute dimensions of the fibres for the analysis of very small samples or for *in vivo* measurements. An added advantage of fibre electrodes is the non-linear diffusion applicable to microelectrodes, which make them less sensitive to hydrodynamic fluctuations and hence occasionally also more sensitive.

On comparing a carbon fibre electrode (8–10 μm in diameter) and a glassy carbon disc thin-layer cell (2 mm in diameter) it was seen that the signal-to-background ratio of the former was roughly 1.6 times higher than that of the latter. Thanks to their ease of construction and low cost, fibre cells can be used as disposable electrodes, thereby avoiding the polishing required by glassy carbon electrodes.[103] Carbon wool electrodes, packed in different shapes (discs, fibres), are somewhat less commonly used in this context.

Metal electrodes (gold, platinum) are also scarcely used as working electrodes. A comparative study of gold fibre, gold-plated platinum and carbon electrodes used in potentiometric and constant-current stripping analysis for Hg^{II} with disc electrodes in a thin-layer cell assembly revealed them to feature increased sensitivity and long-term stability. From a practical point of view, they are also much simpler to handle in relation to leakage problems and electrode changeover. In order to obtain optimal reproducibility, not only carbon and platinum, but also gold fibre electrodes have to be coated with a fresh gold film prior to each electrolysis/stripping cycle. Because this is more readily achieved on gold fibre than on the other two types of fibre, the former is the best suited to routine work. In addition, gold-coated carbon fibre electrodes provide a higher background than gold fibre electrodes in both stripping modes. [104]

With the exception of the determination of mercury or a more noble metal, a simple pretreatment of the working electrode with an Hg^{II} solution at an appropriate potential for the deposition of a thin layer of mercury favours adherence of the metals to be subsequently reduced by formation of an amalgam. This pretreatment is logically unnecessary when a liquid electrode (Hg) is used as working electrode. The resulting simplification of the experimental procedure, however, does not compensate for the greater complexity of the cell to be used, which must include a small reservoir to collect used, wasted mercury drops (Figure 4.19B).

A current promising trend in this context is the use of modified electrodes. The possibilities are as wide as chemistry itself; however, three broad categories can be established according to whether the electrode surface is modified with adsorbed organic substances, inorganic complex ions, or adatoms of metals.[105]

The Ag/AgCl electrode is the reference electrode most frequently used in stripping techniques, followed by the saturated calomel electrode (SCE) and the glassy carbon electrode. A comparative study including a calomel electrode in a thin-layer arrangement, a silver tube (0.7 mm ID) lined internally with AgCl, and a glassy carbon rod electrode (1 mm diameter) mounted perpendicularly to the flow direction on a PVC tube (2 mm ID) showed that, because of its large dimensions, the calomel electrode produces significantly more noise that the other two. On the other hand, the Ag/AgCl electrode cannot be used in media containing Au^{III}, with which silver reacts spontaneously.[105] Another shortcoming of this type of electrode is that the reference potential depends on the chloride concentration of the solution being sucked through the tube. Also, it is impossible to obtain a reference potential in solutions with very low chloride concentrations. Moreover, if the solutions contain high concentrations of bromide or iodide, the reference electrode is transformed into an Ag/AgBr or Ag/AgI electrode.

Another comparative study involving a calomel electrode, a silver tube (8 mm long, 1.0 and 0.6 mm OD and ID, respectively) lined with an AgCl, and an Ag/AgCl electrode—the latter was used both as reference and as counter-electrode—included in the same cell showed background noise produced by the calomel electrode to be ten-fold that caused by the Ag/AgCl tube used as reference and counter-electrode; however, the noise level produced by the Ag/AgCl electrode used as reference only was intermediate between those of the other two electrodes.[103]

Counter-electrodes used in stripping techniques are made of a variety of materials, the most common of which are glassy carbon rods dipped in the out-flow solution stream or a threaded stainless steel block integrated in the flow-cell and usually located at its exit. The ohmic drop is minimized by placing the auxiliary electrode facing the working electrode, which can be readily achieved with thin-layer cells, where the auxiliary electrode is placed in the same or as similar a situation as possible.

4.4.1.4 Instrumentation

The instrumentation used to implement stripping techniques varies in complexity depending on the type of stripping involved. The preconcentration step can be carried out with a simple battery supplying an adequate potential for depositing the unknown analyte(s). It is thus the stripping step proper which usually calls for more complex instrumentation and ultimately imposes the features of the instrument to be used to apply a given method.

The more sophisticated instrumentation is usually required by the controlled-potential stripping modes, particularly when the stripping excitation signal is non-linear (staircase signals, differential pulse scanning, which require the use of polarographic analysers or stripping voltammeters). Laboratory-made digital voltammeters are used to perform electrochemical measurements via microcomputers governed by suitable software. Very often, a master computer programmed both in BASIC and in ASSEMBLY language is used to control the whole system, while potential waveform generation and data acquisition are entrusted to a slave computer communicated to its master via serial or parallel lines.

Chronopotentiometric stripping only requires an appropriate power supply—simpler than a potentiostat—for the measurement step.

Potentiometric stripping calls for the simplest instrumentation in this context. In fact, a straightforward potentiometer can be used as detector, and the only potential complications in this respect arise from the choice of an appropriate data-acquisition scheme or from multi-scanning methods.

4.4.1.5 Experimental Procedure

A continuous stripping process consists of three fundamental steps: (*a*) electrode pretreatment, (*b*) analyte preconcentration, and (*c*) stripping of the deposited substance(s).

The electrode pretreatment involves subjecting it to the deposition potential of Hg^{II} (0.70–0.90 V against the Ag/AgCl reference electrode) by keeping it in contact with a mercury solution until a thin layer is formed to favour the subsequent deposition of the analyte. This step can be implemented in two fashions, namely: (*a*) at the beginning of each working day, by allowing the electrode to be coated for 10–15 minutes; and (*b*) prior to each determination, by injecting a small volume of Hg^{II} at the deposition potential of this metal and subsequently performing a scan up to 0.10 V to strip oxidizable impurities from the film.

Deposition of the analytes takes place as the sample, introduced into the system by injection or aspiration, reaches the electrode, which should have been previously subjected to an appropriate potential. Sample aspiration calls for more strict control of the time during which it is introduced and requires the flow to be changed by a washing solution to effect the stripping. Injected samples are stripped by the carrier itself.

The stripping and measurement steps can be performed with or without halting of the flow, which results in increased precision and sensitivity, respectively. The way in which this step is carried out gives rise to the different variants of this technique, namely: (*a*) by performing a potential scan in the appropriate direction (voltammetric stripping); (*b*) by applying a constant-current intensity (chronopotentiometric stripping); and (*c*) with no electric excitation (potentiometric stripping).

Background information is acquired after each sample is assayed or only at the beginning and/or the end of the day by a similar process applied with no injection.

4.4.1.6 Automation

Continuous stripping methodologies can be automated to varying degrees, the lowest of which only includes the data acquisition and processing steps (real-time data acquisition, digital filtration, peak searching, peak integration, background subtraction). Full automation requires not only a passive detector–computer interface, but also active interfaces allowing the computer to act on the different units of the continuous system (injecting and/or selecting valves, stop/go sequence of the propulsion system) and on the detector to obtain the excitation signal corresponding to each step.

Figure 4.21 illustrates the action on the different units in a general procedure including injection of the Hg^{II} for electrode pretreatment and subsequent injection of the sample, and halting of the propulsion system during the stripping step after a previous rest period. The action on the detector during the electrode pretreatment and preconcentration step is always the same whatever the stripping variant. In voltammetric anodic stripping, the positive potential scan provides the measured *i-t* signal (D), which is obtained by applying a constant current in chronopotentiometric stripping (F) and requires no electric excitation signal in potentiometric stripping (E). The figure also illustrates the overall process involved in obtaining the background signal.

Obtainment of precise results in any stripping technique requires accurate timing of the different stages and a high data acquisition rate which can only be achieved by using an automated system. Unusually low

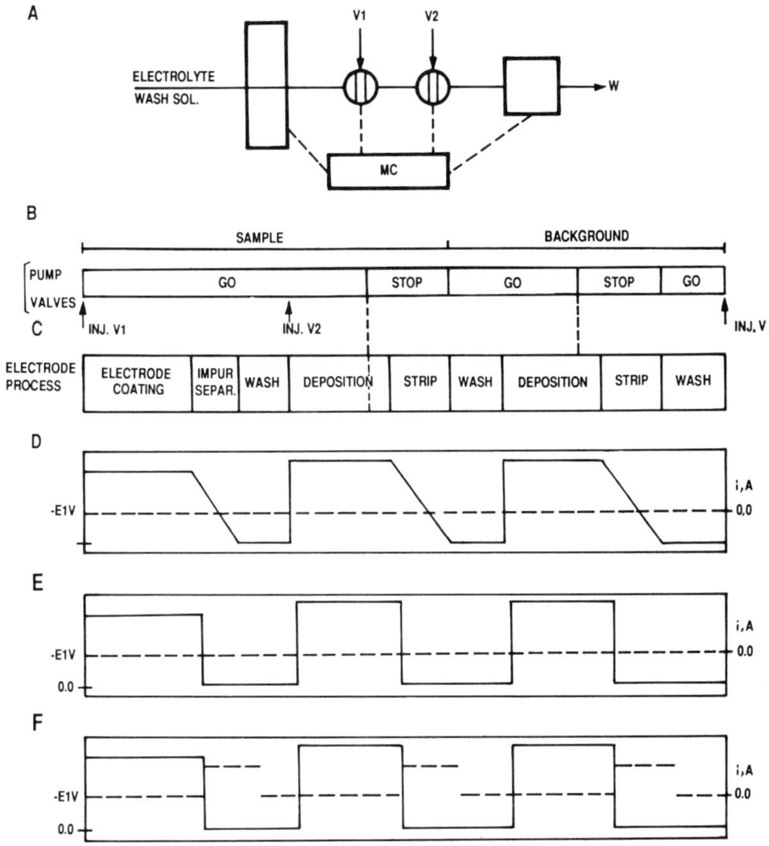

Figure 4.21 *Time sequence of the peristaltic pump and injection valves* (B) *of the FIA system* (A)*. Electrode processes involved* (C) *and excitation signals applied to the electrode in voltammetric* (D)*, potentiometric* (E) *and chronopotentiometric* (F) *stripping techniques*

determination limits call for the use of automated systems in multi-scanning modes, where the stripping and plating cycles alternate so rapidly that the analyte may hardly diffuse from the electrode surface, or for the flow direction to be reversed during the deposition step so that a small volume of sample may be passed successively through the flow-cell in order to increase the analyte concentration at the electrode.

4.4.1.7 Configurations

The configurations used to implement continuous stripping techniques vary basically in the way samples are introduced (by injection or continuous aspiration) and are normally quite simple in any case. Figure 4.22 shows some of the possible configurations used depending on the type of analysis involved.

Figure 4.22A1 depicts a straightforward configuration in which sample and carrier are sequentially aspirated by switching the selecting valve manually and from which data are collected via a recorder. The configuration in Figure 4.22A2 is a fully automated system in which a multichannel valve allows the user to select between electrode pretreatment solutions, sample, standards, carrier, and washing solution. The valve is actuated automatically by the computer. The pump can be stopped during the stripping step. Both the application of the excitation signal to the working electrode (a constant potential for conditioning the electrode followed by a scan for stripping of impurities, a constant deposition potential and a new potential scan for stripping/detection at constant i or E=i=0), data acquisition and processing, and result delivery are also programmed via the computer.

Stripping configurations with sample injection can be manual (Figure 4.22B1) or fully automated (Figure 4.22B2). In the latter, the sampler is furnished with an aspiration probe and contains sample and standards vials, the contents of which are aspirated into the injection valve. The electrode pretreatment solution can be injected sequentially via an additional valve or be included in sampler vials, in which case only a single injection valve is required. The active and passive interfaces to the detector ensure performance of the same level as that of the configuration in Figure 4.22A2.

4.4.2 Voltammetric Stripping

Voltammetry is the oldest of stripping techniques, so its theoretical foundation is well known and solidly established. Unfortunately, equations for stripping voltammetry, solved rigorously for batch and flow systems with steady-state deposition currents,[107-109] are inapplicable to continuous systems in general and FIA systems in particular. The effect of dispersion on the response in stripping voltammetry/flow injection systems has been examined and expressions for the corresponding stripping peak currents have been derived and experimentally tested.[110]

As in other stripping techniques, the current in voltammetric stripping is usually obtained after the plug has passed completely through the detector. For a flow-through mercury film electrode, to which a deposition potential is applied while the entire sample plug is passed under laminar-flow conditions, if tubing of small inner diameter and low flowrates are used, the solution will be homogeneous at the cross-section of the flow.[111] The stripping peak current of a mercury film electrode in the linear-scan stripping mode is given by

$$i_p = 1.116 \times 10^6\, n^2 A C_E l v \qquad (1)$$

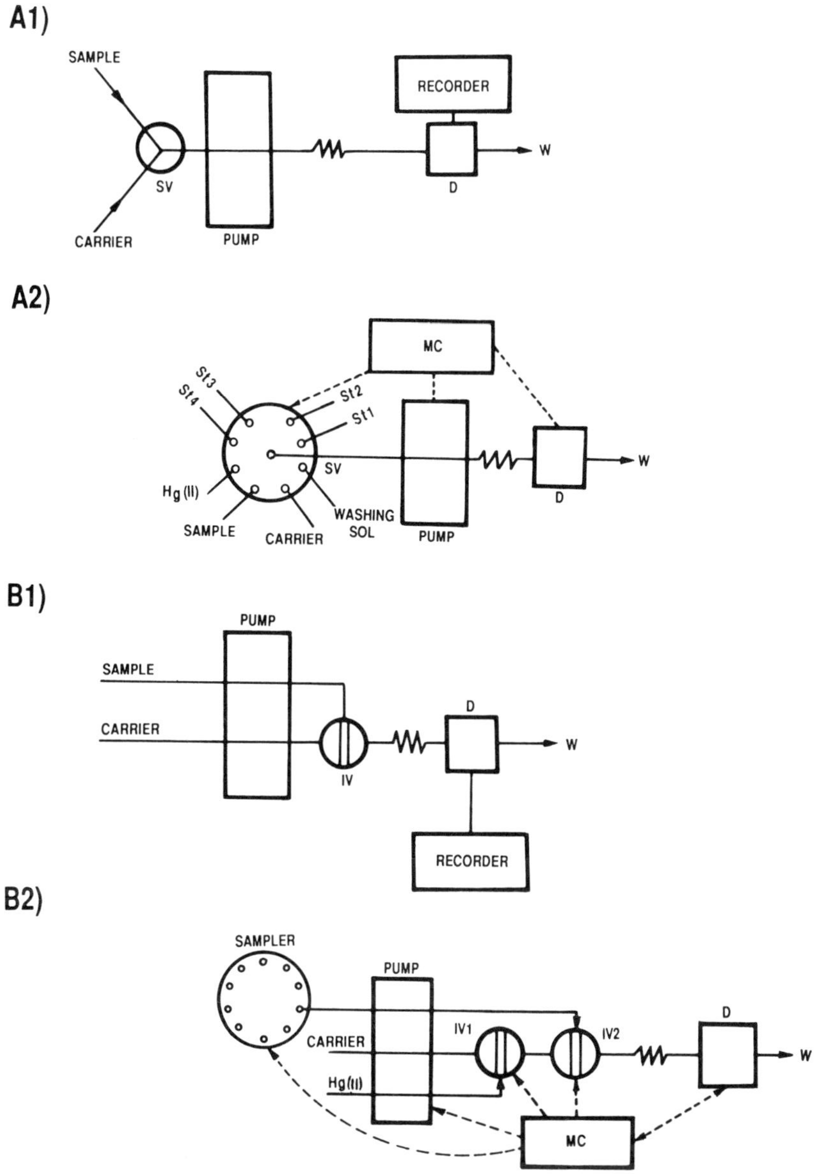

Figure 4.22 *Configurations for implementation of stripping methods.* (A) *With sequential aspiration of the sample.* (A1) *Sequential aspiration of sample and carrier;* (A2) *programmed aspiration of an* Hg^{II} *solution for electrode conditioning, plus sample, carrier, washing solution, and standards.* (B) *With sample injection.* (B1) *Manual injection into a carrier/washing stream.* (B2) *Programmed aspiration of sample and standards from the sampler into the injection valve* (V_2) *after injection of the electrode conditioning solution via* V_1. SV = *selecting valve,* D = *detector,* W = *waste,* MC = *microcomputer*

where n is the number of electrons involved in the process, C_E is the concentration of the metal in the mercury, A and l are the surface area and thickness of the mercury film, respectively, and v is the potential scan rate. For the same electrode used with a differential pulse stripping waveform, i_p is given by[108]

$$i_p = 0.138Q/t_p \tag{2}$$

where Q is the amount of metal deposited and t_p is the pulse width.

According to Faraday's law

$$C_E = Q/nFAl$$

where Q is the area of the current-time peak during the deposition step (passage of the sample plug). Such a peak area can be obtained from

$$Q = \int i(t)\mathrm{d}t$$

where $i(t)$ is the time-dependent deposition current, which in turn is given by[112]

$$i(t) = i_{ss}C(x,d,t)/C_0 \tag{3}$$

i_{ss} being the limiting steady-state current, $C(x,d,t)$ the concentration at any point along the stream (x and d are the axial and radial co-ordinates) and C_0 the initial concentration of the sample. The steady-state deposition current is given by

$$i_{ss} = nFAKC_0U^{\alpha}$$

where K is a constant related to the diffusion coefficient and geometric parameters, U is the fluid flow-rate, and α is a constant dependent on the electrode geometry.

Careful evaluation of the integral $\int i(t)\mathrm{d}t$ leads to the following expression for Q[112]

$$Q = i_{ss}V_s/U$$

where V_s is the sample volume.

A combination of equations (1), (2), and (3) with the above expressions for C_E and i_{ss} allows the stripping peak currents recorded in the linear scan (ls) and differential pulse (dp) stripping modes to be calculated as:

$$i_p(\mathrm{ls}) = 1.116 \times 10^6 n^2 AKC_0U^{(\alpha-1)}V_s v \tag{4}$$

$$i_p(\mathrm{dp}) = 0.138nFAKC_0U^{(\alpha-1)}V_s/t_p \tag{5}$$

Other experimental studies have been aimed at determining the influence of the flow-rate on the metal deposition efficiency; while it can be fitted to a decreasing exponential, function for some metals, no generally applicable expression has been derived so far.[99]

The anodic stripping methods reported so far include the determination of analytes such as As^{III} with a detection limit of 4×10^{-10} M (0.03 μg As l^{-1}) by the normal scanning mode with a stationary gold disc working electrode. The advantages offered by the use of other waveforms for stripping have been shown in the determination of Zn^{II} at a liquid electrode (a static mercury drop electrode) by using a differential pulse scan.[96] A comparison between differential pulse and square wave scans for the determination of lead showed square-wave voltammetry to provide shorter analysis times and better sensitivity than differential pulse voltammetry,[95] which in turn resulted in lower detection limits (as low as 20 pg ml^{-1}) than linear sweep voltammetry (detection limits of a few ng ml^{-1}) in the analysis for lead in 1 ml of undeaerated blood sample. Subtractive anodic stripping voltammetry was first used in flow injection analysis by Wang and Dewald in 1984 with the aim of effectively correcting Faradaic and non-Faradaic background current contributions. In this way, they lowered the detection limit for Cd^{II} to 0.14 ng with a one minute deposition. This mode enables use of undeaerated samples thanks to the effective correction of the oxygen reduction current.[99]

Cathodic stripping voltammetry has been used in continuous systems for the determination of chloride and iodide[100] on a manifold where measurements were made with the aid of a laboratory-made voltmeter in which potential waveform (staircase) generation and data acquisition were done by a single-chip microcomputer acting as the slave of an ABC-80 master computer via which the whole system was automated. One eight-way and three three-way valves were used to switch the flows to the detector. The eight-way valve could also be used as an injection valve, so the manifold allowed introduction of the sample by aspiration and injection. The flexibility provided by the four automatically actuated valves in combination with the different possibilities offered by channel switching make this type of configuration particularly interesting for studying special plating and washing procedures involving compounds strongly adsorbed on the electrode and hence the need to introduce simultaneously several streams for efficient removal of the adsorbed layer from the electrode surface.

Current trends in voltammetric stripping techniques are essentially aimed at the use of chemically modified electrodes.[97, 101, 106] Algae columns are a promising novelty in voltammetric stripping, where they have so far been used for on-line separations. Algae-silica preparations have proved to be highly advisable for use in this field as they combine unique reactivity patterns and the high binding capacity of algal biomass with the hydrodynamic and mechanical features of porous silica for selective and exhaustive removal of interfering constituents to circumvent such common problems as those arising from peak overlap and intermetallic effects.[91]

The iterative reversal of the flow direction[114] was recently used in combination with anodic voltammetric stripping to increase the sensitivity of existing methods by overcoming the best known drawbacks of flow injection stripping methods, namely the limited duration of the preconcentration (deposition), the length of which is determined by the time during which the sample plug resides in the flow-cell. The 'effective' deposition period can be greatly lengthened by passing the sample zone repeatedly through the detector. The increased plating obtained by using small sample volumes results in significantly increased stripping peak currents and hence in substantially lower detection limits. Once the electrode has been conditioned, the sample is injected and, after a given delay time, the flow modulation is initiated by using a preset cycle duration. Following a number of cycles, the forward movement of the carrier is re-established for a few seconds to flush the sample from the system. Then, the solution is stopped and after a short rest period the analyte is stripped from the mercury film by applying a positive-going potential ramp. The prolonged deposition period obtained with ten flow oscillations results in a 13-fold increase in the response and in a detection limit for lead of 7×10^{-9} M compared with 9×10^{-8} M achieved by conventional flow-injection operation.[115]

4.4.3 Potentiometric Stripping

Potentiometric stripping analysis based on the preconcentration of trace metals by potentiostatic deposition on a mercury film electrode and subsequent chemical re-oxidation of the amalgamated analytes under well defined conditions has been adapted to continuous systems in order to shorten analysis times by expediting sample changeover and electrode clean-up between samples. In addition, the chemical re-oxidation (stripping) in a flow system need not necessarily be performed on the sample, unlike in a batch system.

The theoretical foundation of this technique as established by its creators for batch systems was adapted by Hu *et al.*[119] to flow injection systems starting from the Levich equation[120] for the limiting current (i_L) at a flow-through planar electrode:

$$i_L = nkFC_0V_f^x \tag{6}$$

where n is the number of electrons involved in the process, F is the Faraday constant, C_0 is the bulk concentration of the electroactive species, V_f is the flow-rate, k is a complex function of the diffusion coefficient, kinematic viscosity of the fluid, and geometry of the electrode, and x is a constant calculated theoretically to be 1/3 for tubular electrodes and 1/2 for planar electrodes.

For a given set of experimental conditions k is constant, hence

$$i_L = K'V_f^x \tag{7}$$

where

$$K' = nkFC_0$$

The amount of analyte deposited on a glassy carbon electrode over the electrolysis period can be calculated from

$$Q = i_L t_{elect} \tag{8}$$

where t_{elect} is the electrolysis time.

In flow injection analysis, the sample is injected into the flowing stream as a plug. Therefore, the time during which it is exposed to the electrode surface determines the effective plating time, which can thus be defined as

$$t_{elect} = V_s/V_f \tag{9}$$

where V_s is the sample volume and V_f is the flow velocity.

Substitution of t_{elect} into equation (8) yields

$$Q = i_L V_s/V_f \tag{10}$$

or

$$Q = K'V_s V_f^{x-1} \tag{11}$$

According to equation (11), the amount of analyte plated out is inversely proportional to the flow-rate.

By measuring limiting currents at different flow-rates, one can construct a plot of log V_f *vs.* log i_L, the slope of which will be x. Likewise, if the amount of analyte deposited can be experimentally determined, then a plot of log Q *vs.* log V_f will give another straight line of slope $x - 1$.

Buffle *et al.*[121] relate the rate of oxidation of the metal to the oxidant concentration via the equation

$$dN/dt = AD(C'_0 - C'_s)/H \tag{12}$$

where dN/dt is the rate of oxidation, A the electrode surface area, D the diffusion coefficient, H the diffusion layer thickness, C'_0 the bulk oxidant concentration and C'_s the oxidant concentration at the electrode surface.

If the oxidation process is diffusion-controlled and the bulk oxidant concentration is much higher than that at the electrode surface, then equation (12) can be simplified to

$$\mathrm{d}N/\mathrm{d}t = ADC'_0/H \tag{13}$$

The rate dN/dt remains constant as long as the oxidant concentration, C'_0, does not change. The amount of analyte deposited, Q, is related to the time required to oxidize the amalgamated metal, t_{strip} by

$$Q = ADC'_0 t_{\text{strip}} \tag{14}$$

Thus, the time needed to strip a metal from the electrode surface can be used to quantify its concentration. Since

$$K'V_s V_f^{x-1}/nF = ADC'_0 t_{\text{strip}}/H \tag{15}$$

by plotting log t_{strip} *vs.* log V_F one will also obtain a straight line of slope $x-1$.

Potentiometric stripping analysis (PSA) has several salient features that warrant some comment. Thus:

(*a*) Samples need not be deoxygenated since the presence of oxygen facilitates the stripping step. This is of utmost significance to automation as deoxygenation is time-consuming and prone to contamination.

(*b*) No excitation signal need be applied to enact the stripping (chemical stripping), which in turn results in considerably simplified instrumentation.

(*c*) The potential at which re-oxidation takes place depends on the chemical composition of the stripping solution. The re-oxidation potential can be modified by adding an appropriate agent (X) to complex the metal according to

$$\mathrm{M(Hg)} \rightarrow \mathrm{M}^{n+} \xrightarrow{m\mathrm{X}} \mathrm{MX}_m$$

This is of special interest when several metals are deposited and their stripping potentials must be properly resolved.

Figure 4.23 shows typical potentiometric stripping curves obtained after 80 seconds of pre-electrolysis at 1.40 V against the SCE using a 0.5 M HCl solution containing identical amounts of Sn^{II}, Cd^{II}, and Pb^{II} and subsequent stripping in 0.5 M HCl (Figure 4.23A) or 0.2 M NaOH containing 2 M NH_3 (Figure 4.23B). The increased resolution of Sn and Pb arising from complexation by hydroxide ions is outstanding.

(*d*) The stripping event takes place very rapidly and changes in the electrode potential (dE/dt) during the stripping are self-optimizing, *i.e.* dE/dt is quite small on the stripping plateau and larger between

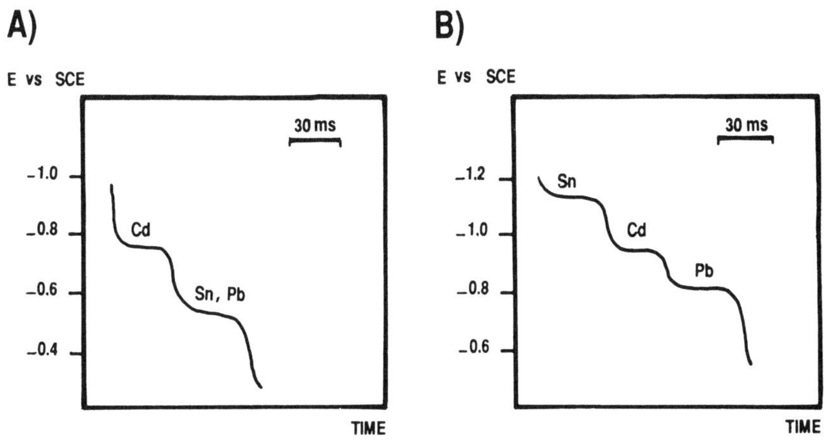

Figure 4.23 *Potentiometric stripping curves obtained in 0.5 M HCl* (A) *and in 0.2 M NaOH/2 M* NH_3 (B)

the metal stripping events. This results not only in decreased analysis times, but also in the need for simpler instrumentation.

The rapidity with which stripping is accomplished makes use of a microcomputer virtually indispensable. The computer can also be used to control the rest of the flow system (valves, sampler, propulsion system), in addition to other tasks such as optimizing the concentration of the stripping solution and controlling the thickness of the mercury film.

The data acquired by the computer can be plotted directly as the primary curve, *i.e.* as E *vs.* t (Figure 4.23), or as the reciprocal differential dt/dE converted from a derivative signal dE/dt against the electrode potential (E). In this last case, the quantitative signal used is the area under the stripping peak, which is normally proportional to the peak height.

Although the features of these techniques require at least data acquisition to be automated, some interesting methods such as that developed by Frenzel and Brater[122] for the simultaneous determination of Zn, Pb, and Cu in tap water, and that of Pb and Cd in acid digests from biological samples make a computer dispensable. In fact, by using a high-speed *x-t* recorder, the authors managed to determine from a few μg l^{-1} to percent levels at a rate of 200 samples h^{-1} using various injected sample volumes (between 14 μl and 1 ml). A fully automated system was used for the determination of Hg^{II} with a detection limit of 0.4 μg kg^{-1} (2 nM) using a gold working electrode.[123]

The determination of Cd and Pb in whole blood by potentiometric stripping with carbon fibre electrodes requires a sample pretreatment involving a single 1 + 19 dilution of two 0.2–0.4 ml sample aliquots with 0.5 M HCl, and a standard addition to one of the aliquots. The computer-controlled flow system used automatically executes a pre-programmed number of cycles on each sample pair before delivering the final result. Each cycle, which has a duration of 80 and 135 seconds for Pb^{II} and Cd^{II}, respectively, involves displaying the derivative stripping signal on a printer or plotter and integrating the stripping peaks obtained. Lead is stripped with 0.5 M HCl, which eliminates the interference of copper, though at the cost of a relatively high stripping rate compared with the 5 M $CaCl_2$ used to strip cadmium.[124] The flow-cell used consists of a silver chloride-lined silver tube which acts both as reference and as counter-electrode, and a disposable carbon fibre working electrode mounted on a PVC tube, which has a useful lifetime of 50–200 cycles.

The increased sensitivity obtained by halting the flow and optimizing some variables typical of the manifold used was studied by Mannino in his determination of Pb^{II} and Sn^{II} in fruit juices and soft drinks using a fully automated continuous-flow system.[125] After plating in a strong acidic medium, an appropriate stripping solution was used to separate Pb and Sn, which were subsequently determined. In an experiment conducted by operating at the maximum speed of the peristaltic pump (1.3 ml min^{-1}) during plating and by stopping the flow during stripping, the Pb and Sn signals increased by a factor of about ten with respect to those obtained when the stripping flow-rate was maintained at that of the plating. The flow was stopped 20 seconds before the stripping.

Nakayama *et al.*[126] reported a special methodology combining voltammetric and potentiometric stripping for the automatic determination of iodine species in natural water. The continuous system used for this purpose included two flow-cells for analyte preconcentration and detection, respectively (Figure 4.24). Iodide was electrochemically oxidized to iodine, which was concentrated quantitatively at a carbon wool electrode housed in a preconcentration cell (C_1 in Figure 4.24). After interfering ions had been removed from the cell, iodine was eluted with a reductant and subsequently determined at a polished Ag_3SI electrode in the detection cell. Iodate was determined after chemical reduction to iodide. The resulting amperometric current was proportional to the iodide or iodate concentration in the original solution. The sensitivity and detection limit thus achieved were 0.4–0.5 nA μg^{-1} I^- and 5 ng of I^-, respectively. Iodine species were accurately determined as iodide and total iodine (iodide + iodate) in less than 50 ml of sea water sample.

4.4.4 Chronopotentiometric Stripping

This variant of potentiometric stripping is implemented by applying a constant current intensity to the working electrode as excitation signal during the stripping step. Despite its limited development, this technique offers some disadvantages over ordinary PSA (*e.g.* the need for simpler instrumentation as a result of the application of the excitation signal characteristic of this mode). In addition to the typical *E-t* recordings obtained by stripping the analyte under galvanostatic conditions, this variant also allows reciprocal derivative constant current measurements

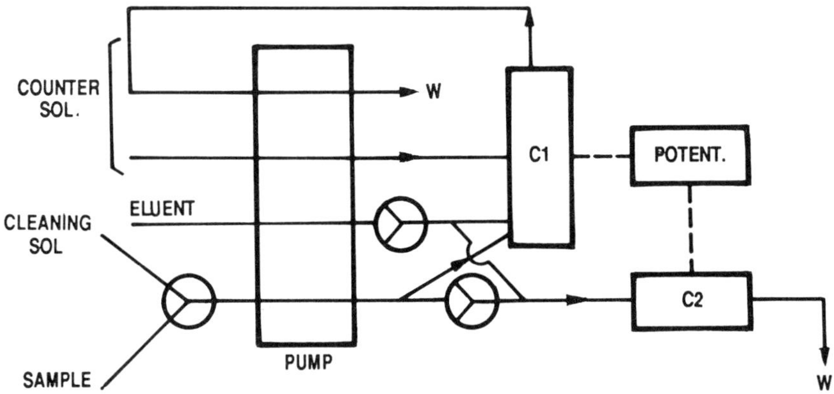

Figure 4.24 *Automatic configuration for the determination of iodine species in natural waters.* C_1 *and* C_2 = *preconcentration and detection cell*

to be made and hence higher sensitivity and resolving power to be achieved.[127]

Jagner *et al.* conducted studies on automated continuous systems by using the two modes of anodic potentiometric stripping (with and without application of the excitation signal to the working electrode). In one study they constructed a fully automated configuration by using a novel flow-cell and a personal computer-based system in which the thin-layer cell used was designed for Teflon-embedded glassy carbon electrodes placed in a spring-regulated holder. This system was absolutely leak-free and easy to install. The personal computer was interfaced to hardware components such as a potentiostat, an amplifier, a peristaltic pump, a six-inlet valve for insertion of different solutions, an autosampler, and a strip-chart recorder. User-accessible commands allowed the electrolysis, stripping current, pump flow-rates, valves, stripping transient recording, derivation, filtration, and integration to be controlled, and unattended operations to be accomplished via a number of user-defined programs. Through a software interface to a high-level language (BASIC), results from any number of analytical runs could be further evaluated by standard addition, from calibration curves or by statistical methods.[128] Later studies with this system involved constructing carbon fibre flow electrodes and comparing them with glassy carbon disc layer cells,[103] as well as constructing other electrodes from various materials (gold, platinum and carbon) and comparing their performance.[102] In contrast with the potentiometric stripping signal, the magnitude of the constant-current stripping signal is dependent on the surface area of the working electrode.

Continuous-flow reductive stripping chronopotentiometry has been used for the determination of selenium in biological materials (acid-digested mussels and NBS bovine liver) with excellent results.[129]

4.4.5 Coulometric Stripping

Coulometric stripping has so far been chiefly used to estimate film thicknesses. After the working electrode has been conditioned, the potential is stepped from −0.700 V, at which coating is done, to +0.700 V, and the current response resulting from the oxidation of the mercury on the electrode surface is integrated. The film thickness is calculated by assuming an even distribution of the mercury.

Some attempts at the practical application of this technique to quantitative analyses have been made by using a novel design of cell compartment and/or long electrolysis times in order to achieve thorough depletion of the metal ions. New electrode materials such as carbon felt or reticulated vitreous carbon can be used to construct porous flow-through electrodes that yield up to 100% conversion under appropriate conditions.

The chief advantage of this coulometric methodology is the avoidance of errors arising from variations in the electrolysis time, plating potential, mass transport, and temperature. The only constraint to its application is determining the time during which the potential should be applied in order to ensure complete deposition. Such a time can be quite short provided large electrode areas and small sample volumes are used and mass transfer is reasonably effective. In addition, the amount of metal deposited can be calculated from Faraday's law. To achieve complete deposition of the metal, the flow can be stopped while the sample is inside the flow-cell. Only small sample volumes should be used in order to ensure that all the sample material is within the active cell volume. Stripping is obviously done as the solution is set into motion again. The amount of metal deposited is calculated from the sample concentration and the injected volume.[130]

4.4.6 Stripping Following Non-electrolytic Preconcentration

Ordinary stripping analysis is generally inapplicable to organic compounds, most of which, unlike metals, cannot be deposited at an electrode. However, in view of the inherent sensitivity afforded by analyte accumulation prior to voltammetric measurements, much endeavour has been devoted to the development of alternative preconcentration methods involving no Faradaic process; this search has given rise to the so-called 'adsorptive stripping analysis',[131] in which a spontaneous adsorption process occurring at a stationary electrode is purposely used as an effective preconcentration step for trace measurements of major species that cannot be accumulated by electrolysis. This step can be based on two types of phenomenon: (*a*) interfacial accumulation at an ordinary electrode; and (*b*) a specific reaction (covalent co-ordination, ion-exchange) at a modified electrode. Judging by the wide range of applications of the adsorptive concept developed over the last five years, it appears to be the most universal stripping approach. In addition to the use of chemically modified electrodes as a means of selectively preconcentrating organic and inorganic analytes, rich chemistry can be used for preconcentration at these molecularly tailored surfaces, which in turn opens up exciting prospects for travel electroanalysis. These non-electrolytic preconcentration strategies add new dimensions to stripping analysis by extending its scope of application to a wide range of organic and inorganic analytes.

Stripping of the preconcentrated species can be accomplished by applying the electrode a potential scan as one of different possible waveforms (linear-scan differential pulse, square wave, staircase, or normal pulse), or a constant-current excitation signal.

The nature of the species to be determined ultimately determines its controlled interfacial accumulation over a given period at a given potential and convection rate onto the working electrode. Thus, the adsorbed species can be of different nature, namely:

(*a*) Complexes for measurement of metal traces. Detection limits at the subnanomolar concentration level have been estimated for transition metals, actinides, and rare earth elements following short (3–5 minutes) accumulation periods and using ligands such as dimethylglyoxime, di-*o*-hydroxyazo dyes, catechol, oxine, *etc.* This adsorptive step has also proved to be extremely useful in distinguishing between different oxidation states. This ability has been attributed to the specificity of some of the co-ordination reactions towards some oxidation states of the metals. Thus, chelation with Solochrome Violet RS has proved to be useful for selective measurements or Fe^{III} in the presence of Fe^{II}.[132]

(*b*) Measurement of traces of organic compounds. Trace levels of organic compounds have been measured by ordinary stripping techniques in only a few instances involving film-formation cathodic stripping schemes based on reactions with electrolytically generated mercury ions. Adsorptive stripping analysis has been shown to be highly suitable for measuring organic compounds with surface-active properties. Organic molecules are adsorbed from aqueous solutions onto the electrode surface mainly because of their hydrophobic character. As a rule, the less water-soluble an organic compound is, the more strongly it is adsorbed. Neutral organic compounds are adsorbed most firmly on uncharged electrodes.

Non-electrolytic preconcentration is optimally accomplished by using chemically modified electrodes (CMEs), which allow the determination of species with slow electron transfer kinetics (electrodes modified with electrocatalytic moieties), offer increased selectivity and stability (electrodes coated with permselective polymer layers) and/or favour preferential accumulation of the analytes onto bound surface functionalities. The preconcentrating agent used can be included in the electrode in various ways, thus giving rise to two basic types of CEM:

(*a*) *Polymer electrodes*. Polymer layers are commonly produced by spin-coating or dip-coating methods. The film is then held on the electrode surface by a combination of chemisorptive and solubility effects. The polymer layer can also be generated by electrodeposition or by inducing the polymerization of monomers by other means.

(*b*) *Modified carbon paste electrodes*. A number of preconcentrating agents can be readily inserted into carbon paste without the need to devise individualized attachment schemes for each modifier. The effective surface coverage of these electrodes can be altered by carrying the weight of the modifier added to the paste mixture. In addition, the surface can be removed and renewed for subsequent re-use with a loss of reproducibility of only 5–10%.

Preconcentration at a CME can be achieved by using ligand-containing complexing electrodes and by electrostatic ion-exchange binding if the electrode coating is made of an appropriate polyelectrolyte or, otherwise by incorporating an ion exchanger into the carbon paste matrix.

The most serious problem posed by this stripping alternative is the almost universal need to renew the electrode surface after each analysis, which is rather tedious in batch procedures. This problem can be overcome by using a continuous configuration—preferably with sample injection—in which the electrode surface is automatically regenerated by simply switching a valve to select the most suitable medium for each step.

4.4.7 Applications

The broad scope of application of stripping techniques in different areas is clearly reflected in the examples described below, which also testify to their brilliant analytical performance.

Eskilsson *et al.*[133] developed a determination of metals such as Ni^{II} and Co^{II} after potentiostatic concentration of their dimethylglyoximates in a well-stirred solution at a glassy carbon electrode coated with a thin mercury film. The analytes were then reduced in a 5 M $CaCl_2$ medium by applying a constant current, simultaneously with a microprocessor recording the *E-t* curve of the working electrode. Finally, the mercury film was removed by mild oxidation with an iodine/iodide solution, and the glassy carbon surface was cleaned with ethanol and sodium hydroxide prior to the next analytical cycle. The detection limits for the one sigma level after one minute of potentiostatic adsorption were nine and eleven for Ni^{II} and Co^{II}, respectively. Both analytes were determined in reference samples of bovine liver and sea water sediments after acid digestion.

The speciation of Fe^{III} and Fe^{II} in wines was accomplished on the basis of the specific complexation of the ferric ion by Solochrome Violet Red (SVR); total iron was determined with catechol by transferring the dye solution to the electrolytic cell and applying the accumulation potential (−0.45 V for SVR and −0.10 V for catechol, both against the Ag/AgCl electrode) to a fresh mercury drop for a few seconds after degassing. Following preconcentration, stirring was stopped and, after equilibration for 15 seconds, a differential pulse voltammogram was recorded from 0.0 to 1.0 V. After the background stripping potential was recorded, an appropriate volume (50 μl) of the dilute sample solution was added to the cell solution and the voltammogram was recorded. The high sensitivity of the method (smaller than a few p.p.b. of the analyte) requires the sample to be highly diluted in order to overcome the potential interference of surface-active organic species such as tannic acid, which might compete for sites on the mercury surface.[132]

Electroinactive species such as chloride and cyanide adsorbed on pH electrodes alter the rate of surface oxide formation following a positive potential step. Thus, triple step potential waveforms similar to those successfully used for pulsed amperometric detection of electroactive adsorbates (*e.g.* alcohols, carbohydrates, and amino-acids) can be also successfully applied to electroinactive adsorbates injected into a stream of electrolyte. Depending on the waveform used, the total anodic current at the detection peak can be greater or smaller than the baseline signal corresponding to the oxide formation in the absence of the adsorbate. The sensitivity thus achieved is quite high (0.2 ng of chloride), so the method can be used for determinative purposes.[134]

The determination of organic compounds that usually undergo no reversible electrode processes is the most interesting application of stripping with non-electrolytic preconcentration. One typical example is the determination of chlorpromazine in a 0.01-fold excess of non-adsorbable solution species with similar redox potentials. The increased selectivity achieved is a result of combining the intimate contact between the surface-bound species and the electrode surface with the medium-exchange procedure in a single-channel flow injection configuration. The system operation was optimized in order to achieve the desired balance between high sensitivity, a fast assay rate, and minimal carry-over. By using a flow-rate of 0.3 ml min^{-1}, injection rates of 24 samples h^{-1}, and detection limits of a few nanograms can be readily achieved. Reproducible quantitation of chlorpromazine in urine was accomplished with no sample treatment by applying a procedure in which the injected sample was preconcentrated at a carbon paste electrode by applying a potential of + 0.3 V for 60 seconds, after which surface-bound chlorpromazine was measured by applying a differential pulse anodic potential ramp to + 0.9 V. Keeping the electrode at + 0.9 V for 60 seconds allowed the carbon paste to be cleaned of all chlorpromazine and redyed the system for the next injection.[135]

Preconcentration and quantitation of doxorubicin, a drug used in cancer chemotherapy, was also accomplished by using a flow injection method based on the adsorption of the drug onto a carbon paste electrode, medium exchange, and application of a differential voltammetric pulse to the adsorbing surface. The determination was carried out by injecting the desired sample volume into the flowing buffer stream and allowing it to pass over the carbon paste electrode. The flow rate was 0.1 ml min^{-1} and the electrode must be maintained at −0.30 V during adsorption. After the response from the working electrode was stabilized, a differential potential voltammetric scan was recorded from an initial potential of +0.30 V (pulse amplitude = 50 mV; potential scan rate = 2 or 5 mV s^{-1}). A linear response was obtained for concentrations above 10^{-6} M, and the drug could be determined in urine by direct injection of the sample with no preliminary conditioning. The sensitivity, linear determination range, and selectivity of this flow injection method were much better than those of its batch counterpart, in which adsorption was done in a stationary sample solution. Each preconcentration experiment was started with a fresh electrode surface.[136]

4.5 Continuous Precipitation–Filtration–Dissolution

4.5.1 General Considerations

Methodologies based on the formation and/or dissolution of a precipitate have been widely used in classic analytical chemistry, both for qualitative (systematic separations into groups, identification reactions[137-139]) and for quantitative purposes (gravimetries[140]). However, the popularity of precipitation reactions has declined in the last few decades, mainly because of: (*a*) their inherent limitations (scarce sensitivity and selectivity, narrow scope of application, slowness and intensive involvement of the operator in the process); (*b*) the spectacular advances in analytical instrumentation; and (*c*) the difficulty involved in implementing precipitation and dissolution reactions in automatic systems, whether discrete or continuous.

Filtration has been used in automatic analysers for different purposes, the most common of which is the use of filters of different designs to remove suspended particles from the sample or the reagents, particles that might give rise to spurious analytical results or even reversible or irreversible deterioration of one or several modules of the analyser.[141] A somewhat less common application involves obtaining a precipitate and determining the analyte(s) in the supernatant liquid. The problems derived from the filter regeneration or the removal of accumulated particles can be overcome by changing the filter or, more ingeniously, by implementing an automatic washout cycle involving passage of a washing solution in the opposite direction to the sample to sweep any residues remaining on the filter.

A more complex and also less frequently used alternative to the separation of suspended solid in analysers relies on microcentrifugation, whereby the liquid and suspended precipitate are placed in the bottom of a centrifuge and the supernatant is collected from the top by aspiration. An intermediate washing operation is required to remove the precipitate via the same conduit as the sample.

Early attempts at incorporating precipitation reactions into air-segmented continuous analysers[142] met with little success. It was not until a few years ago that automatic continuous unsegmented analytical methodologies based on FIA principles succeeded in implementing precipitate formation, washing and dissolution.[143, 144] All of such methodologies use atomic absorption spectrophotometers for detection. As far as their scope of application is concerned, they have been used for the indirect determination of organic and inorganic anions, the preconcentration–determination of metal traces and the removal of interferents.

Continuous precipitation–filtration–dissolution (PFD) processes can be best described as continuous liquid–solid systems in which the second, solid phase is generated and disappears *in situ*, *i.e.* it is not a permanent part of the system unlike in the sorption, ion-exchange, or stripping systems described above. They bear some resemblance to leaching insofar as they involve the selective dissolution of the analyte—the newly formed precipitate in this case.

Turbidimetric and nephelometric methods based on the monitoring of precipitates obtained *in situ* are in a way precedents of these systems; however, the precipitates are kept in suspension and finely divided for passage through the continuous detector.[145–148] The chief difference with the systems described in this section lies in the absence of a continuous filtration unit and in the precipitate dissolution. The determination of sulphide based on the precipitation of CdS in a rather complex manifold including two peristaltic pumps, an exchange microcolumn, and an atomic absorption spectrometric detector can also be situated in this context.[39]

4.5.2 Basic Units

The automatic continuous systems used to implement precipitation reactions consist of a series of common essential elements such as filters, valves and detectors, which are described separately below.

4.5.2.1 Filters

The filter is the key part of the analyser insofar as it is intended to retain quantitatively the precipitate formed in the reaction coil. It should be of a material as resistant as possible to the solutions to be circulated through it—usually plastic or stainless steel, but also occasionally paper (*e.g.* Whatman filters). Figure 4.25 shows four types of filtering modules commonly used in continuous PFD configurations.

Most of the PFD applications reported to date rely on the use of cylindrical filters originally conceived for cleaning purposes in HPLC. The model depicted in Figure 4.25A is a removable screen-type stainless steel filter of 0.5 μm pore diameter, 580 μl inner volume, and *ca.* 3 cm^2 filtration area. Cylindrical filters with inner volumes of 400–800 μl and pore sizes of

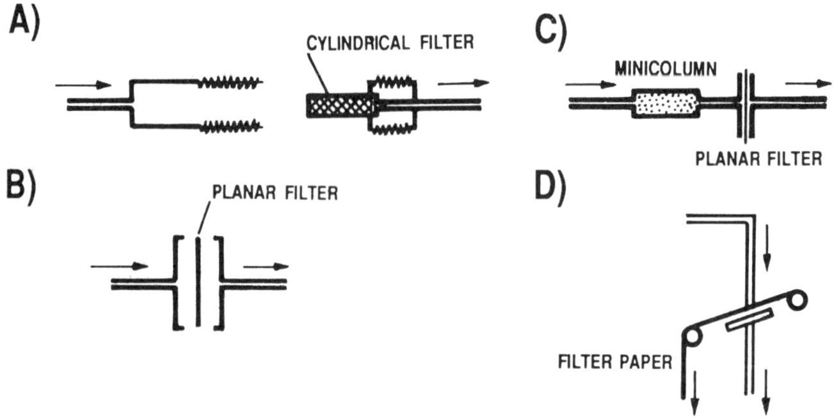

Figure 4.25 *Types of filters used in continuous analysers with precipitate formation and dissolution.* (A) *Cylindrical,* (B) *Planar,* (C) *Moving filter paper,* (D) *Combination of a minicolumn and a planar filter*

0.2–2 μm provide basically the same performance. However, the smaller the chamber inner volume or filtration area is, the more often the column and the filter should be cleaned.

Planar filters of different designs (Figure 4.25B) have also been used for this purpose. One of them, made of stainless steel, consists of a planar circular filter rather than a cylinder, which results in a considerably smaller filtration area (*ca.* 7 mm^2). Disposable membrane filters of 0.45 μm pore size and 3, 13, and 25mm diameters incorporated into Telman syringes, and straightforward filter papers have also been used in this context.

A combined filter consisting of a 5 cm × 3 mm ID PTFE minicolumn packed with 35–44 mesh polystyrene granules and a PTFE membrane (0.2 μm pore size, 5 mm ID) was also recently reported (Figure 4.25C). This combined device allows graded collection of the precipitate while eliminating excessive back-pressure, which might result in leakage.

The continuous filtration system originally used in Technicon continuous segmented analysers features a peculiar design (Figure 4.25D) and operation. The flowing liquid stream from a conventional manifold is dropped onto a belt of filter paper to continuously remove the precipitate. The filter paper is then recycled to the pump from the underside of the filter. The system can only be used with aqueous solutions as volatile solvents tend to evaporate in the process.

In configurations where measurements are based on the decrease in the concentration of a cation-reagent in the carrier (Figures 4.26A and 4.27A), *i.e.* with no precipitate dissolution, the precipitate gradually builds up on the filter as successive samples or standards are introduced into the system unless a moving filter paper is used. In order to avoid overpressure and clogging problems, the filter should be regenerated periodically by removing the accumulated precipitate. This can be done manually by replacing the planar or cylindrical filter after stopping the continuous analyser. Alternatively, a solvent stream can be passed to dissolve the precipitate after a preset number of determinations. This clean-up operation can be expedited by irradiating the filter (steel) with ultrasounds while the solvent is circulated through it. The nature of the precipitate formed decisively influences the capacity of the analyser to withstand multiple injections without the need to replace or clean the filter. Gelatinous (*e.g.* Fe_2O_3) and curdy precipitates (*e.g.* AgCl) are scarcely compact and therefore allow readier passage of the flowing solutions and hence call for less frequent filter changes or clean-up (only after *ca.* 250 injections of sample or standards). On the other hand, crystalline precipitates (*e.g.* CaC_2O_4) are more compact and less permeable, so they tend to occupy the filter pores and clog the filter outlets after only *ca.* 20 consecutive injections.

4.5.2.2 Valves

PFD systems use virtually the same types of valve as other analytical flow systems.

Injection valves are used both to inject samples containing the analyte(s) (normal FIA configurations) or cation-reagent (reversed FIA manifolds).

Selecting valves are normally positioned in front of or behind the propulsion unit. Three-way valves are typically used before the propulsion system to switch between the solutions to be aspirated (blank or standard, washing solution or solvent). Four-way valves are employed in the reaction zone or between the reaction-precipitation coil and the filter in order to select the reaction mixture stream or additional streams (washing or precipitate solvent) to be successively passed through the filter. They also allow the stream that is not passed through the filter to be driven to waste. Occasionally, one selecting valve is used on each side of the continuous filter.

4.5.2.3 Precipitation Coils

Precipitation coils are most often helically coiled pieces of Teflon tubing typically 0.3–0.7 mm in inner diameter. The coil length is key to the quality of the FIA peaks obtained and hence of the analytical results. It should ensure both a minimal residence time, which depends on the reactor length and the flow-rate, and complete precipitation of the analyte. On the other hand, too long reaction coils result in passage of an excess of cation-reagent into the reaction zone in normal FIA manifolds and consequently in the obtainment of broader, less negative FIA peaks and ultimately in decreased sensitivity and reproducibility. In practice, coil lengths typically range between 60 and 300 cm.

4.5.2.4 Detector

Insofar as most of the precipitation reactions used for analytical purposes involve metal ions, whether as reagents or as analytes, atomic absorption spectrometers are not surprisingly the most commonly used in this context.

The use of an AAS detector coupled on-line to a PFD system poses two problems arising from incompatibilities between the two. The differences between the flow-rate of the streams circulated through the FIA manifold and the nebulizer aspiration rate can be compensated for by connecting the filter and the detector via a T-piece, which will allow the continuous aspiration of air or water as required. However, one incompatibility arises from the different concentration of the cation-reagent required to ensure complete and rapid precipitation in the FIA manifold—particularly in the determination of organic compounds—and the optimal concentration for ensuring a linear detector response. If the latter condition is not met, a water dilution stream should be intercalated between the filter and the nebulizer to ensure that the analyte concentration is suited to each zone of the analyser while maintaining the aforementioned compatibility between the FIA and nebulizer flow-rates.

It is significant that, despite the lack of technical hindrances to the joint use of an inductively coupled plasma atomic emission spectrometer and a continuous precipitation system, no applications of this potentially productive hyphenated methodology have been reported to date.

Continuous air-segmented analysers based on precipitation reactions use ordinary detectors such as colorimeters or photometers rather than atomic detectors. However, none has so far proved to be as efficient as the above-described unsegmented systems.

4.5.3 Continuous Configurations

These configurations can be classified according to various criteria such as the type of FIA manifold used (normal and reversed), whether or not the precipitate is dissolved, and the aim (indirect determinations, preconcentration–dissolution, *etc.*). However, we shall deal with them here according to the way in which the liquid sample is introduced, *i.e.* by injection or by continuous aspiration.

4.5.3.1 With Sample Injection

Figure 4.26 shows the scheme of FIA configurations with sample injection involving the sequential formation and filtration of a precipitate (A), its washing (B), and its dissolution and transport of the dissolved ingredients to an atomic spectroscopic detector (C). The figure also shows the typical signal obtained in each of the above stages, as well as the time programming of the valves used. In short, there are two types of contin-

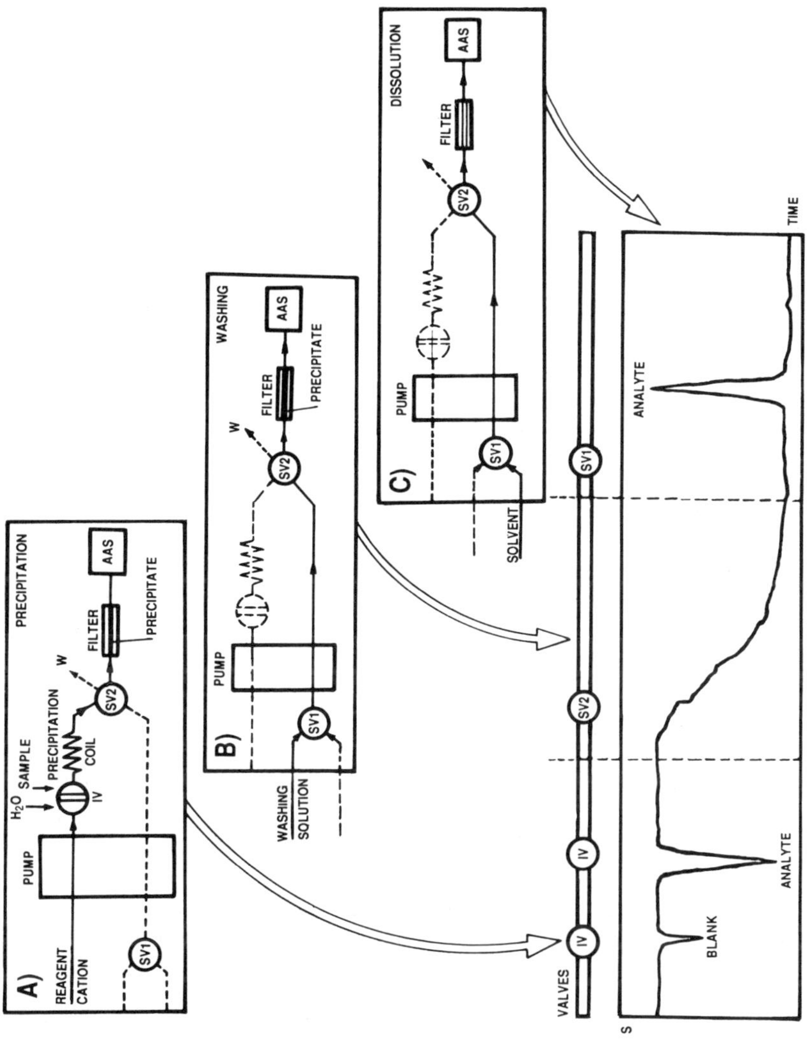

Figure 4.26 *Scheme of a continuous precipitation–washing–dissolution system with sample injection for the determination of anionic species also showing the FIA peaks obtained in each operation.* IV = *injection valve.* SV = *selecting valve*

uous precipitation configurations with sample injection, *viz.* with and without precipitate dissolution.

Continuous analysers with sample injection in which the precipitate formed is not dissolved prior to detection are depicted schematically in Figure 4.26A. A cation-reagent stream is continuously introduced into the system and reaches the detector, where it yields a constant, tall signal that can be taken as the baseline. Two microvolumes of water and the liquid sample containing an ion are injected sequentially via the same valve. The first injection gives rise to a small negative peak due to the dilution of the reagent—which yields the detector signal—in the carrier which is used as a blank signal. Injection of an identical volume of sample results in precipitate formation in the reaction coil and in a concomitant decrease in the cation-reagent concentration in the reaction zone of the carrier; this yields a negative peak, the height of which will increase with the concentration of the analyte-anion in the sample. The difference between the heights or areas of the two peaks will be proportional to such a concentration. This is thus an indirect way of determining anionic species by atomic absorption spectroscopy. As noted earlier, a diluting stream is occasionally inserted between the filter and the detector in order to avoid problems arising from differences in the cation-reagent concentrations required by the FIA manifold and the detector. The precipitate formed in each determination or calibration gradually builds up on the filter, from which it should be removed after a given number of injections that depends on the nature of the precipitate as stated above.

By incorporating another channel including two selecting valves, before (SV_1) and after (SV_2) the propulsion system, into the previous configuration, the precipitate can be washed and dissolved in an automatic fashion. In a first step (Figure 4.26B), SV_1 allows the continuous introduction of a washing solution that is driven to the detector via SV_2 and the filter while the manifold stream is sent to waste. This operation is intended to wash the precipitate and flush the cation-reagent from the conduits connecting SV_2 and the detector. As a result, the signal decreases gradually to a new baseline corresponding to a zero cation-reagent concentration. In the last step (Figure 4.26C), and by switching SV_1, an appropriate solution is inserted to dissolve rapidly the precipitate accumulated on the filter. The resulting plug reaches the detector and yields a positive FIA peak proportional to the concentration of the analyte-anion in the injected liquid sample.

4.5.3.2 With Sample Aspiration

There are two general alternatives to the continuous introduction of the sample by aspiration according to whether the cation-reagent is also introduced by aspiration or by injection (Figure 4.27).

Figure 4.27A depicts a reversed FIA manifold equipped with a selecting valve located prior to the propulsion unit and intended to select the car-

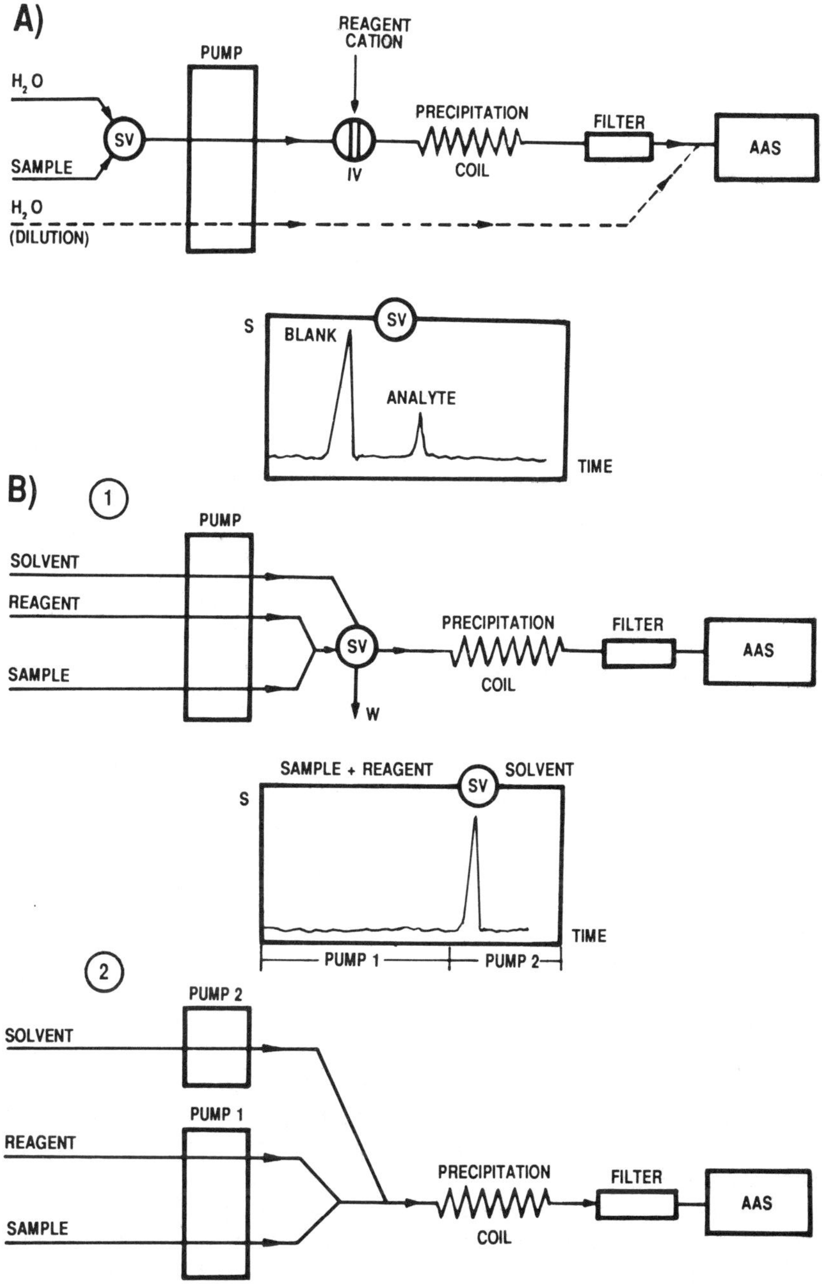

Figure 4.27 *Schemes of continuous precipitation systems with continuous aspiration of the sample.* (A) *With injection via IV of the reagent or precipitant (for the determination of anions).* (B) *With continuous aspiration of the precipitating reagent (for the determination of metal traces):* (1) *with a selecting valve* (SV)*;* (2) *with intermittent pumping*

rier. In the first step, a microvolume of the cation-reagent solution is injected into a continuous water stream that drives it directly to the atomic absorption spectroscopic detector, where it yields a positive FIA peak that is used as a blank. In the second step, the selecting valve is switched to introduce a sample stream into the system. A reagent injection identical with that performed in the previous step gives rise to the formation of a precipitate with the anion-analyte in the sample which is retained on the filter while the excess reagent yields a small positive FIA peak which decreases with increasing analyte concentration. This in turn is calculated by subtracting the two peaks. If required, a diluting stream is used between the filter and the detector. As in the previous case, the filter should be cleaned after a given number of injections that depends on the nature of the precipitate.

Figure 4.27B shows two alternatives to the simultaneous introduction of sample and reagent by aspiration. Both rely on the precipitate dissolution and its direct transport to the detector, so no filter cleaning is needed. They were conceived for the determination of metal traces with a base or anion that yields no signal on passage through the detector. Thus no blank need thus be used to obtain a reference peak. The sample volume introduced (of the order of a few ml) is controlled via the flow-rate and the aspiration time. The configuration shown in Figure 4.27B.1 uses a selecting valve after the peristaltic pump. In the first step, the sample and reagent streams are merged prior to the reaction coil, where a precipitate is continuously formed and retained on the filter. In the second step, the selecting valve inserts a solvent stream to rapidly dissolve the precipitate and transport the resulting plug to the detector, where it yields a positive peak proportional to the amount of analyte originally introduced into the system. In the configuration depicted in Figure 4.27B.2, the selecting valve is replaced with an additional peristaltic pump (pump 2) that operates in an intermittent fashion like the main pump (pump 1). In the first step, pump 1 works while pump 2 is at rest, and the opposite holds in the second step. The signal obtained is identical with that provided by the previous system. However, reagent consumption is somewhat lower.

A critical comparison of the systems with sample injection and aspiration shown in Figures 4.26 and 4.27, respectively, allows one to draw the following conclusions: (*a*) analysers with sample injection feature lower sample consumption (of the order of a few microlitres); (*b*) those with continuous sample introduction and reagent injection are more sensitive on account of the inherent acknowledged advantages of these reversed FIA configurations; and (*c*) those with sample and reagent aspiration allow analytes to be preconcentrated by factors of 50–1000 and hence the sensitivity to be indirectly enhanced.

4.5.4 Applications

Despite their recent advent, PFD systems have so far been used to develop a number of interesting analytical methods for a variety of purposes, some of which are discussed below.

4.5.4.1 Determination of Inorganic Anions

This is the most immediate and natural application of PFD analysers relying on indirect atomic spectroscopic measurements. Thus, chloride has been determined in water by using silver ion as precipitant in normal (Figure 4.26A) and reversed (Figure 4.27A) FIA configurations furnished with the filter depicted in Figure 4.25.[149] The rFIA method provided a lower detection limit (0.3 μg ml^{-1}) than that afforded by the nFIA method (3 μg ml^{-1}). The sample throughput of the former was also greater (200 *vs.* 50 samples h^{-1}). The results obtained by the two methods were consistent with those provided by the classical potentiometric method for the determination of chloride.

It is worth emphasizing the substantially increased selectivity achieved by carrying out precipitation in a continuous fashion rather than batchwise. Some of the methods described below are also more selective than their manual counterparts. Figure 4.28 shows the selectivity factor, *viz.* the ratio between the maximum tolerated concentration of each continuous method and the corresponding manual method, for methods with (Figure 4.26B) and without (Figure 4.26A) dissolution of the AgCl precipitate formed in the determination of chloride traces. As can be seen: (*a*) a large number of ions have a selectivity factor greater than unity, which is good proof of the higher selectivity of FIA methods; (*b*) the

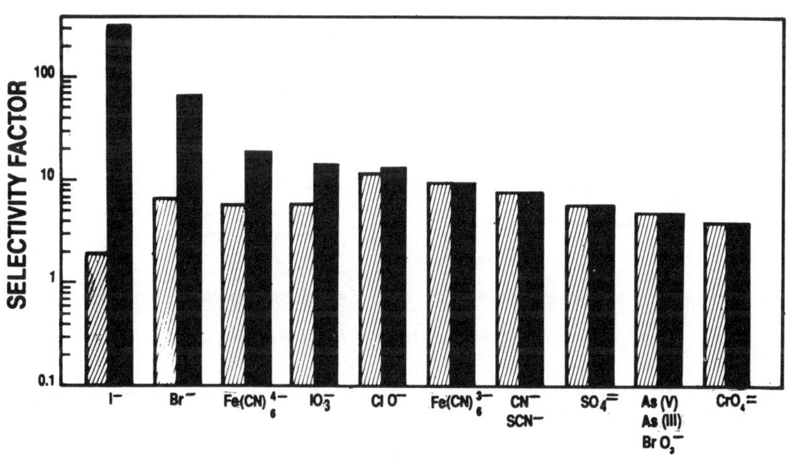

Figure 4.28 *Comparison of the selectivity level afforded by precipitation methods with and without precipitate dissolution in the determination of chloride ions with silver ions as reagent and ammonia as solvent. The selectivity factor is given by the ratio between the maximum tolerated concentration of the automatic method and that of its manual counterpart*

automatic method based in the dissolution of the precipitate into a stream of 6 M ammonia features a somewhat higher selectivity level and a substantially increased factor for iodide, bromide, and ferrocyanide ions (note the logarithmic scale on the y-axis).

The manifold shown in Figure 4.29 allows the simultaneous determination of chloride–iodide mixtures at the ng ml^{-1} level in foodstuffs (milk, cheese, bread, coffee, eggs) and wines.[150] It relies on the selective dissolution of the AgCl precipitate in the presence of AgI. Two analytes are precipitated in the first step. The negative peak obtained, once subtracted from the blank, will be proportional to the total analyte concentration. On introduction of a stream of 6 M ammonia into the system, the AgCl is rapidly dissolved and a positive peak proportional to the concentration of chloride alone is obtained. That of iodide is then calculated by difference. The method features a relative standard deviation less than 2.5% and a sampling frequency of 10 h^{-1}. Up to 100 samples can be assayed before the filter requires any cleaning.

The configuration in Figure 4.27A allows the determination of sulphate over the range 2–20 μg ml^{-1} using Pb^{II} as injected cation-reagent. The relative standard deviation of this selective method is 3% and its sample throughput 60 samples h^{-1}.[151]

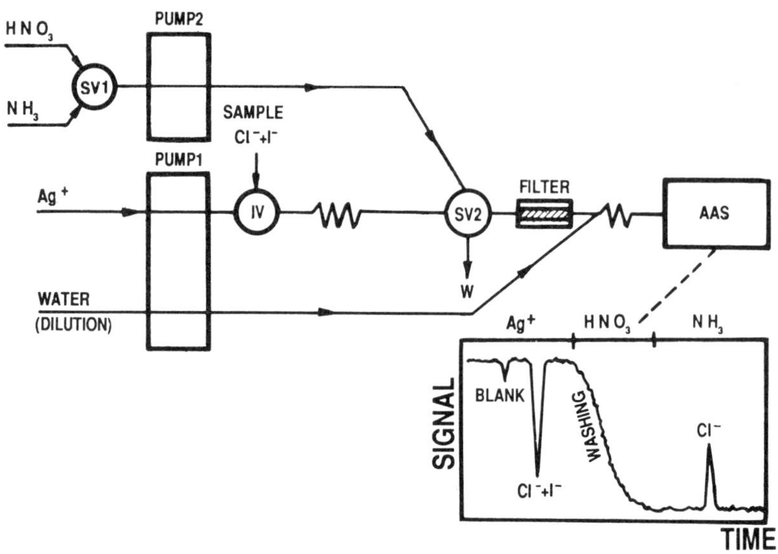

Figure 4.29 *Determination of chloride-iodide mixtures with a continuous automatic analyser involving the formation, washing, and selective dissolution of silver precipitates.* IV = *injection valve.* SV = *selecting valve. (For details, see text).*
(Reproduced from reference 150 with permission of Elsevier Science Publishers)

4.5.4.2 Determination of Drugs

Manual indirect determinations of organic compounds by atomic absorption spectroscopy after a separation step involving metal ions are usually time-consuming and scarcely reproducible.[152] Application of automatic continuous PFD methods in this context results in remarkable improvements in some analytical features as shown below.

The configuration in Figure 4.27A was used for the determination of five sulphonamides in various pharmaceutical preparations (capsules, tablets, syrups, ophthalmic solutions) and urine using Ag^{I} or Cu^{II} as precipitant.[153] The sulphonamides were determined over the range 2.5–3.5 $\mu g\ ml^{-1}$ with a relative standard deviation of 1.5–3% at a sampling rate of 100–150 samples h^{-1}. The sensitivity achieved was roughly 100 times higher than those of the corresponding manual methods. Excipients and diluents were tolerated in amounts over 250 times as large, compared with the 4–12-fold amounts tolerated by the manual methods. Logically, the automatic method was much faster and involved far less human participation than typical manual precipitation methods. The use of copper rather than silver ions offered two advantages, namely higher sensitivity, which allowed the method to be applied to the determination of sulphonamides in biological fluids, and the need for no diluting stream to suit the reagent concentration to each zone of the analyser.

Cobalt II was used as precipitant in the FIA configuration shown in Figure 4.27A for the determination of local anaesthetics (lidocaine, tetracaine, and procaine) over the range $1.0–11.0 \times 10^{-5}$ in pharmaceuticals with a relative standard deviation of 0.6% at a rate of 100 samples h^{-1}.[154]

A determination for chlorhexidine in pharmaceutical preparations was recently developed by using the configuration depicted schematically in Figure 4.30. The process involves formation of a precipitate with Cu^{II}, filtration, washing, and dissolution. The nFIA manifold in question is slightly different from that shown in Figure 4.26 as it uses three switching valves to sequentially introduce streams of ammonia (SV_1, washout) and nitric acid (SV_2, dissolution) and divert the stream emerging from the filter to waste (SV_3) before the solvent (nitric acid) is introduced into the detector, which is continuously fed with an HNO_3 stream in order to obtain a stable baseline. The method has a linear calibration range of 5.0–20 p.p.m. and a relative standard deviation of 3.6%. Samples can be assayed at a frequency of up to 10 h^{-1}.[155]

4.5.4.3 Preconcentration–Determination of Metal Traces

In this type of application, the analyte is directly determined at the atomic spectroscopic detector used. The most immediate application in this respect is the determination of metal traces in various matrices whenever direct measurements are unfeasible so preconcentration is required.

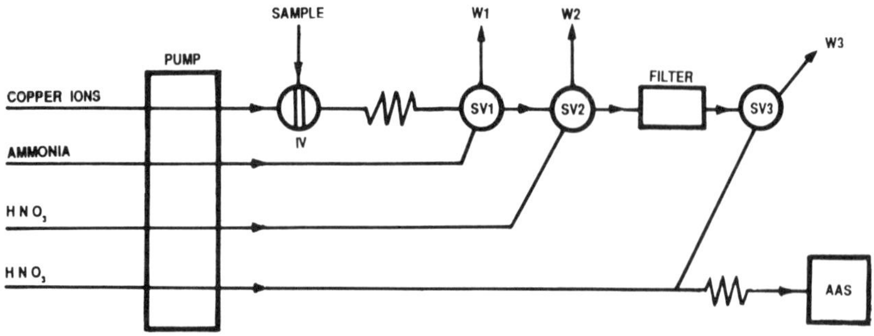

Figure 4.30 *Flow injection configuration for the determination of chlorhexidine based on its precipitation with copper ion, washing of the precipitate with ammonia and dissolution into nitric acid.*
(Reproduced from reference 155 with permission of Pergamon Press)

This can be accomplished with one of the manifolds shown in Figure 4.27B.

The configuration in Figure 4.26B.1 was used for the determination of lead at the ng ml^{-1} level in tap water using ammonia and nitric acid streams to form and dissolve the precipitate, respectively.[156] The determination limit and r.s.d. thus achieved were 1.5 ng ml^{-1} and *ca.* 3%, and the sample throughput was between 5 and 15 samples h^{-1}, depending on the sample volume aspirated (10–250 ml), which in turn was dictated by the concentration level at which the analyte was present in the sample. It is interesting to note that the hollow-cathode lamp used, which results in poor sensitivity in the manual method, performed reasonably well. The maximum concentration factor achieved was 700. The automatic method was highly selective; in this context it is worth mentioning the negative interference of Fe^{III}, which was unpredictable because of the high selectivity of the AAS technique towards lead. After much study, the perturbation was found to be of physical origin: the precipitated iron basic salt occluded some of the lead on the filter, so the dissolution and subsequent transport of the lead to the detector was slow and erratic and thus resulted in smaller signals. In any case, this interference was overcome by adding some tartrate to the ammonia carrier in order to avoid precipitation of the iron basic salt.

The configuration in Figure 4.31 was recently used to develop two automatic continuous methods based on the formation, filtration and dissolution of insoluble chelates of copper[157] and cobalt[158] present at trace levels in silicate rocks. Both yielded excellent results. The most salient feature of these determinations is their high selectivity; in fact, the sample matrix contained some ions such as Fe^{III} and Al^{III} at concentrations 100 to 1000 times higher than those of the analytes. The enhanced selectivity was a result of the precipitation reaction used, namely chelation with a highly selective ligand (rubeanic acid for copper and α-nitroso-β-naphthol for

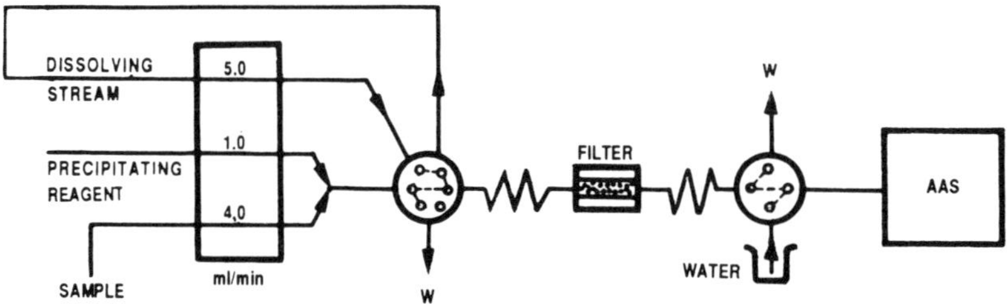

Figure 4.31 *Configuration for the determination of traces and subtraces of copper and cobalt in silicate rocks by precipitation–filtration–dissolution of a metal chelate using two selecting valves located before and after the filter. (For details, see text)*

cobalt) under specific experimental conditions that excluded precipitation of other species. The dissolving stream was also carefully chosen as it should ensure rapid dissolution of the precipitate retained on the filter. The reducing properties of rubeanic acid in the case of copper compelled the use of an oxidant (potassium dichromate) in the dissolving carrier. Dissolution of the cobalt chelate was aided by using an organic solvent (ethanol) and heating the filter at 55 °C. Both configurations were of closed-loop type in order to avoid wasting the reagent in the precipitation/filtration operation. The sampling frequency achieved was 5–10 h^{-1}, depending on the concentration level of the analyte.

This alternative to continuous preconcentration surpasses the more generally used ion-exchange minicolumns in some aspects. Thus, it does not suffer from the low selectivity of chelating resins, which retain metals other than that of interest; in addition, it requires no second phase, which reduces costs, nor does it give rise to the problems associated with resin swelling.

4.5.4.4 Interference Removal

Recently, Tyson *et al.*[159–161] applied continuous precipitation coupled on-line to atomic absorption spectroscopy in order to eliminate the interferences posed by substances accompanying the analyte in the sample. Thus, they resorted to the precipitation of calcium with oxalate to isolate this alkaline earth from synthetic rock-sample solutions containing aluminium, silicon and phosphorus as interferents. The precipitating stream used contained ammonia and ammonium oxalate, while the dissolving stream consisted of 5% HCl. The enrichment factor obtained in the determination of calcium at concentrations between 0 and 1 ng ml^{-1} was about 650.[160] By using a similar configuration and the same chemical system but also a filter combined with a packed column plus a membrane (Figure 4.25C), they succeeded in separating 10 mg ml^{-1} Ca^{II} from 200 mg ml Al^{III} at low flow-rates and room temperature.[161]

4.5.5 Role of kinetics

The inherent chemical and physico-chemical kinetics of precipitation processes have traditionally been a decisive factor in opting for developing or implementing precipitation methodologies. Time is known to be highly influential not only to the complete formation of the precipitate, but also on its physical features and impurification.

Conversion from an ordinary manual precipitation methodology to an automatic continuous methodology implies a major kinetic transformation that in turn occasionally involves remarkable advantages, so much so that some former shortcomings are turned into valuable assets.

Automatic continuous precipitation methods are highly selective thanks to the well known phenomenon of kinetic discrimination which occurs during the short time over which the analytical process takes place. The typically short reaction time minimizes precipitation of foreign substances since the system is optimized for the analyte. In addition, the short time of contact between the precipitate formed and the other components significantly reduces or even eliminates problems arising from adsorption and post-precipitation.

One other substantial advantage of automatic continuous precipitation methodologies is that the precipitate is dissolved only a few seconds after it is formed, thereby preventing ageing. The dissolution kinetics are significantly aided by the precipitate being finely divided. Otherwise, the FIA peak obtained is rather broad and distorted, which prevents accurate data processing. This can be best avoided by using an efficient, rapid dissolving reagent.

4.6 On-line Leaching

Leaching or lixivation, whereby a solid sample is treated with a solution that extracts one or more of the species it contains, has so far scarcely been applied in continuous systems.

In the solid–liquid separation techniques described above, the solid phase used was part of the system in order to facilitate retention (separation and preconcentration) of the analyte present in the unknown sample solution, or, alternatively was generated *in situ* from the analyte and reagent (*e.g.* in precipitation techniques). In leaching techniques, the sample itself makes the solid phase, from which the analyte is transferred to the liquid phase in a more or less selective fashion. Therefore, the second phase is also the liquid phase and the separation process takes place in the opposite direction to any other solid–liquid separation technique, *i.e.* from the solid to the liquid phase. As the analyte is usually more concentrated in the solid than in other techniques, leaching is not intended for preconcentration purposes, but rather for the separation of species not readily soluble in the leaching agent, which should be quite selective in this respect.

This technique bears some resemblance to the procedures for introduction of gas samples dealt with in Chapter 2: the sample makes the solid (gas) phase rather than being created within the system or added to form the separation interface.

Continuous leaching normally requires some form of additional energy, usually electronic or ultrasonic,[162] for expedient, boosted performance.

4.6.1 Leaching with the Aid of Electric Current

This leaching mode is only applicable to metals in alloys. The sample, usually a cylinder, makes the anode of an electric assembly that also includes a cathode and a stabilized current source capable of providing the required amount of electricity over short, precise intervals of time.

The leaching cell, an electrochemical dissolution unit (Figure 4.32) basically consisting of a Teflon block with a groove on one of its sides, one hole on each end for entry and exit of the leaching solution (Figure 4.32A) and a central hole for electric connection to the cathode, is the key unit of a leaching system. The cathode in turn is circular in shape (Figure 4.32B) and is located within the cell; it has an oval hole that defines the space between it and the cathode and the anode (sample) and hence the electrolytic cell proper.

The leaching technique aided by electric energy has been applied to the determination of minor components such as aluminium and molybdenum in steels by using a leaching cell similar to that depicted in Figure 4.32 and continuous configurations with and without injection of the solution containing the analyte extracted into a stream of derivatizing reagent.

Figure 4.33A shows the flow diagram of the determination of aluminium. The current pulses applied between the sample and cathode (100, 200, 400, 600, and 1200 mA) for electrolysis times of 5, 10, 15, and 20 seconds caused dissolution of an amount of sample proportional to the precisely delivered number of coulombs. The well defined, reproducible zone established was pushed forward by the electrolyte solution. The

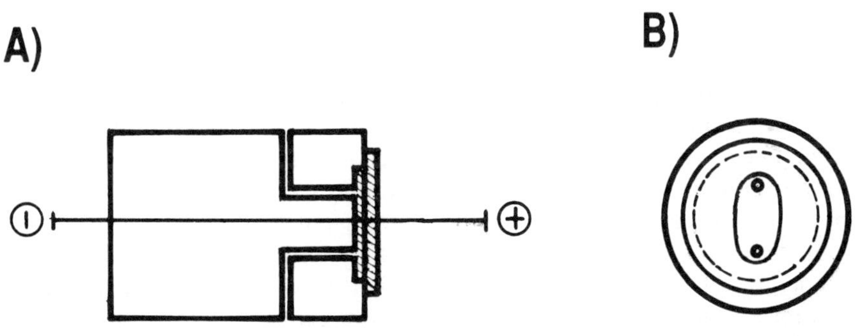

Figure 4.32 *Side* (A) *and top view* (B) *of a typical electrolytic unit*

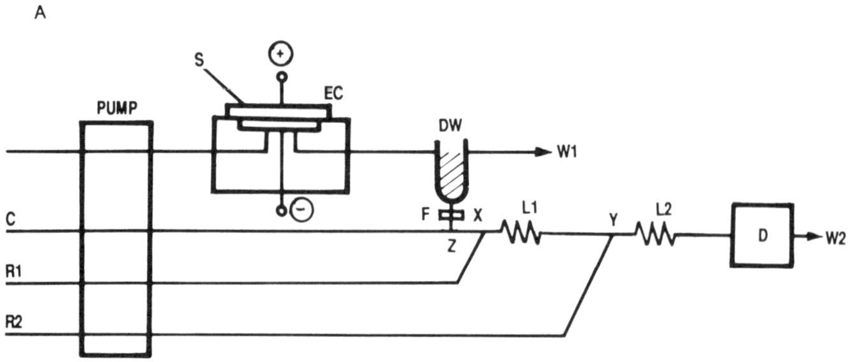

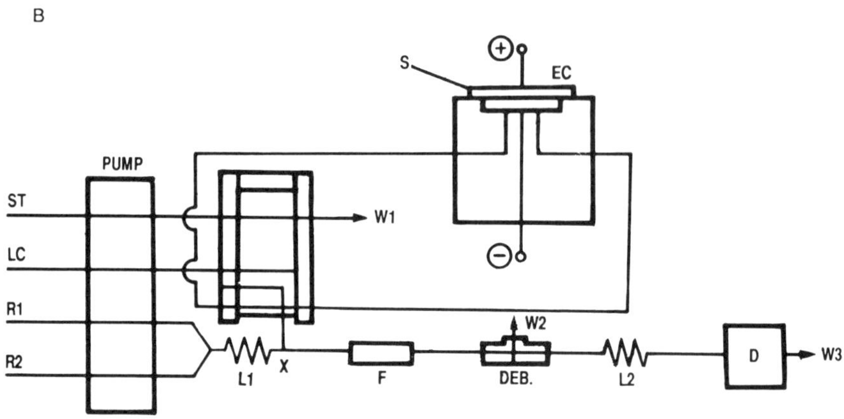

Figure 4.33 *Configurations for the on-line leaching–reaction–detection of aluminium* (A) *and molybdenum* (B). *In* (A), *the sample, S, is placed in the electrolysis chamber* (EC), *through which the leaching carrier,* LC, *is circulated.* DW = *debubbling well;* F = *filter;* L_1 *and* L_2 = *reaction coils;* R_1 *and* R_2 = *cyanine/ascorbic acid and hexamine streams;* W = *waste. In* (B), *the leached analyte is inserted at point* X *into a mixed stream of* Sn^{II} (R_1) *and thiocyanate* (R_2). ST = *standard;* DEB = *debubbler*

gases generated in the electrolytic process were efficiently removed on passage of the sample through a debubbling well (DW). Most of the degassed zone was aspirated into the manifold across a Whatman no. 1 filter placed between the bottom of the well and the merging point Z prior to the derivatizing reaction. The colour-forming, reducing agent (thiocyanate) was merged with the sample zone at point X. The final pH was adjusted by merging the analytical stream with another of hexamethylenetetraamine buffer. After passing through coil L_2, the processed sample reached the spectrophotometer flow-cell, where the colour produced was monitored at 535 nm and a typical FIA peak was obtained as a result. The performance of this system was found to be particularly dependent on:

(*a*) The nature and pH of the electrolyte solution. An acid strength smaller than that provided by 0.5 M HCl was found to be inadequate to keep the aluminium and iron ions in solution, whereas too high an HCl concentration made adjustment of the final pH for reaction development rather difficult. On the other hand, ionic strengths smaller than 0.5 M ensure application of the current pulse in a short time.

(*b*) The current intensity and pulse duration. An increase in the electrolytic current resulted in a proportional increase in sensitivity. However, the current could not be increased indefinitely (beyond 600 mA) because of the formation of large amounts of solids—chiefly carbides—that clogged the filter in a very short time. The electrolysis time was found to play a similar role to the injected volume in FIA; however, rather long times resulted in the amount of aluminium dissolved per time unit approaching that leaving the electrolytic cell, thereby giving rise to a stationary signal. Thus, aluminium can be determined in steels in proportions between 0.01 and 0.13 (w/w) at a sampling rate of 20–40 samples h^{-1}.[163]

Molybdenum was determined by the spectrophotometric thiocyanate method in alloys after leaching of the sample and electrolytic dissolution of the analyte in a design similar to that used for the above-described determination of aluminium (Figure 4.33B), from which the debubbling well and the filtering device were removed in order to endow the system with greater stability and reproducibility. The filter used (F) was a sintered glass cylinder (40–60 μm pore size, 10 mm length, 4 mm ID) encapsulated in Tygon tubing. The debubbler (D), made of Perspex, avoids the large dead volume of a debubbling well while substantially decreasing the analyte dispersion. In the system depicted in Figure 4.33B, the sample is placed in the electrolytic dissolution chamber. Sample changeover is effected with the manifold in the situation shown in the Figure in order to avoid leakage by stopping the electrolyte flow through the dissolution chamber. After commutation, a direct current pulse is applied and the dissolved metal ions are pushed by a stream (LC) towards the detector. The electrolysis current (200 mA applied for 8 seconds) results in the separation of some solid particles from the sample which are retained at F. The small gas bubbles formed are collected on the filter as large bubbles that are readily removed by the debubbler. The electrolytic medium must be acidic in order to avoid the formation of oxide films on the sample surface—1 M KCl/0.1 M HCl was found to provide optimal results. The calibration graph, constructed from a reference steel standard, was linear over the range 0.7–2.7% (w/w) molybdenum, and the r.s.d. was 2.3% for an analyte content of 2.3%. The sample throughput achieved was 40 samples h^{-1}.[164]

Compared to the configuration shown in Figure 4.33A, that in Figure

4.33B has the advantage of allowing injection of standard solutions into the sample zone and hence application of the standard-addition method. However, electrically-aided leaching also has some disadvantages such as its limited scope of application: only metal samples liable to be made into the cathode of the electrolytic system are viable candidates. In addition, appropriate standards are difficult to obtain and slight compositional differences between these and the samples may drastically alter the amount of analyte dissolved by a given amount of electricity. This pitfall is aggravated by the unavoidable need to use solid standards imposed by the non-quantitative dissolution of the analyte. Additional problems occasionally arise from the formation of gasses and the release of solid particles during the electrolytic process.

4.6.2 Leaching with the Aid of Ultrasonic Energy

Ultrasonic energy is a still virtually unexplored tool of a potential high value to analytical chemistry.[162] So far, ultrasounds have proved to be of great use in accelerating homogeneous and heterogeneous reactions in FIA systems, with improvements in process yields of up to 200–300% (*e.g.* in catalysed reactions taking place in homogeneous media).[165]

Ultrasounds can be effectively and conveniently used to facilitate the selective dissolution of one or more components of the sample provided this is finely divided.

The leaching cell typically used for this purpose (Figure 4.34) consists of a methacrylate piece of 20 μl inner volume at each end of which is placed a Technicon filter (0 4F 824 UK) to retain the sample. The main block in turn consists of two identical pieces of mass 4 g that allow the sample to be conveniently weighed in one and are then fitted by means of

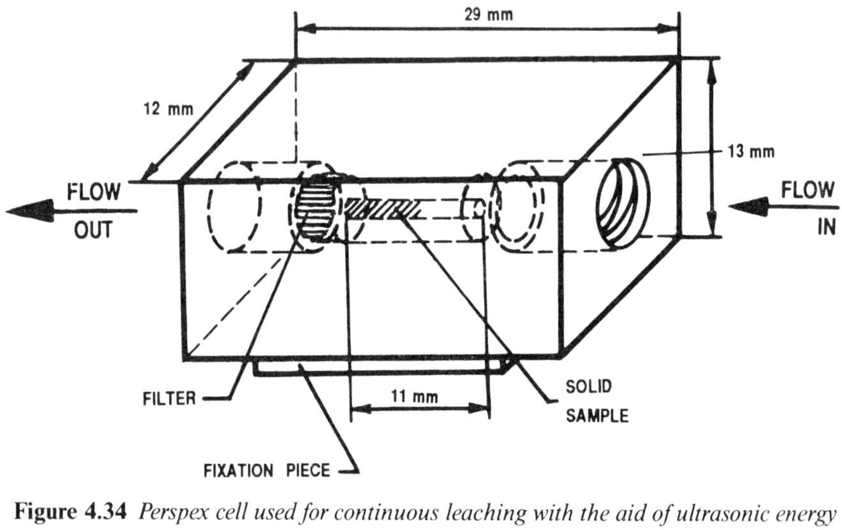

Figure 4.34 *Perspex cell used for continuous leaching with the aid of ultrasonic energy*

screws at the ends through which the block is connected to the flow system. Once loaded, the cell is subjected to ultrasonic irradiation while the leaching solution is circulated through it.

The design of the configuration required to implement continuous leaching with the aid of ultrasounds is a function of the leaching kinetics; if this is reasonably fast, then the leaching solution can be passed at a low flow-rate through the ultrasound-irradiated cell to separate an adequate amount of analyte for measurement. On the other hand, if the leaching kinetics are rather slow, the leaching solution must be passed repeatedly through the electrolytic cell.

Figure 4.35 illustrates the different operational alternatives of this leaching variant. In the case of fast leaching kinetics, application of ultrasounds to the sample cell while the leaching solution is circulated (Figure 4.35A) has the same effects as injection in a normal FIA system: the analyte is 'inserted' into the leaching carrier stream to be transported to the detector, with or without an intervening derivatization reaction.

With slow leaching kinetics, repeated passage of the leaching solution through the sample cell can be accomplished in two ways, namely:

(*a*) By using a closed circuit and a peristaltic pump to propel the same portion of leaching agent recurrently through the cell (Figure 4.35B). After a preset time required for the analyte to be adequately enriched, a switching valve is used to insert it into the flow system for detection, following or not a derivatizing reaction.[166]

(*b*) By using a programmable peristaltic pump to iteratively change the flow direction at preset intervals so that a small volume of leaching agent is circulated through the sample in both directions (Figure 4.35C).[167] The leaching efficiency thus achieved is higher than that provided by the previous alternative because the sample is prevented from compacting in the cell zone close to the exit by the repeated reversal of the flow direction, which results in a more effective action of the leaching agent. Once the programmed number of reversal cycles has been completed, the leaching carrier is driven to the detector.

Figure 4.36 shows a recirculating leaching system coupled on-line to a continuous flow configuration for development of the leaching and derivatizing reaction/detection steps used for the determination of various analytes in agricultural samples.

Iron was determined in plant material by complexation of the dissolved analyte with 1,10-phenanthroline. The overall process consisted of the two steps illustrated in Figure 4.36. The leaching step took place in the units drawn in solid lines in Figure 4.35A, namely the selecting valve (SV), the debubbler—which released the air introduced with the solid into the sample cell, and the peristaltic pump (P_1), which propelled the leaching solution through the solid sample. In order to optimize the circuit volume (360 μl) and hence analyte dilution, the connecting and pump tubing were of a small bore (0.3 and 0.5 mm ID, respectively), and

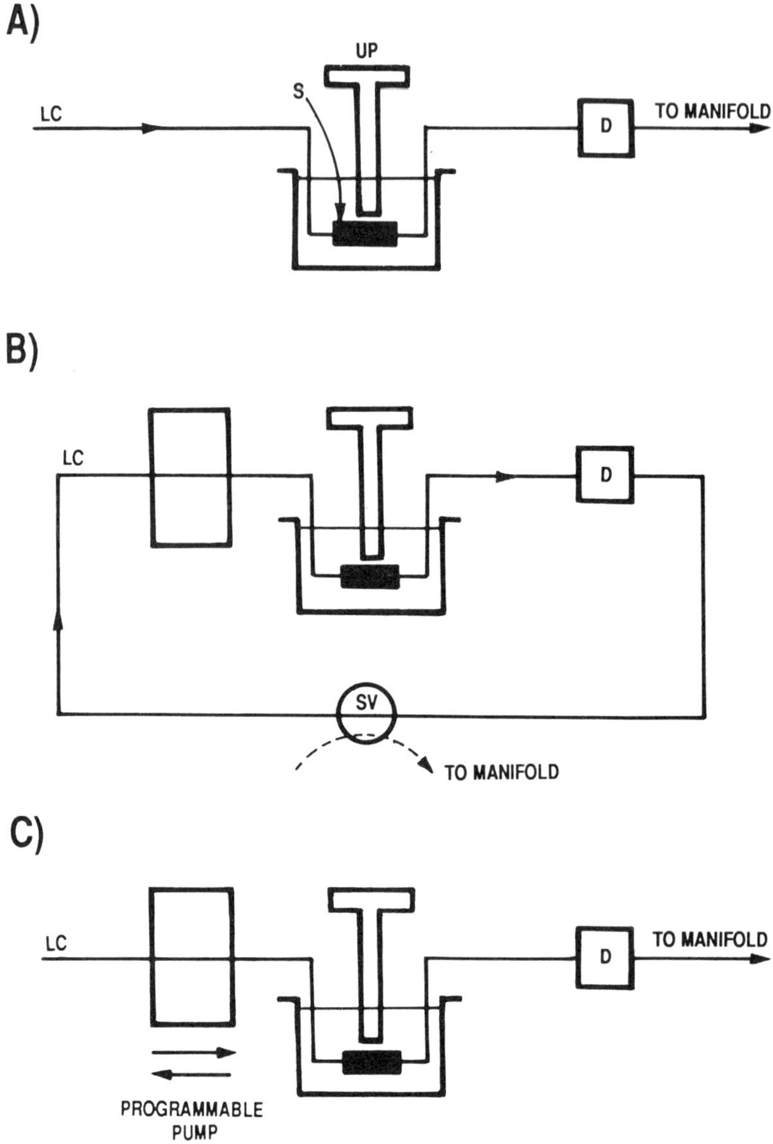

Figure 4.35 *Different ways of accomplishing continuous leaching with the aid of an ultrasonic probe* (UP) *depending on whether the process has a slow* (A) *or fast kinetics* (B,C). LC = *leaching carrier,* D = *debubbler,* SV = *switching valve. (For details, see text)*

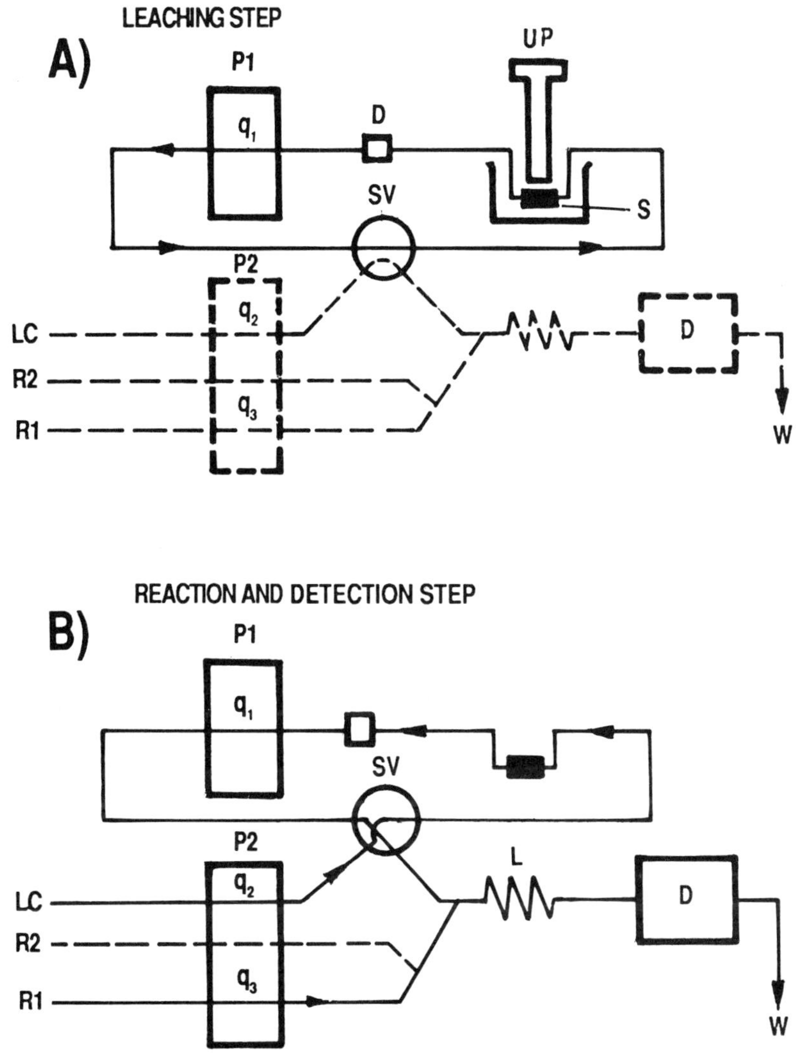

Figure 4.36 *Manifold for the determination of total iron in plant materials or boron in soils by on-line leaching under ultrasonic irradiation.* P = *pump,* SV = *switching valve,* D = *detector,* W = *waste,* R_1 *and* R_2 = *reagents,* LC = *leaching carrier*

the leaching solution in the degassing arm only reached a small height. The reaction and detection step involved the whole manifold in Figure 4.36B, *i.e.* the units in the leaching circuit plus the peristaltic pump (P_2), which was used to propel the leaching and reagent solutions at the same flow rate, $q_2 = q_3 = 0.4$ ml min^{-1}. At this stage, P_2 was used to propel the carrier first to the sample and then to the debubbler; both flow-rates (q_1 and q_2) must be equal in order to avoid entry of air into the system or accumulation of the leached analyte in the debubbler. The leaching circuit and the rest of the manifold were connected via a switching valve. An amount of dried, ground plant material between 3 and 5 mg was introduced into the sample cell for each analysis. With the switching valve in the leaching position, the leaching flow was halted and the sample cell was inserted into the circuit, which was then plunged and fixed to the sonication reservoir—this in turn was placed in a water bath at 60 °C. Next, the leaching stream was circulated through the solid sample for five minutes under ultrasonic irradiation. After this interval had elapsed, SV was switched for the photometric determination of the extracted iron after derivatization. One blank signal per sample was thus obtained, but using a reagent stream containing no ligand. The iron contained in the solid sample was quantified by comparing the analytical sample (*viz.* that resulting from subtracting the blank from the sample signal) with a calibration curve obtained by filling the leaching circuit with a standard solution of Fe^{III}. The performance of the method was assessed by using samples of beech leaves and spruce needles that were subjected to interlaboratory analyses prior to their use as standards. The analyte content was determined by three different procedures, namely: (*a*) hot attack with a 3:1:1 mixture of nitric, sulphuric, and perchloric acid for decomposition of organic matter, followed by determination of iron by AAS; (*b*) the novel procedure described above; and (*c*) as in (*b*), but in the absence of ultrasounds. Procedures (*a*) and (*b*) provided quite consistent results, while (*c*) resulted in incomplete partial extraction.[168]

The determination of boron in solids by continuous leaching of the sample dispenses with the time-consuming reflux heating required by the manual method.[169] The experimental procedure involves accurate weighing of between 5 and 10 mg of finely powdered and homogenized soil in the sample cell and insertion into the manifold. The sample was leached for 30 seconds by circulating 0.1 HCl through the cell under ultrasonic irradiation at 80 °C. The valve was then switched to inject the sample into the carrier stream, which was then merged with the reagent stream (azomethine H), after which the absorbance was monitored at 416 nm. The ultrasonic source was switched off during the injection process to avoid unwanted noise. Any air bubbles resulting from the ultrasonic treatment and the high temperature used were effectively eliminated by the debubbler. The calibration graph was obtained by using boron standards containing between 0.20 and 2.25 μg ml^{-1} B in the same volume used in the leaching step. The graph was quite linear and the correlation coefficient was 0.9997. The relative standard deviation obtained in five replicate injections of mid-range B concentrations was 2.6%, and the sampling frequency achieved was 25 h^{-1}. The results obtained for the various ground, homogenized soil samples assayed were consistent with those provided by the recommended curcumin manual method.[170] Insofar as boron dissolution in the manual procedure is reportedly quantitative, the consistency of the results obtained by the continuous leaching method reveals that it must provide quantitative dissolution as well.

Continuous leaching with iterative reversal of the flow direction was used for the simultaneous determination of phosphate and sulphate in plant material. The configuration used for this purpose (Figure 4.37) consisted of a single, programmed peristaltic pump and two ordinary injection valves for insertion of the derivatizing reagents for sulphate ($BaCl_2$ solution) and phosphate (MoO_4^{2-}/Sn^{II} solutions) after the programmed number of reversal cycles had been completed. Detection was done by means of a diode array detector that allowed measurements to be carried out at different wavelengths (between 400 and 500 nm for sulphate, and between 600 and 700 nm for phosphate) where the absorbance differences between the reaction products of the two analytes were maximal. The procedure included a solution of nitric–perchloric acid as leaching carrier and was applied to beech leaf and spruce needle standards. The results obtained were consistent with those provided by the reference manual method used for contrast; this demonstrates that the two analytes were extracted quantitatively. In addition, the leaching method provided excellent reproducibility with minimal consumption of sample (4–5 mg) and reagents and sample throughputs between 25 and 40 samples h^{-1} compared with 1–2 hours required for complete application of the manual method to each analyte.[171]

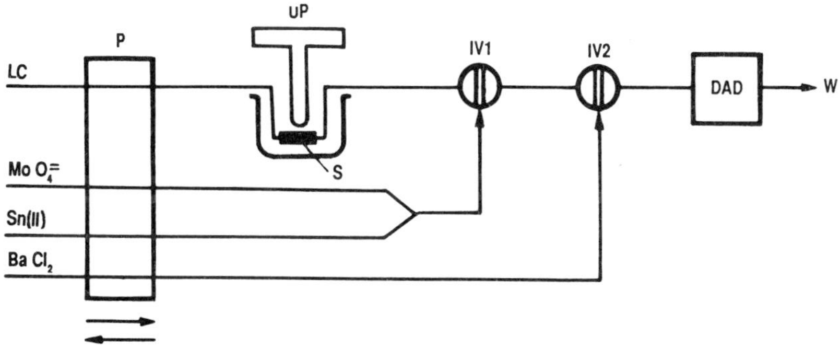

Figure 4.37 *Configuration with iterative reversal of the flow direction for the simultaneous determination of sulphate and phosphate in plant materials with the aid of ultrasonic irradiation.* P = *programmable peristaltic pump,* UP = *ultrasound probe,* S = *sample,* IV = *injection valve,* DAD = *diode array detector,* W = *waste*

Ultrasonic energy as a means of boosting leaching offers some advantages over electric current, namely:

(*a*) A broader scope of application including virtually any type of powdered sample (plant materials, soils, filtration tails, sorbents, sediments, active pharmaceutical principles, micronutrients, food additives and pollutants, *etc.*), whatever its shape or size.

(*b*) Quantitative extraction of the analyte, which facilitates standardization with solution matrices similar to that of the sample. In any case, quantitativeness should be rigorously checked in every case to avoid significant errors as it depends on the type of sample and particle size in question.

However, ultrasounds also have some disadvantages such as the need to use ear muffs and use powdered samples, and the occasional occurrence of blank signals arising from the simultaneous undesirable partial leaching of other components of the sample.

Continuous leaching poses two major problems, namely:

(*a*) Differences between samples of the same type (*e.g.* in particle size or surface composition) may give rise to irreproducible results.

(*b*) Reference materials with different analyte concentrations are scarce and/or unaffordable by routine control laboratories.

Automatic continuous processing of solid samples is a predictably fertile research field insofar as it focuses on the most difficult to automate step of the analytical process (sampling and sample treatment). The problem posed by the diversity of samples encountered is aggravated by the slowness of this step, particularly when dealing with solid samples, manual processing of which takes too long a time that cannot be offset by the rapid, highly automated development of the subsequent analytical steps (measurement and transducing of the analytical signal and data acquisition and processing).

References

1. H.F. Walton, 'Ion-Exchange Chromatography', J. Wiley, New York, 1976.
2. Z. Fang, J. Ruzicka, and E.H. Hansen, *Anal. Chim. Acta*, 1984, **164**, 23.
3. M.A. Marshall and H.A. Mottola, *Anal. Chem.*, 1985, **57**, 729.
4. T. Miller, *Adv. Instrum.*, 1980, **35**, 21.
5. W. Lei, P.K. Dasgupta, and J.L. López, *Anal. Chem.*,1989, **61**, 499.
6. M. Valcárcel and M. Gallego, 'Separation Techniques', Chapter 5 in 'Flow Injection Atomic Spectroscopy', ed. J.L. Burguera, M. Dekker, New York, 1989.
7. M. Valcárcel, 'Flow Injection Analysis', Chapter 11 in 'Sample Introduction in Atomic Spectroscopy', ed. J. Sneddon, M. Dekker, New York, 1990.
8. T. Kumamaru, H. Matsuo, Y. Okamoto, and M. Ikeda, *Anal. Chim. Acta*,1986, **181**, 271.
9. S. Olsen, L.C.R. Pessenda, J. Ruzicka, and E.H. Hansen, *Analyst (London)*, 1983, **108**, 905.

10. P. Hernández, L. Hernández, and J. Losada, *Fresenius' Z. Anal. Chem.*, 1986, **325**, 299.
11. G. Knapp, K. Müller, M. Strunz, and W. Wegscheider, *J. Anal. At. Spectrom.*, 1987, **2**, 611.
12. F. Malamas, M. Bengtsson, and G. Johansson, *Anal. Chim. Acta*, 1984, **160**, 1.
13. M. Bengtsson, F. Malamas, A. Torstebsson, O.R.Egnell, and G. Johansson, *Mikrochim. Acta* ,1985, **III**, 209.
14. Y. Zhan, P. Riby, A.G. Cox, and C.W. McLeod, *Analyst (London)*,1988, **113**, 125.
15. S. Hirata, Y. Umezaki, and M. Ikeda, *Anal. Chem.*, 1986, **58**, 2602.
16. X. Wang and R.M. Barnes, *J. Anal. At. Spectrom.*, 1989 **4**, 509.
17. S. Hirata, K. Honda, and T. Kumamaru, *Anal. Chim. Acta*, 1989, **221**, 65.
18. J. Ruzicka and A. Arndal, *Ibid*, 1989, **216**, 243.
19. Z. Fang and B. Welz, *J. Anal. At. Spectrom.*, 1989, **4**, 543.
20. Z. Fang, S. Xu, and S. Zhang, *Anal. Chim. Acta*, 1984, **164**, 41.
21. S.D. Hartenstein, J. Ruzicka, and G.D. Christian, *Anal. Chem.*, 1985, **57**, 21.
22. A.G. Cox, C.W. McLeod, D.L. Miles, and J.M. Cook, *J. Anal. At. Spectrom.*, 1987, **2**, 553.
23. I.G. Cook, C.W. McLeod, and P.J. Worsfold, *Anal. Proc.*, 1986, **23**, 5.
24. H. Wada, K. Asakura, G.V. Rattaiah, and G. Nakagawa, *Anal. Chim. Acta*, 1988, **214**, 439.
25. H. Bergamin F., B.F. Reis, A.O. Jacintho, and E.A.G. Zagatto, *Anal. Chim. Acta*, 1980, **117**, 81.
26. X. Wu and W. Qi, *Anal. Chim Acta*, 1988, **214**, 279.
27. T. Yamane, K. Watanabe, and H.A. Mottola, *Anal. Chim. Acta*, 1988, **207**, 331.
28. C.M. Sakamoto-Arnold and K.S. Johnson, *Anal. Chem.*, 1987, **59**, 1789.
29. L. Risinger, *Anal. Chim. Acta*, 1986, **179**, 509.
30. R.H. Gaddy and R.S. Dorsett, *Anal.Chem.*, 1968, **40**, 429.
31. T. Urbányi and A. O'Connell, *Anal.Chem.*, 1972, **44**, 565.
32. O.F. Kamson and A. Townshend, *Anal. Chim. Acta*, 1983, **155**, 253.
33. J.M. Wang, T.S. Wei, and Y.M. Chen, *J. Chin. Chem. Soc.*, 1986, **33**, 109.
34. J.L. Burguera, M. Burguera, and A. Townshend, *Anal. Chim. Acta*, 1981, **127**, 199.
35. S. Motomizu and M. Onoda, *Anal. Chim. Acta*, 1988, **214**, 289.
36. A.G. Cox, I.G. Cook, and C.W. McLeod, *Analyst (London)*, 1985, **110**, 331.
37. E.B. Milosavljevic, L. Solujic, J.H. Nelson, and J.L. Hendrix, *Mikrochim. Acta*, 1985, **III**, 353.
38. E.B. Milosavljevic, J. Ruzicka, and E.H. Hansen, *Anal. Chim. Acta*, 1985, **169**, 321.
39. B.A. Petersson, Z. Fang, J. Ruzicka, and E.H. Hansen, *Anal. Chim. Acta*, 1986, **184**, 165.
40. E.B. Milosavljevic, L. Solujic, J.L. Hendrix, and J.H. Nelson, *Analyst (London)*, 1989, **114**, 805.
41. F. Cañete, A. Ríos, M.D. Luque de Castro, and M. Valcárcel, *Analyst (London)*, 1987, **112**, 263.
42. S. Tesfalidet and K. Irgum, *Anal. Chem.*, 1989, **61**, 2079.
43. M. Valcárcel and M.D. Luque de Castro, *Analyst (London)*, 1990, **115**, 699.
44. K. Yosimura and H. Waki, *Talanta*, 1985, **32**, 345.
45. K. Yosimura, *Anal. Chem.*, 1987, **59**, 2922.

46. F. Lázaro, M.D. Luque de Castro, and M. Valcárcel, *Anal. Chim. Acta*, 1988, **214**, 217.
47. M. de la Torre, F. Fernández, F. Lázaro, M.D. Luque de Castro, and M. Valcárcel, *Talanta*, in press.
48. J. Ribereau-Gayon, E. Peynaud, P. Suddrau, and P. Ribereau, Trait d'Enologie. Sciences et Techniques du Vin, Tome I, Dunod, Paris, 1976, pp. 270.
49. F. Lázaro, M.D. Luque de Castro, and M. Valcárcel, *Anal. Chim. Acta*, 1989, **219**, 231.
50. M. de la Torre, M.D. Luque de Castro, and M. Valcárcel, *Analyst (London)*, 1991, **116**, 81.
51. D.A. Russell and R. Narayanaswamy, *Anal. Chim. Acta*, 1989, **220**, 75.
52. Danhua Chen, M.D. Luque de Castro, and M. Valcárcel, *Anal. Chim. Acta*, 1990, **234**, 345.
53. Danhua Chen, M.D. Luque de Castro, and M. Valcárcel, *Talanta*, in press.
54. P. Linares, M.D. Luque de Castro, and M. Valcárcel, *Anal. Chim. Acta*, 1989.
55. C.H. Lochmller, K.R. Lung, and M.R. Cushman, *J. Chromatogr. Sci.*, 1985, **23**, 429.
56. K.K. Unger, N. Becker, and P. Roumeliotis, *J. Chromatogr.*, 1976, **125**, 115.
57. H. Engelhardt and G. Ahr, *Chromatographia*, 1981, **14**, 227.
58. G.E. Berendsen and L. De Galan, *J. Chromatogr.*, 1980, **196**, 21.
59. G.V. Lisichkin and G.V. Kudryavtsev, *Zh. Anal. Khim.*, 1983, **38**, 1684.
60. K.K. Unger, 'Porous Silica', Elsevier, Amsterdam, 1979.
61. R.J. Reddingius, G.J. De Jong, U.A. Th. Brinkman, and R.W. Frei, *J. Chromatogr.*, 1981, **205**, 77.
62. C.E. Werkhoven-Goewie, U.A.Th. Brinkman, R.W. Frei, C. de Ruiter, and H. de Vries, *J. Chromatogr.*, 1983, **276**, 349.
63. C.E. Werkhoven-Goewie, C. de Ruiter, U.A. Th. Brinkman, R.W. Frei, G.J. de Jong, C.J. Little, and I. Stahel, *J. Chromatogr.*, 1983, **255**, 79.
64. H. Nielsen, *J. Chromatogr.*, 1986, **381**, 63.
65. W. Roth, *J. Chromatogr.*, 1983, **278**, 347.
66. M.W.F. Nielen, E. Sol, R.W. Frei, and U.A.Th. Brinkman, *J. Liq. Chromatogr.*, 1985, **8**, 1053.
67. K. Zech and R. Huber, *J. Chromatogr.*, 1986, **353**, 351.
68. D. Ishii, K. Hibi, K. Asai, M. Nagaya, K. Mochizuki, and Y. Mochida, *J. Chromatogr.*, 1978, **156**, 173.
69. M. Krejsi, K. Slais, D. Kourilova, and M. Vespalcova, *J. Pharm. Biomed. Anal.*, 1984, **2**, 197.
70. W.Th. Kok, U.A.Th. Brinkman, R.W. Frei, and H.B. Hanekamp, *J. Chromatogr.*, 1982, **237**, 357.
71. R.P.W. Scotto and P. Kucera, *J. Chromatogr.*, 1979, **185**, 27.
72. W. Roth and K. Beschke, *J. Pharm. Biomed. Anal.*, 1984, **2**, 289.
73. 'Selective Sample Handling and Detection in High-Performance Liquid Chromatography', Part A, ed., R.W. Frei and K. Zech, Elsevier, Amsterdam, 1988.
74. 'Selective Sample Handling and Detection in High-Performance Liquid Chromatography', Part B, ed., K. Zech and R.W. Frei, Elsevier, Amsterdam, 1989.
75. M. Valcárcel and M.D. Luque de Castro, 'Automatic Methods of Analysis', Elsevier, Amsterdam, 1988.

76. M.W.F. Nielen, G. Koomen, R.W. Frei, and U.A. Th. Brinkman, *J. Liq. Chromatogr.*, 1985, **8**, 315.
77. M.W.F. Nielen, U.A. Th. Brinkman, and R.W. Frei, *Anal. Chem.*, 1985, **57**, 806.
78. H.Y. Mohammed, H. Veening, and D.A. Dayton, *J. Chromatogr.*, 1981, **226**, 471.
79. R. Huber, unpublished results.
80. H. Nielsen, *J. Chromatogr.*, 1986, **381**, 63.
81. H.P. Fiedler and H.P. Reuschenbach, *Chromatographia*, 1984, **19**, 246.
82. W. Voelter, Th. Kronbach, K. Zech, and R. Huber, *J. Chromatogr.*, 1982, **239**, 475.
83. M.W.F. Nielen, H.E. van Ingen, A.J. Valk, R.W. Frei, and U.A.Th Brinkman, *J. Liq. Chromatogr.*, 1987, **10**, 617.
84. J. de Jong, M.W.F. Nielen, R.W. Frei, and U.A.Th. Brinkman, *J. Chromatogr.*, 1986, **381**, 431.
85. R.P.W. Scott, *J. Chromatogr. Sci.*, 1980, **18**, 49.
86. M. Novotny, *Anal. Chem.*, 1981, **53**, 1294A.
87. F.J. Yang, *J. High Res. Chromatogr. Chrom. Commun.*, 1983, **6**, 348.
88. 'Microcolumn High-Performance Liquid Chromatography', ed. P. Kucera, Elsevier, Amsterdam, 1984.
89. 'Small Bore Liquid Chromatography Columns: Their Properties and Uses', ed. R.P.W. Scott, J. Wiley, New York, 1984.
90. M.W.F. Nielen, R.C.A. Koordes, R.W. Frei, and U.A.Th Brinkman, *J. Chromatogr.*, 1985, **330**, 113.
91. W.W. Kubiak and J. Wang, *Anal. Chem.*, 1989, **61**, 468.
92. M.B. Glauert, *J. Fluid Mech.*, 1956, **1**, 625.
93. D.T. Chin and C.H. Tsang, *J. Electrochem. Soc.*, 1978, **125**, 1461.
94. W. Matuszewski, M. Trojanowicz, and W. Frenzel, *Fresenius' Z. Anal. Chem.*, 1988, **332**, 148.
95. H. Gunasingham, K.P. Ang, C.C. Ngo, and P.C. Thiak, *J. Electroanal. Chem.*, 1986, **198**, 27.
96. C. Yarnitzky and E. Ouziel, *Anal. Chem.*, 1976, **48**, 2024.
97. J. Wang and M. Ariel, *Anal. Chim. Acta*, 1978, **99**, 89.
98. H.B. Hanekamp, W.H. Voogt, P. Bos, and R.W. Frei, *Anal. Chim. Acta*, 1980, **118**, 81.
99. J. Wang and H.D. Dewald, *Anal. Chem.*, 1983, **55**, 933.
100. J. Wang and M. Ariel, *J. Electroanal. Chem.*, 1977, **85**, 289.
101. S.D. Brown and B.R. Kowalski, *Anal. Chim. Acta*, 1979, **107**, 13
102. E.B.T. Tay, S.B. Khoo, and S.G. And, *Analyst (London)*, 1989, **114**, 1271.
103. 'Stripping Analysis: Principles, Instrumentation and Applications', ed. J. Wang, VCH, Deerfield Beach, Florida, 1985.
104. H. Huiliang, C. Hua, D. Jagner, and L. Renman, *Anal. Chim. Acta*, 1987, **193**, 61.
105. H. Huiliang, D. Jagner, and L. Renman, *Anal. Chim. Acta*, 1987, **201**, 1.
106. K.Z. Brainina, A.V. Tchernyshova, N.Y. Stozhko, and L.N. Kainyshevskaya, *Analyst (London)*, 1989, **114**, 173.
107. W.D. Ellis, *J. Chem. Educ.*, 1973, **50**, A131.
108. T.R. Copeland and R.K. Skogerboe, *Anal. Chem.*, 1974, **46**, 1257A.
109. J. Wang, *Environ. Sci. Technol.*, 1982, **16**, 104A.
110. J. Wang and H.D. Dewald, *Anal. Chim. Acta*, 1984, **162**, 189.

111. P.L. Meschi and D.C. Johnson, *Anal. Chim. Acta*, 1981, **124**, 303.
112. P.L. Meschi, D.C. Johnson, and G.R. Luecke, *Anal. Chim. Acta*, 1981, **124**, 315.
113. D. Jagner and K. Aren, *Anal. Chim. Acta*, 1978, **100**, 375.
114. A. Ríos, M.D. Luque de Castro, and M. Valcárcel, *Anal. Chem.*, 1981, **60**, 1540.
115. J. Wang, H. Huiliang, and W. Kubiak, *Electroanal. Chem.*, 1990, **2**, 127.
116. D. Jagner, *Anal. Chem.*, 1978, **50**, 1924.
117. D. Jagner, *Anal. Chem.*, 1979, **51**, 342.
118. L. Anderson, D. Jagner, and M. Josefson, *Anal. Chem.*, 1982, **54**, 1371.
119. A. Hu, R.E. Dessy, and A. Granli, *Anal. Chem.*, 1983, **55**, 320.
120. V.G. Levich, 'Physicochemical Hydrodynamics', Prentice-Hall, Englewood Cliffs, New Jersey, 1962.
121. J. Buffle, M. Pelletier, and D. Monnier, *J. Electroanal. Chem.*, 1973, **43**, 185.
122. W. Frenzel and P. Bratter, *Anal. Chim. Acta*, 1986, **179**, 389.
123. D. Jagner, M. Josefson, and K. Aren, *Anal. Chim. Acta*, 1982, **141**, 147.
124. L. Almestrand, D. Jagner, and L. Renman, *Anal. Chim. Acta*, 1987, **193**, 71.
125. S. Mannino, *Analyst (London)*, 1984, **109**, 905.
126. E. Nakayama, T. Kimoto, and S. Okataki, *Anal Chem.*, 1985, **57**, 1157.
127. X. Ruan and H. Chang, *Talanta*, 1988, **35**, 861.
128. L. Renman, D. Jagner, and R. Berglund, *Anal. Chim. Acta*, 1986, **188**, 137.
129. H. Eskilsson and C. Haraldsson, *Anal. Chim. Acta*, 1987, **198**, 231.
130. G. Schulze, M. Husch, and W. Frenzel, *Mikrochim. Acta*, 1984, **I**, 191.
131. J. Wang, 'Electroanalytical Chemistry', Vol. 16, ed. A.J. Bard, M. Dekker, New York, 1989.
132. J. Wang and S. Mannino, *Analyst (London)*, 1989, **114**, 643.
133. H. Eskilsson, C. Haraldsson, and D. Jagner, *Anal.Chim. Acta*, 1985, **175**, 79.
134. J.A. Polta and D.C. Johnson, *Anal. Chem.*, 1985, **57**, 1373.
135. J. Wang and B.A. Frelha, *Anal. Chem.*, 1983, **55**, 1285.
136. E.N. Chaney Jr. and R.P. Baldwing, *Anal. Chim. Acta*, 1985, **176**, 105.
137. D.T. Burns, A. Townshend, and A. Carter 'Inorganic Reaction Chemistry. Reactions of the Elements and their Compounds', Vol 2, Parts A and B. Ellis Horwood, Chichester, 1981.
138 D.T. Burns, A. Townshend, and A. Catchpole, 'Inorganic Reaction Chemistry. Systematic Chemical Separation'. Ellis Horwood, Chichester, 1980.
139. F. Burriel, F. Lucena, S. Arribas, and J. Hernández-Méndez 'Química Analítica Cualitativa'. Paraninfo, Madrid, 1989.
140. L. Erdey, 'Gravimetric Analysis' Parts I and II. Pergamon Press, Oxford, 1965.
141. J.F. van Staden. *Fresenius' Z. Anal. Chem.*, 1982, **312**, 438.
142. J.M. Skinner and A.C. Docherty, *Talanta*, 1967, **14**, 1393.
143. P. Martínez, M. Gallego, and M. Valcárcel, *Anal. Chem.*, 1987, **59**, 69.
144. M. Valcárcel and M. Gallego, *Trends Anal. Chem.*, 1989, **8**, 34.
145. F.J. Krug, H. Bergamin F., E.A.G. Zagatto, and S.S. Jogensen, *Analyst (London)*, 1977, **102**, 503.
146. J.F. van Staden and W.D. Basson. *Lab. Pract.*, 1980, **29**, 1279.
147. A. Hughes and P. Worsfold. *Anal. Proc.*, 1985, **22**, 16.
148. J. Martínez-Calatayud and P. Campins. *Talanta*, 1986, **33**, 625.
149. P. Martínez, M. Gallego, and M. Valcárcel, *J. Anal. At. Spectrom.*, 1987, **2**, 211.
150. P. Martínez, M. Gallego, and M. Valcárcel, *Anal. Chim. Acta*, 1987, **193**, 127.
151. J. Zorro, M. Gallego, and M. Valcárcel, *Microchem. J.*, 1989, **39**, 71.

152. S.S.M. Hassan, 'Organic Analysis Using Atomic Absorption Spectrometry'. Ellis Horwood, Chichester, 1984.
153. R. Montero, M. Gallego, and M. Valcárcel, *J. Anal. At. Spectrom.*, 1988, **3**, 725.
154. R. Montero, M. Gallego, and M. Valcárcel, *Anal. Chim. Acta*, 1988, **215**, 241.
155. J. Martínez-Calatayud and V. García-Mateo, *J. Pharm. Biomed. Anal.*, in press.
156. P. Martínez, M. Gallego, and M. Valcárcel, *Analyst (London)*, 1987, **112**, 1233.
157. R.E. Santelli, M. Gallego, and M. Valcárcel, *Anal. Chem.*, 1989, **61**, 1427.
158. R.E. Santelli, M. Gallego, and M. Valcárcel, *J. Anal. At. Spectrom.*, 1989, **4**, 547.
159. C.E. Addeyinwo and J.F. Tyson, *Anal. Chim. Acta*, 1988, **214**, 339.
160. C.E. Addeyinwo and J.F. Tyson, *Anal. Proc.*, 1989, **26**, 375.
161. C.E. Addeyinwo and J.F. Tyson, *Anal. Proc.*, 1989, **26**, 58.
162. P. Linares, F. Lázaro, M.D. Luque de Castro, and M. Valcárcel, *J. Autom. Chem.*, 1988, **10**, 88.
163. H. Bergamin F., F.J. Krug, E.A.G. Zagatto, E.C. Arruda, and C.A. Coutinho, *Anal. Chim. Acta*, 1986, **190**, 177.
164. H. Bergamin F., F.J. Krug B.F. Reis, J.A. Nobrega, M. Mesquita, and I.G. Souza, *Anal. Chim. Acta*, 1988, **214**, 397.
165. P. Linares, F. Lázaro, M.D. Luque de Castro, and M. Valcárcel, *Anal. Chim. Acta*, 1987, **200**, 51.
166. A. Ríos, M.D. Luque de Castro, and M. Valcárcel, *Anal. Chem.*, 1985, **57**, 1803.
167. A. Ríos, M.D. Luque de Castro, and M. Valcárcel, *Anal. Chem.*, 1988, **60**, 1540.
168. F. Lázaro, M.D. Luque de Castro, and M. Valcárcel, *Anal. Chim. Acta*, 1991, in press.
169. L. Lotti and G. Galoppini, 'Análisis Químico Agrario', Alhambra, Madrid, 1986.
170. D. Chen, F. Lázaro, M.D. Luque de Castro, and M. Valcárcel, *Anal. Chim. Acta*, 1989, **226**, 221.
171. F. Pérez-Roldán, unpublished results.

CHAPTER 5

Evaluation of Non-chromatographic Continuous Separation Techniques

5.1 Introduction

In the three previous chapters have been described a wide variety of non-chromatographic continuous separation techniques classified according to the type of interface (liquid–gas, liquid–liquid, or liquid–solid) across which mass transfer, the very foundation of the separation process, takes place. No other type of interface was dealt with either because implementation of the corresponding separation technique was technically unfeasible (*e.g.* solid–solid or gas–gas) or because only a few specific examples have so far been reported (*e.g.* gas–solid).[1] The first chapter was aimed at placing non-chromatographic separation techniques within the broad setting of current separation techniques in analytical chemistry.

This final chapter is concerned with those non-chromatographic continuous separation techniques which, on account of their peculiar features, cannot be placed in any of the groups described in the previous chapters, yet are gaining significance day by day in analytical chemistry, namely field-flow fractionation and capillary electrophoresis. Both are non-chromatographic in nature and entail use of a carrier and/or analyte stream to accomplish the pursued separation. It also discusses briefly and critically the continuous separation techniques dealt with in the previous chapters, which are also compared with the other separation techniques. Finally, the most salient trends to be expected in this field are also outlined.

5.2 Field Flow Fractionation

Under this denomination are classed a series of relatively novel techniques based on the coupling of a fractionated sample (solute) concentration distribution and fluid flow non-uniformities inside a narrow channel that causes differential migration of the solute species and thus separation.[2] A lateral physical or chemical field (*e.g.* a temperature gradient, an electrical, magnetic, gravitational, or centrifugal field, a chemical potential gradient, *etc.*) of constant or variable intensity acts through a channel formed by two planar, parallel, polished surfaces (Figure 5.1A). Separation takes place along the channel that links the sample injection inlet to its outlet, which is connected to the detector. On interaction with the field, the particles or molecules concerned are squeezed on the channel walls. The concentration gradient thus created induces a countercurrent diffusion flow. After some time, equilibrium is reached and the solute is exponentially distributed along the channel. The liquid flow velocity at a cross-section of the channel will vary with the distance to the channel walls. This velocity profile arises from the viscosity effects associated with flow processes and is shaped parabolically under laminar flow conditions (Figure 5.1B).

Classical field flow fractionation (FFF) techniques rely on the establishment of an exponential solute concentration profile along the channel, the maximal concentration occurring at the accumulation wall of the channel in relation to the velocity profile of the moving fluid (Figure 5.1B). On the other hand, the new FFF techniques (focusing techniques)[3] are based on the migration of the solute under the influence of driving forces aided by the velocity profile to a position where the field intensity

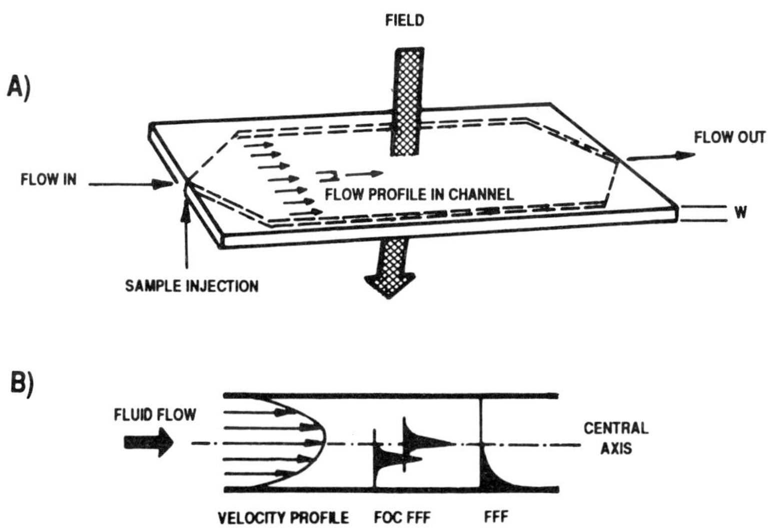

Figure 5.1 (A) *Operational principle of field flow fractionation.* (B) *Typical parabolic velocity profile and distribution of the solute along the channel in focusing and classical FFF techniques*

can be classified according to various criteria. A basic classification establishes two broad groups, namely classical and focusing FFF techniques, which differ in the driving force involved, its intensity depending or not on the co-ordinate of the direction along which its action is exerted. In focusing FFF techniques, the driving force is a function of the position in the focusing direction and must be zero at the equilibrium point. The intensity of the driving force across the channel remains constant in this type of FFF technique, which in turn can be classified according to the nature of the applied field and its intensity (see Table 5.1).

5.2.3.1 Classical Field Flow Fractionation Techniques

Thermal field flow fractionation (thermal FFF) is the earliest of FFF techniques and also one of the easiest to implement.[5,6] The channel to be used is relatively easy to construct as it only requires two metal blocks with highly polished surfaces to be squeezed together with a intervening spacer. The upper block is heated electrically, while the lower block is cooled with water. The channel shape proper is cut into the spacer. The temperature gradient between the walls causes non-selective thermal diffusion of the solute species, thereby giving rise to accumulation of the solute, usually near the cold wall.

Sedimentation field flow fractionation is also one of the earliest FFF techniques[7-10] and relies on the effect of gravitational or centrifugal force. The channel is usually coiled around the interior basket of a centrifuge rotor.

Electrical field flow fractionation (electrical FFF) is a more sophisticated variant of the previous techniques.[11] The altogether homogeneous field is induced by electrical current across the channel. The channel walls are formed by two semi-permeable membranes that allow passage of small ions and separate the channel space from the electrode compartment.

Flow field flow fractionation (flow FFF) is the best-known technique in this group. The solvent flow, perpendicular to that of the basic medium in the channel, creates an external field, the driving force behind the process.[12-14]

Pressure field flow fractionation (pressure FFF)[15] is very similar to flow FFF. The basic difference between the two lies in the fact that while the flow field is applied externally across the channel bound by two planar, parallel membranes in flow FFF, in pressure FFF the lateral flow across the walls of a circular cross-section capillary is started by an internal pressure drop in a liquid pumped along the semi-permeable capillary.

In *magnetic field flow fractionation* (magnetic FFF)[16-18] the channel is formed by a coiled Teflon circular cross-section capillary or by two parallel planes onto which a magnetic field is applied.

Concentration field flow fractionation (concentration FFF) is the sole technique of this type that can exploit the concentration gradient of a sol-

Table 5.1 *Classification of FFF techniques*

Field forces External	Internal	External + internal	*Field intensity*
Classical FFF		*Focusing FFF*	
Thermal FFF Sedimentation FFF Electrical FFF Flow FFF Magnetic FFF Concentration FFF	Pressure FFF Shear FFF	SFFFFF Isoelectric focusing FFF	Normal
Steric FFF			High

vent mixture across the channel to introduce effective chemical forces or a chemical field. This technique has only been developed in the theoretical plane so far.

In *shear field flow fractionation* (shear FFF), an internal shear force field is induced in the annular space between two concentric cylinders that rotate with respect to each other.[20] This technique is quite suitable for the separation of large macromolecules and globular particles.

5.2.3.2 Focusing Field Flow Fractionation Techniques

Only two of the FFF techniques developed so far respond to the operational principle behind focusing FFF techniques, namely:

(*a*) *Sedimentation–flotation focusing field flow fractionation* (SFFFFF). In applying this technique, particles or macromolecules are allowed to settle or float as a result of the density gradient established in a liquid phase depending on the differences between the local density of the liquid and that of the particles or macromolecules concerned. At *quasi*-equilibrium, the solute species are focused on a fine layer in the position where the density of the environment equals that of the solute. The focused zones are separated longitudinally as a result of the differential axial velocity of the movement induced by an appropriate shape of the fluid flow velocity profile.[21]

(*b*) *Isoelectrical focusing field flow fractionation* (IEFFFF) relies on the fact that the electrophoretic mobility of amphoteric macromolecules is a function of the pH and becomes zero at the isoelectric point. Therefore, on application of an electric field, the amphoteric solute is focused on the position of the isoelectric point. At equilibrium, the transport is counteracted by ordinary diffusion.[22]

Only some of these techniques can be implemented on commercially available instrumentation. Figure 5.3 illustrates the scope of application of such instrumentation and how the selectivity achieved varies with the molecular weight of the analytes to be separated.[23] For comparison, the Figure also shows the corresponding selectivity achieved by size-exclusion chromatography.

5.2.4 Applications

Field flow fractionation techniques have so far been used to develop methods of analysis for macromolecules and particulate samples, and for the physico-chemical characterization of some substances.

The mass property of the retained product is related to the retention ratio, R, and the value of parameter λ*.[24] In most classical FFF techniques, the mass property used is the size of the macromolecules or particles to be separated, expressed as molecular mass or effective diameter. In some of the classical FFF techniques, λ and the size of the fractionated solute are not related through a straightforward expression. Thus, in thermal FFF, λ is a function of the thermal diffusion coefficient, which in turn is dependent on the diffusion coefficient. However, the diffusion coefficient is an inverse function of the molar mass or another dimension of the solute molecule. In electrical FFF, λ is a function of the ratio between the diffusion coefficient and the electrophoretic mobility. In this case, λ is rather a complicated function of the size of the solute molecule. This parameter is always inversely proportional to the molecular mass or the size of the retained solute molecules, so, in classical FFF techniques, retention (R) is an inverse function of extensive properties of the solute.

The situation in focusing FFF is rather different. Let us consider SFFFFF by way of example. The position of the focused zone in the density gradient is determined by the solute density (*i.e.* by an intensive property). Extensive properties (*e.g.* size) affect the equilibration rate and hence the width of the focused zone in this case.

The feasibility of applying these techniques to particular problems thus relies on the above reasoning. The difference between analytical applications and those aimed at obtaining physico-chemical measurements lies rather in how the results are used than in the fundamental procedure employed. Analytical FFF methods can be developed on the basis of the direct relationship with the retention parameters (R or λ) or by including one of the calibration procedures typically used in analytical chemistry. Both kinds of principle have been used in the methods reported to date.

* $\lambda = R^0T/Fw$, where F is the effective force of the field that acts on one mole of the solute, R^0 is the gas constant, T the absolute temperature, and w the thickness of the fractionation channel.

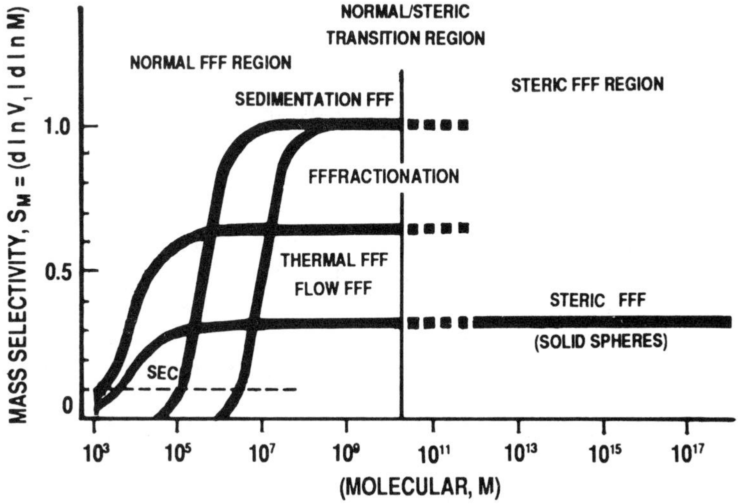

Figure 5.3 *Mass selectivity range afforded by commercially available instrumentation. A representative value for size-exclusion chromatography is shown for comparison*
(Reproduced from reference 22 with permission of the American Chemical Society)

Below are described some illustrative examples of the application of different FFF techniques.

The earliest work on thermal FFF was aimed at checking theoretical hypotheses (the relationship between λ and the thermal diffusion factor).[25, 26] Subsequent improvements in the features of the separation channel such as the incorporation of a metering valve at its outlet[27] allowed the temperature range of the channels and hence the separation potential to be expanded; on the other hand, the use of a pressurized system[28] enabled work at temperatures above the boiling points of the solvents at the atmospheric pressure, thereby allowing low molar-mass substances such as crude oils, asphalts[28] and asphaltanes[29] to be efficiently separated. Also, the thinner the channel, the higher the resolution achieved.[30] Janca and Klepárnik used thermal FFF to determine the average molecular masses of standard polystyrene samples with molar masses between 3600 and 867 000. The results were obtained at three temperature increments (10, 15, and 40°C), and the average molar masses calculated from the fractograms were quite consistent with the data supplied by the manufacturer and those obtained by molecular size-exclusion chromatography. Polydispersity indices were calculated from the fractogram with no corrections for the zone width.[31]

Sedimentation FFF has reached the highest degree of development among field flow fractionation techniques. In addition to its proven ability for the analysis of particles such as those of some polymer latexes,[32] inorganic particulates and natural and synthetic emulsions,[33] the possibility of characterizing biopolymers and particles of biological interest[34] has increased its interest dramatically.

Electrical FFF techniques have been chiefly devoted to protein fractionation. Various types of channels including flexible[35] or rigid membranes[36] and circular channels[37] have proved to be rather suitable for these applications. The influence of experimental variables (sample and buffer concentrations, pH, and solute–solute interactions) on this type of separation is decisive. As can be seen from Figure 5.4A, there are significant differences between the fractogram of the sample and those of individual proteins.[38] This technique is confronted with some problems on application to denatured proteins and latexes:[35] denatured proteins are not retained under some working conditions; however, polyethylene latexes are strongly adsorbed on application of an electric field, the adsorbed fraction increasing with increasing field intensity.

Flow FFF has also been used for the separation and characterization of proteins, as well as to investigate their structural transformations. Figure 5.4B shows the fractograms yielded by human and rat plasma, the diffusion coefficients for various components of which were calculated from retention data.[39] Other applications such as the fractionation of water-soluble polymers[40] and the separation of polystyrene in organic solvents[41] testify to the potential of this technique—in the latter case, the logarithm of the molecular mass and the diffusion coefficient are linearly related over a wide range of molecular masses (from 20 to 400 to 1 800 000).[42]

Few applications of pressure FFF have so far been reported. One example is the fractionation of blue dextran and human plasma in a semi-permeable circular capillary of 0.2 mm ID reported by Lee *et al.*[43]

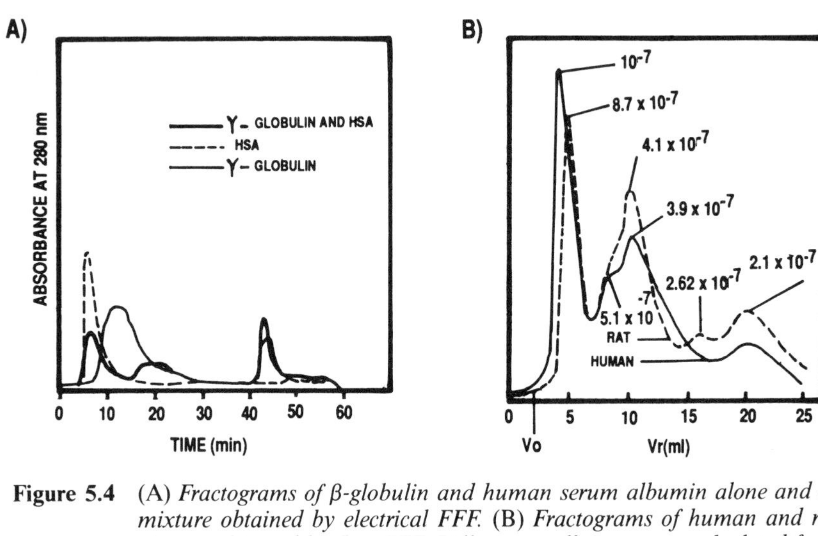

Figure 5.4 (A) *Fractograms of β-globulin and human serum albumin alone and in mixture obtained by electrical FFF.* (B) *Fractograms of human and rat plasma obtained by flow FFF. Diffusion coefficients were calculated from individual peaks*

Work on magnetic FFF has to date been devoted to the retention of bovine serum albumin in the presence and absence of nickel nitrate[44] and that of ferric oxide particles suspended in acetonitrile.[45, 46]

Steric FFF has been applied both in its gravitational and centrifugal variants, basically in theoretical studies, but also occasionally in nature (*e.g.* the characterization of fine coal particles,[47] and that of human, sheep, rabbit, and horse blood cells[48]).

As far as focusing FFF methods are concerned, only a few applications of SFFFFF have been reported to date despite their promising potential. The best results in this respect were obtained by using a trapezoidal cross-section fractionation channel, the thickness and permeability of which varied along the same direction as the density gradient. In addition to the separation of particles of different densities in a density gradient medium (Percoll),[49] this technique has been applied to the separation of polystyrene and polyglycol methacrylate latexes that were injected into the channel and eluted with Percoll as carrier fluid. Figure 5.5 shows the fractogram of this separation, which involved spectrophotometric detection. As can be seen, separation between the particles of the two types of latex is outstanding.

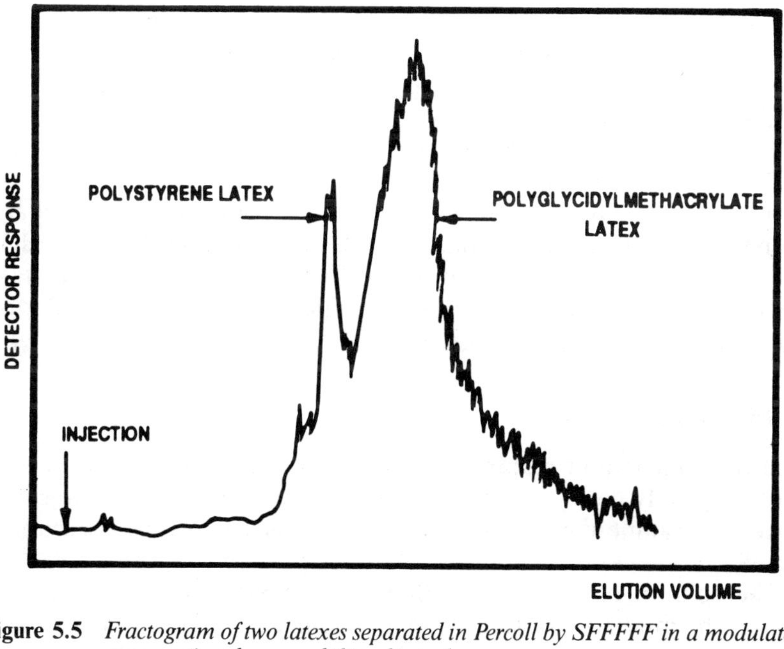

Figure 5.5 *Fractogram of two latexes separated in Percoll by SFFFFF in a modulated cross-sectional permeability channel*

5.3 Isotachophoresis

Capillary electrophoresis, commonly also referred to as 'isotachophoresis', like other electrophoretic techniques, is based on the application of an electrical field that induces migration of charged particles, thereby creating sharp boundaries between the separate zones of the sample components.[51, 59] Such zones remain in contact as they migrate at the same rate, the migration sequence being determined by the relative electrophoretic mobilities of the separated species. Isotachophoresis is the only electrophoretic technique that can be implemented in a continuous fashion; it involves the typical steps of a continuous hydrodynamic separation technique: sample injection, separation, and detection of the components.

Figure 5.6A illustrates the different stages of the overall process:

(1) The column (typically 300 mm long and 0.5 mm ID) is filled with two electrolytes with charge of the same sign but higher (leading electrolyte) and lower mobility (terminating electrolyte) than the species to be isolated.

(2) The sample is injected into the column.

(3) The high-voltage power supply used is switched on, thereby creating an electrical field (2–30 kV, 5–500 μA) between the leading and terminating electrolyte. A membrane prevents the sample zone from reaching the electrodes.

(4) The separated zones reach the detection system sequentially.

Figure 5.6B shows the typical recording obtained, namely a baseline adjusted with the leading electrolyte, signals corresponding to the different analytes and a new baseline established by the tailing electrolyte in the integral mode (*a*), or the classical signals similar to chromatographic peaks in the differential mode (*b*).

The position and profile of the zones that reach the detector are defined by different characteristic principles of the isotachophoretic system such as the uniform migration velocity of each zone, the self-sharpening effect of the zone boundary and the concentrating effect, as well as by the properties of specific zones (*e.g.* field strength, conductivity, temperature) and the electrolytes (composition, concentration and pH), optimization of which results in easier, improved detection.

5.3.1 Basic Units

The instrumentation required to implement the isotachophoretic technique typically consists of the following basic units: column, injection system, separation channel, detection cell, electrodes and electrolyte chambers, detection system, and high-voltage power supply.

The column of the isotachophoregraph holds the electrolytes and houses the migration process. It should be made of a material with adequate conductivity and resistance to electric discharges. There are two

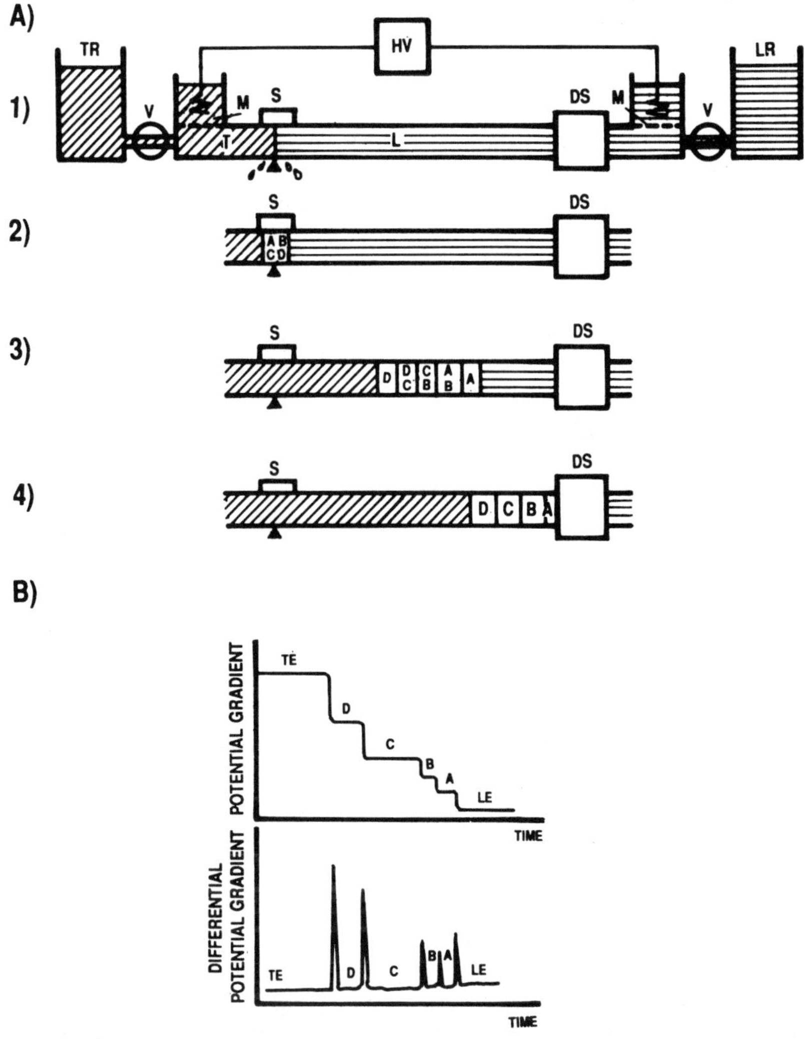

Figure 5.6 (A) *Scheme of an isotachophoremeter and analytical steps involved:* (1) *loading of column with the electrolytes;* (2) *injection of sample;* (3) *separation step;* (4) *detection step.* (B) *Integral and differential potential recordings.* TR, LR = *tailing and leading electrolyte reservoir, respectively;* V = *switching valve;* S = *sample injection system;* D = *detection system;* HV = *high-voltage power supply;* A–D = *analytes*

basic column designs: modular,[52, 53] in which each of the other parts of the instrument is independent so that it can be freely combined (high versatility); and compact,[54] in which the units make up a single block.

The injection system is usually composed of a microsyringe for small (0.1–10 μl)[52] and a sampling valve for large sample volumes (up to one ml).[52, 55] The latter provides pulse-free injections with higher reproducibility.

The separation channel is the essential part of the column as it is there that the electromigration process proper takes place. It should be made of a material with a high electric resistance, good heat conductivity, and a low zeta potential, and be readily connectable to the other parts of the column. This function is usually served by capillaries with inner diameters below 1 mm to take advantage of the stabilizing wall effect and offset the effects of transversal temperature profiles in the zones. When analyses call for larger column hold-ups, a single capillary of uniform diameter is usually inadequate. This shortcoming can be circumvented in two ways: (*a*) by volume coupling, *i.e.* by using two capillaries serially connected via an interchangeable piece of capillary[56] or by providing the separation channel with an auxiliary inlet in the central zone of the capillary;[56] and (*b*) by column coupling,[58,59] *i.e.* by previously isolating and removing interfering bulk substances in a wide channel furnished with a telltale detector and an auxiliary electrode. Multi-column systems[60] represent a further development of the latter alternative. Passage of the current through this unit produces heats, so it is usually immersed in a thermostated system (usually of Peltier type).

The electrodes are usually made of platinum and placed within chambers filled with an appropriate electrode and connected to the separation channel through a semi-permeable membrane. The chambers, the electrolyte reservoirs, and the membranes should all be made of a material resistant to the solvents to be used.

The measuring cell should have as small an inner volume as possible to minimize zonal interference.

The detector should provide a rapid response (within a few tenths of a second) and minimal background noise and drift; in addition, the signal should not be affected by phenomena extraneous to the detection cell. The detectors used in isotachophoresis can be classified into universal and selective. The pioneer among the former, the thermocouple, measures zone heating[61] and is scarcely sensitive and responsive. Conductivity detectors of the contact[53] and non-contact types,[53] and potential gradient detectors, whether conventional[64,65] or multi-channel,[66] are other types of universal sensors commonly used in isotachophoretic techniques. Conventional[67] and dual-wavelength UV absorbance detectors[68] are the universal detectors most commonly used with capillary electrophoresis, followed by the much less frequent fluorescence,[69] radiometric,[70] and polarographic detectors.[71]

The high-voltage power supply used provides the direct driving current, which is automatically stabilized at the preset value (5–500 μA and up to 30 kV).

Automation trends have also reached isotachophoresis. Injection systems and valves are currently actuated by microprocessors or computers via active interfaces, while isotachophoregrams are electronically evaluated by using modified chromatographic integrators or on-line microcomputers.[72] Multi-channel detection allows the separation process to be monitored and the analysis to be evaluated automatically after equilibrium is reached.[66]

5.3.2 Analytical Aspects of Isotachophoresis

The analytes assayed by isotachophoresis can be determined over rather broad ranges as very concentrated components are diluted in the course of the separation process, while trace components are concentrated. Thus, the determination ranges for concentrated and diluted components can easily extend between 10^{-4} and 10^{-1} M for the former, and from 10^{-10} to 10^{-11} M for the latter. The time required for a complete analysis may vary between 30 and 40 minutes.

This technique allows implementation of two types of analysis (qualitative and quantitative), in addition to preparative methods.

Qualitative isotachophoretic analysis is based on the identification of each of the stages recorded in the isotachophoregram by visual contrast with that of a known substance (standard) whose presence in the sample is suspected. Each stage is characterized by two parameters, *viz.* its order and height. The characterization process is complicated by the occurrence of mixed zones, whether transient[59] or stable.[64]

Dubious characterizations can only be confirmed by applying a micropreparative procedure[61] or some specific sample pretreatment.[63]

As a rule, the amount of injected sample used in an isotachophoretic analysis is a compromise between those providing the highest sensitivity and sampling frequency.

Isotachophoresis is particularly suitable for separation/quantitation purposes. Currently available instrumentation allows the simultaneous separation/determination of cations and anions in one run. Quantitative isotachophoretic analysis relies on the fact that the electrical charge generated on passage of the sample zone through the detector is independent of the magnitude of the driving current, the instrumental parameters and the concentration of the leading electrolyte; hence, it can be used to develop an absolute method requiring no calibration.[73] The best way of assessing the amount of electrical charge involved is by coulometry;[74] however, commercially available instrumentation does not usually carry this type of detector. Nevertheless, isotachophoretic quantitation procedures normally include calibration, which can be performed by any of the techniques typically used in column chromatography (internal or external standard, standard addition, *etc.*).[75]

In addition to batch micropreparation,[76] continuous preparative isotachophoresis can be performed with continuous injection of the sample and simultaneous removal of the separated substances.[77, 78] This procedure is implemented on a free-flow system where a flat slit formed by two well cooled glass plates lying 0.5 mm apart serves as the separation channel.

5.3.3 Advanced Isotachophoretic Techniques

Research endeavours in isotachophoresis have been chiefly aimed at widening its scope of application, accomplishing better separations, increasing sensitivity, and extending the concentration ratios of major components to minor components.

Counter-flow isotachophoresis enables prolongation of the effective migration path of the zones, thereby increasing the separation ability of the system without the need to modify the separation channel or increase the applied voltage.[79] Basically, it involves introducing a stream of the leading electrolyte into the column in the opposite direction of zone migration (Figure 5.7A).

Continuous sampling is resorted to when the concentration of the analytes in the sample is so low that the usual injected volume (1–30 μl) results in zones too short for quantitative evaluation (Figure 5.7B).[80]

The *concentration cascade* technique allows the separation efficiency to be increased by using two different concentration levels of the leading electrolyte (Figure 5.7C).[57]

The *volume coupling* technique is also aimed at increasing separation capacity while preserving the usual voltage and sensitivity levels. It involves modifying the separation channel by means of an interchangeable transversal section; the larger it is, the larger the volume of electrolyte between the detector and the injection port and hence the greater the separation capacity will be (Figure 5.7D).

Column coupling (or switching) is the most advanced isotachophoresis mode and provides substantially increased selectivity. It relies on the use of a separation channel consisting of two capillaries: one for pre-separation and the other for the final separation and detection (Figure 5.7E).[58, 59]

5.3.4 Scope of Application

Routine isotachophoretic analyses call for three basic requisites to be met, namely: (*a*) the sample should contain detectable amounts of the compounds to be assayed; (*b*) the sample components should be separated in as short a time as possible; and (*c*) while liquid samples of high enough conductivity can be assayed directly, those of low conductivity are prone to overheat on injection, which can be avoided by adding to them an adequate amount of the leading or terminating electrolyte or by using a very small injected volume.

Isotachophoretic separation-determination procedures have been applied in virtually every area of social interest, particularly in clinical analyses involving body fluids (very often in blood plasma or serum and urine, somewhat less frequently in tissue homogenates and cerebrospinal fluid, and very rarely in saliva, semen, bile, aqueous humor and urinary calculi), with excellent results. The analytes thus determined include a

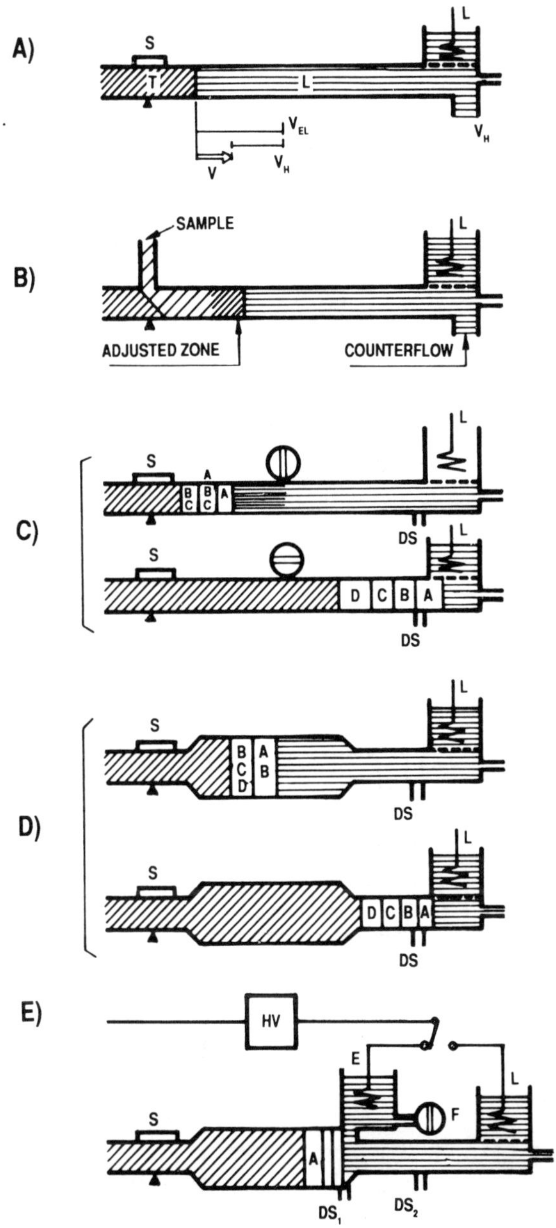

Figure 5.7 *Advanced isotachophoretic techniques:* (A) *Counter-flow,* (B) *continuous sampling,* (C) *concentration cascade,* (D) *volume coupling,* (E) *column coupling.* V_{EL} = *isotachophoretic migration velocity;* V_H = *hydrodynamic flow velocity;* V = *resulting travelling velocity of the zones;* T, L = *tailing and leading electrolyte, respectively;* DS = *detection system;* S = *sample injection system;* HV = *high-voltage power supply;* A–D = *electrolytes;* F = *electrolyte inlet;* DS_1, DS_2 = *tell-tale detector and analytical detector, respectively*

wide variety of substances such as carboxylic acids,[81] amino-acids and peptides,[82] purines, pyrimidines and organic bases,[83] nucleosides and nucleotides,[84] proteins[85] and inorganic anions,[86] among others.

Drugs such as antibiotics,[87] peptides,[88] and alkaloids,[89] and their degradation products have also been successfully analysed isotachophoretically. Figure 5.8 shows a normal and a differential lipidic recording obtained in an analysis for ephedrine and procaine.

Isotachophoresis has also been applied to the analysis for taste additives in foodstuffs,[89] as well as in environmental analyses on air,[91] water[92] and soil samples,[93] and in agricultural [94] and industrial analyses.[95]

The electrolytes and concentrations most frequently used in isotachophoretic analyses are as follows:

(*a*) Separation–determination of anionic species. Leading electrolyte: 5–10 mM HCl + β-phenylalanine, 1–7 mM HCl + glycylglycine and 5–10 HCl + Ammdiol. Tailing electrolyte: 5–10 mM caproic, propionic, or acetic acid, or 10 mM GABA, EACA or β-alanine + $Ba(OH)_2$.

(*b*) Separation and analysis for cationic species. Leading electrolyte: 5–20 mM potassium acetate + acetic acid. Tailing electrolyte: 20 mM GABA, acetic acid, or β-alanine.

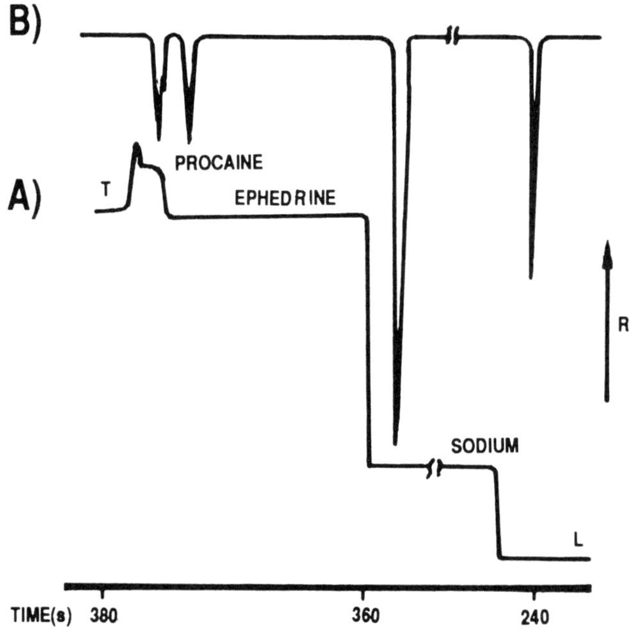

Figure 5.8 *Test for the cationic composition of the pharmaceutical preparation Solutan*™. (A) *Integral and* (B) *differential recording*
(Reproduced from reference 88 with permission of Elsevier Science Publishers)

5.4 Scope of Non-chromatographic Continuous Separation Techniques

Non-chromatographic continuous separation techniques come into play in some stage of the preliminary operations of the analytical process. Their intrinsically dynamic nature makes automation easier than with other analytical alternatives, which in turn results in a number of assets such as a high compatibility with continuous analysers, increased precision and sample throughput, reduced sample and reagent consumption and lower analytical costs, to name a few. Thus, it is hardly surprising that the number of papers reported on this topic has soared in the past few years[96,97] and that modules for implementation of a continuous separation technique to be fitted to available chromatographs, atomic absorption spectrophotometers, *etc.* have started to be marketed. To the authors' minds, this is the best proof that this topic has a solid future despite the fact that, with few exceptions (*e.g.* the incorporation of dialysers into continuous unsegmented analysers marketed by Technicon and Skalar), developments in this area have so far been of scientific rather than applied nature. The extensive development of continuous unsegmented analysers (FIA, CCFA)[99-101] has helped in the maturation process of continuous separation techniques, which are today regarded as a useful and appealing tool for solving problems associated with the preliminary operations and of great aid for such basic purposes as sample clean-up and preconcentration, which in turn are indirect means of increasing the sensitivity and selectivity of the analytical process.

The varied foundations of the wide range of non-chromatographic continuous separation techniques available today allow them to solve a variety of analytical problems in such areas as environmental pollution, clinical and pharmaceutical chemistry, and food and industrial analysis, among others. The description of each continuous separation technique in this book ends with a section dealing with its analytical applications in the above-mentioned areas. As a whole, they appear to have a promising future as they make a useful, convenient alternative to the determination of analytes in a variety of matrices (biological fluids, liquid and solid foods, vegetable and animal materials, alloys, *etc.*).

The technique of choice in each case is dictated by the physico-chemical features of the analytes and their potential interferents, as well as by the analytical requirements and the characteristics of the continuous detection system to be used. Thus, if the analytes or their reaction products are volatile, the gas–liquid systems described in Chapter 2 are the most suitable. On the other hand, solid samples are best dealt with by automatic leaching (Chapter 4), while for analytes with much smaller ionic or molecular sizes than their potential interferents—quite a frequent occurrence in clinical chemistry—dialysis should ideally be applied, which is described in detail in Chapter 3. For preconcentration, the continuous sorption systems dealt with in Chapter 4 offer a variety of solutions for a host of analytes, both ionic (ion-exchange) and molecular (*e.g.* bonded silica phase). Indirect determinations of organic and inorganic anions by atomic absorption spectrophotometry are good candidates for application of a wide variety of continuous separation techniques such as liquid–liquid extraction, precipitation and ion-exchange.[102, 103] Some techniques are more versatile than others; thus, liquid–liquid extraction can be applied to the determination of a broad range of analytes, whereas gas diffusion is only applicable to volatile species and others that can be volatilized upon chemical reaction, and electrochemical stripping is virtually solely suitable for metal species (some anions and organic species, however, can indeed be determined by cathodic and adsorptive stripping, respectively).

Not all of the continuous separation techniques described in this book are equally suitable for use in routine laboratories. The best options are always those featuring (*a*) simplicity in the continuous configuration required for implementation, (*b*) ease of handling, (*c*) reduced human participation, (*d*) a short start-up time, (*e*) smooth functioning for a long time or a large series of samples without the need for part replacement, and (*f*) repeatable, reproducible results. All these requisites should be met as far as possible by any continuous separation technique to be routinely applied in laboratories through commercial instrumentation.

5.5 Batch *versus* Non-chromatographic Continuous Separation Techniques

Comparing these two technical alternatives to implementing a given separation process on the basis of the same principles and type of interface requires distinguishing between several possibilities that are commented on below.

Whenever application of a given separation technique in batchwise fashion is unfeasible, *i.e.* it cannot be routinely used in an analytical control laboratory, whether or not it can be applied for research purposes such as the elucidation of chemical, physico-chemical, and biochemical mechanisms, or the calculation of equilibrium constants, which is the case with ion-exchange, dialysis, solid–liquid extraction, and gas diffusion among others, the corresponding continuous alternative comes to the rescue by providing efficient solutions to specific analytical problems as shown in Chapters 2–4 of this book. The dynamics introduced by the flowing streams of one or two phases is the key to their practical implementation.

When both the batch and the continuous version of a given separation technique are technically suitable for routine analyses (*e.g.* liquid–liquid extraction, hydride generation, electrochemical stripping, precipitation), the continuous alternative offers the following advantages: (*a*) easier incorporation as modules into automatic analysers, (*b*) reduction of hazards involved in the manipulation of toxic reagents and solvents by the manual operator, (*c*) increased precision resulting from the minimization of errors associated with the so-called 'human factor', (*d*) substantially increased sample throughput, (*e*) dramatically reduced sample and reagent consumption, and (*f*) lower analytical costs. These assets have fostered the development of a host of continuous separation methodologies by adaptation of the corresponding well established existing batch methodologies, which are often—though not always—used as guidelines in designing and optimizing experimentally implementable methods.

Thus, continuous methodologies offer clearly higher levels of such basic analytical features as precision and rapidity than their batch counterparts; in addition, they occasionally enable practical application of the technique concerned.

Other basic analytical features such as sensitivity and selectivity are also improved in converting a batch method to a continuous method, mainly because of chemical and physico-chemical kinetic factors. In batch modes, contact times between phases are usually quite long and the mass transfer process through the interface normally reaches equilibrium; moreover, the interface area can be increased to a great extent in order to favour mass transfer, as is the case with the formation of emulsions in liquid–liquid extraction as implemented in separatory funnels. On the other hand, in continuous separation processes, the very dynamics of the systems involved prevents attainment of thermodynamic equilibrium in most cases. This not only affects the separation process proper, but also may lead to incomplete development of complementary reactions intended to yield separable products in unsegmented continuous-flow systems, which is one of the essential features of FIA.[99, 101] As an immediate result, sensitivity can be somewhat lower (between 5 and 15%) than that afforded by the corresponding batch alternative. This drawback is largely offset by the increased selectivity resulting from the so-called 'kinetic discrimination'.[104] Potentially interfering substances acting either on the preliminary chemical reactions or on mass transfer have a very short time to exert a significant action as the system is optimized for the analyte rather than the accompanying species. Systematic studies on this topic in relation to liquid–liquid extraction,[105] gas diffusion,[106] and precipitation[107] testify to the exactness of this assertion, which is of paramount significance as it allows separation techniques to be applied to real analytical problems as described in the preceding chapters.

5.6 Chromatographic *versus* Non-chromatographic Continuous Separation Techniques

Many non-chromatographic continuous separation techniques such as dialysis, gas diffusion, precipitation, electrochemical stripping, leaching, and hydride generation have no chromatographic counterpart. Despite the formal similarities between some techniques of the two types (*e.g.* dialysis and exclusion chromatography, both of which rely on ionic or molecular size differences, electrochemical stripping and ionic chromatography, which allow the individual determination of ions in complex mixtures on the basis of a rather different principle), no practical conclusions can be drawn on comparing these two broad categories of continuous separation techniques as comparisons would be mostly artificial.

Neither can any significant conclusions be drawn on comparing planar chromatographic alternatives (*e.g.* paper, thin-layer, high-performance thin-layer) with non-chromatographic separation techniques as the two differ significantly in the way they are implemented (discrete and continuous flow) or in the type of detection used (batch and continuous).

On the other hand, column chromatographies (*e.g.* GC and HPLC) bear interesting similarities to some continuous separation techniques, either in their foundation, the type of continuous system used for implementation, or both.

Thus, liquid–liquid extraction as implemented in a continuous fashion in the configurations described in Chapter 3 relies basically on the same principle as liquid partition chromatography, which in turn is quite similar to that of bonded-phase liquid chromatography, even though the two are implemented in rather a different way. The number of equilibrations involved in the non-chromatographic mode is much smaller, so the efficiency (number of theoretical plates) of the chromatographic alternative is much greater.

The continuous configurations used to implement sorption processes (*e.g.* ion exchange), which are described in detail in Chapter 4, bear some formal resemblance to a liquid chromatograph. Both use the same modules (solvent reservoirs, propulsion system, injector, packed column and continuous separation system), although the working pressure is rather different. In addition, the foundation of the separation process proper is the same insofar as the interface involved is of the same type, and so is the active solid material used and the type of analyte that takes part in the mass transfer between the phases concerned. The basic difference lies in the fact that retention and elution are performed simultaneously on a group of analytes in the low-pressure, non-chromatographic mode, while they are done by steps in its high-pressure chromatographic counterpart, which enables analyte discrimination. The reason behind this difference in separation efficiency is very simple: the higher pressure used in the chromatographic mode allows the use of a finely divided solid phase, so the active surface area (interface) offered is much larger than in the non-chromatographic mode. Hence the analytical purpose of the two techniques is quite different. While chromatographic techniques are aimed at separation of analytes from one another and from the interferences, their non-chromatographic counterparts are chiefly devoted to clean-up and preconcentration and are better suited to processing a large number of samples in a short time and providing the overall analyte concentration unless they are connected on-line to a chromatograph.

5.7 Trends

As noted elsewhere in this chapter, non-chromatographic continuous separation techniques are a promising reality in the context of analytical chemistry. Once the hurdles hindering extensive application in routine control laboratories have been overcome, the magical 'barrier' that separates the basic development of every novel analytical methodology and its massive use for solving real analytical problems will have also been broken. That was the position of gas and liquid chromatography about three and two decades ago, respectively. Breaking such a barrier and reaching a well established status requires analytical instrumentation manufacturers to bet firmly on these techniques. In this respect, Perkin-Elmer recently marketed two automatic continuous hydride-generation and amalgamation modules for incorporation into atomic absorption instruments. Also, some firms such as Gilson have started commercialization of automatic dialysers for use as pre-column modules in HPLC.

This consolidation process can be aided by introducing the technique to the analytical community in general and to those in charge of control laboratories in particular. This was the primary reason that prompted us to write this book. On the other hand, basic research aimed at miniaturization, the development of new modes, and finding new materials with suitable properties for separation purposes and new types of energy facilitating the mass transfer process should be encouraged, as should contributions of further evidence of the great potential of these techniques for solving present analytical problems (*e.g.* from within analytical biotechnological on-line process control systems).

Modes currently in an incipient stage of development but with a seemingly high potential (*e.g.* some involving continuous configurations where separation is integrated with detection, and others relying on automatic continuous leaching systems that enable direct introduction of undissolved, undisaggregated solid samples) should be investigated more deeply with a view to future consolidation.

Non-chromatographic continuous separation techniques lend themselves particularly readily to on-line combination with other analytical systems and are therefore a promising source for hyphenated methodologies, one of the most salient current trends in analytical chemistry. Development of such methodologies can be approached in different ways, namely:

(*a*) By combining two or more non-chromatographic continuous separation techniques, whether on-line or by integrating two separation processes in a single separation unit, with the potentially highly interesting associated complementary advantages.

(*b*) By on-line coupling of non-chromatographic analytical separation techniques with others such as gas (pre-column arrangement) or liquid chromatography (pre- and post-column arrangements) for automatic clean-up, preconcentration or derivatization, which are of such great interest to chromatography.

(*c*) By on-line coupling to currently widely used instrumental techniques (*e.g.* atomic or mass spectrometry) for the automation of pretreatment operations (*e.g.* preconcentration, indirect determinations, volatilization, solvent changeover), which are indispensable to the obtainment of good analytical results and preserving the instrument integrity.

(*d*) By co-ordinate coupling of continuously operating systems with others functioning in a discrete fashion with a view to overall automation. This requires designing appropriate interfaces, which is no doubt the limiting factor of this development. Two cases in point are the joint use of a continuous separation system and an electrothermal-vaporization atomic absorption spectrophotometer, and the combination of a typically discrete analytical technique such as lyophilization with a continuous analyser.

References

1. S.M. Ramasamy and H.A. Mottola, *Anal. Chem.*, 1982, **54**, 283.
2. J. Janca 'Field-Flow Fractionation. Analysis of Macromolecules and Particles', M. Dekker, Inc., New York, 1987.
3. J. Janca, *Makromol. Chem. Rapid Commun.*, 1982, **3**, 887.
4. J. Janca, *Makromol. Chem. Rapid Commun.*, 1983, **4**, 267.
5. G.H. Thompson, M.N. Myers, and J.C. Giddings, *Sep. Sci. Technol.*, 1967, **2**, 797.
6. G.H. Thompson, M.N. Myers, and J.C. Giddings, *Anal. Chem.*, 1969, **41**, 1219.
7. J.C. Giddings, *Sep. Sci. Technol.*, 1966, **1**, 123.
8. H.C. Berg and E.M. Purcell, *Proc. Natl. Acad. Sci. USA*, 1967, **58**, 862.
9. H.C. Berg, E.M. Purcell, and W.W. Stewart, *Proc. Natl. Acad. Sci. USA*, 1967, **58**, 1286.
10. H.C. Berg and E.M. Purcell, *Proc. Natl. Acad. Sci. USA*, 1967, **58**, 1821.
11. K.D. Caldwell, L.F. Kesner, M.N. Myers, and J.C. Giddings, *Science*, 1972, **176**, 296.
12. J.C. Giddings, F.J. Yang, and M.N. Myers, *Science*, 1976, **193**, 1244.
13. J.C. Giddings, F.J. Yang, and M.N. Myers, *Anal. Chem.*, 1976, **48**, 1126.
14. H.L. Lee and E.N. Lightfoot, *Sep. Sci. Technol.*, 1976, **11**, 417.
15. H.L. Lee, J.F.G. Reis, J. Dohner, and E.N. Lightfoot, *AIChE J.*, 1974, **20**, 776.
16. T.M. Vickrey and J.A. García-Ramírez, *Sep. Sci. Technol.*, 1980, **15**, 1297.
17. T.C. Schunk, J. Corse, and M.F. Burke, *Sep. Sci. Technol.*, 1984, **19**, 653.
18. J. Corse, T.C. Schunk, and M.F. Burke, *Sep. Sci. Technol.*, 1985, **19**, 1073.
19. J.C. Giddings, F.J. Yang, and M.N. Myers, *Sep. Sci. Technol.*, 1977, **12**, 381.
20. J.C. Giddings and S.L. Brantley, *Sep. Sci. Technol.*, 1984, **19**, 631.
21. J. Janca, *Makromol. Chem. Rapid Commun.*, 1982, **3**, 887.
22. J. Chmelik and J. Janca, Proc. Natl. Conf. Chem. Soc., Banska Stiavnica, 1984, pp. 38.

23. K.D. Caldwell, *Anal. Chem.*, 1988, **60**, 959A.
24. J. Janca, M. Janicek, D. Pribylova, and M. Klesnil, *Anal. Instrum.*, 1986, **15**, 149.
25. J.C. Giddings, M.E. Hovingh, and G.H. Thompson, *J. Phys. Chem.*, 1970, **74**, 4291.
26. J.C. Giddings, K.D. Caldwell, and M.N. Myers, *Macromolecules*, 1976, **9**, 106.
27. S.L. Brimhall, M.N. Myers, K.D. Caldwell, and J.C. Giddings, *J. Polym. Sci. Polym. Phys. Ed.*, 1985, **23**, 2443.
28. J.C. Giddings, L.K. Smith, and M.N. Myers, *Anal. Chem.*, 1975, **47**, 2389.
29. M. Martín and R. Reynaud, Collect. Colloq. Semin. Inst. Fr. Pet., 40, Charact. Huiles Lourdes Residus Pet., 1984, p. 263.
30. J.C. Giddings, M. Martín, and M.N. Myers, *J. Chromatogr.*, 1978, **158**, 419.
31. J. Janca and K. Kleparnik, *Sep. Sci. Technol.*, 1981, **16**, 657.
32. J. Janca, D. Pribylova, K. Bouchal, V. Tyrackova, and E. Zurkova, *J. Liq. Chromatogr.*, 1986, **9**, 2059.
33. L.E. Oppenheimer and T.H. Mourey, *J. Chromatogr.*, 1984, **298**, 217.
34. A. Fox, L.E. Schallinger, and J.J. Kirkland, *J. Microbiol. Methods*, 1985, **3**, 273.
35. L.F. Kesner, K.D. Caldwell, M.N. Myers, and J.C. Giddings, *Anal. Chem.*, 1976, **48**, 1834.
36. J.C. Giddings, G.C. Lin, and M.N. Myers, *Sep. Sci.*, 1976, **11**, 553.
37. J.F.G. Reis and E.N. Lightfoot, *AIChE J.*, 1976, **22**, 779.
38. A.S. Chiang, E.H. Kmiotek, S.M. Langan, P.T. Noble, J.F.G. Reis, and E.N. Lightfoot, *Sep. Sci. Technol.*, 1979, **14**, 453.
39. J.C. Giddings, F.J. Yang, and M.N. Myers, *Anal. Biochem.*, 1977, **81**, 395.
40. J.C. Giddings, G.C. Lin, and M.N. Myers, *J. Liq, Chromatogr.*, 1978, **1**, 1.
41. S.L. Brimhall, M.N. Myers, K.D. Caldwell, and J.C. Giddings, *Polym. Mater. Sci. Eng.*, 1984, **50**, 48.
42. J.C. Giddings, F.J. Yang, and M.N. Myers, *Sep. Sci. Technol.*, 1977, **12**, 499.
43. H.L. Lee, J.F.G. Reis, J. Dohner, and E.N. Lightfoot, *AIChE J.*, 1974, **20**, 776.
44. T.M. Vickrey and J.A. García-Ramírez, *Sep. Sci. Technol.*, 1980, **15**, 1297.
45. T.C. Schunk, J. Corse, and M.F. Burke, *Sep. Sci. Technol.*, 1984, **19**, 653.
46. J. Corse, T.C. Schunk, and M.F. Burke, *Sep. Sci. Technol.*, 1985, **19**, 1073.
47. K.A. Graff, K.D. Caldwell, M.N. Myers, and J.C. Giddings, *Fuel*, 1984, **63**, 621.
48. K.D. Caldwell, Z.Q. Cheng, P. Hradecky, and J.C. Giddings, *Cell Biophys.*, 1984, **6**, 233.
49. J. Janca and V. Jahnova, *J. Liq. Chromatogr.*, 1983, **6**, 1559.
50. F.M. Everaerts, J.L. Beckers, and P.E.M. Verheggen, Th., 'Isotachophoresis. Theory, Instrumentation and Applications', Elsevier, Amsterdam, 1976.
51. P. Bocek, M. Deml, P. Gebauer, and V. Dolnik, 'Analytical Isotachophoresis', ed. B.J. Radola, VCH publishers, 1988.
52. F.M. Everaerts and P.E.M. Verheggen, Th., *J. Chromtogr.*, 1970, **53**, 315..
53. F.M. Everaerts, J.L. Beckers, and P.E.M. Verheggen, Th., 'Isotachophoresis. Theory, Instrumentation, and Applications', Elsevier, Amsterdam, 1976, pp. 1–418.
54. P. Bocek, M. Delm, and J. Janak, *J. Chromatogr.*, 1975, **106**, 283.
55. D. Kamiansky and P. Havasi, 'Isotachophoresis. Basic course, Advanced course', ed. P. Bocek, Institute of Radioecology and Applied Nuclear Techniques, Spisska Nova Ves, Czecoslovakia, 1984, pp. 122–145.
56. P.E.M. Verheggen, Th., and F.M. Everaerts, *J. Chromatogr.*, 1982, **249**, 221.

57. P. Bocek, M. Delm, and J. Janak, *J. Chromatogr.*, 1978, **156**, 323.
58. P.E.M. Verheggen, Th., F.E.P. Mikkers, D.M.J. Kroonenberg, and F.M. Everaerts, *Anal. Chem. Symp. Ser.*, 1980, **5**, 41.
59. F.M. Everaerts, P.E.M. Verheggen, Th., and F.E.P. Mikkers, *J. Chromatogr.*, 1979, **169**, 21.
60. V. Dolnik, M. Delm, and P. Bocek, *J. Chromatogr.*, 1985, **320**, 89.
61. A.J.P. Martín and F.M. Everaerts, *Anal. Chim. Acta*, 1967, **38**, 233.
62. B.P. Konstantinov and O.V. Oshurkova, *Dokl. Akad. Nauk SSSR*, 1963, **148**, 1110.
63. J. Vacik, J. Zuska, and I. Muselasova, *J. Chromatogr.*, 1985, **320**, 233.
64. T. Haruki and J. Akiyama, *Anal. Lett.*, 1973, **6**, 985.
65. M. Delm, P. Bocek, and J. Janak, *J. Chromatogr.*, 1975, **109**, 49.
66. E. Schumacher, W. Thormann, and D. Arn, *Anal. Chem. Symp. Ser.*, 1981, **6**, 33.
67. L. Arlinger and H. Lundin, *Protides Biol. Fluids*, 1973, **21**, 667.
68. J.C. Reijenga, P.E.M. Verheggen, Th., and F.M. Everaerts, *J. Chromatogr.*, 1983, **267**, 75.
69. J.C. Reijenga, P.E.M. Verheggen, Th., and F.M. Everaerts, *J. Chromatogr.*, 1984, **283**, 99.
70. D. Kaniansky, P. Rajec, A. Svec, P. Havasi, and F. Macasek, *J. Chromatogr.*, 1983, **258**, 238.
71. D. Kaniansky, P. Havasi, J. Marak, and R. Sokolik, *J. Chromatogr.*, 1986, **366**, 153.
72. J.C. Reijenga, W. van Iersel, G.V.A. Aben, P.E.M. Verheggen, Th., and F.M. Everaerts, *J. Chromatogr.*, 1984, **292**, 217.
73. ITP 84, 4th International Symposium on Isotachophoresis, Symposium Abstracts. Institute of Analytical Chemistry, Czechoslovak, Academy of Sciences, ed. M. Svoboda, X. Svobodova, and J. Bacik, P. Bocek, Brno, 1984, p. 43.
74. P. Bocek, M. Delm, J. Stikarovsky, and J. Janak, 'Progress in Chromatography' (2nd Danube Symposium), Carlsbad, Czechoslovak Chemical Society, Prague, 1979, pp. 42.
75. J. Novak, 'Quantitative Analysis by Gas Chromatography', M. Dekker, New York, 1975.
76. Shimadzu Capillary-Type Isotachophoretic Analyzer IP-2A, Shimadzu Corp., Kyoto, 1985.
77. Z. Prusik, *J. Chromatogr.*, 1974, **91**, 867.
78. H. Wagner and V. Mang, *Anal. Chem. Symp. Ser.*, 1981, **6**, 41.
79. F.M. Everaerts, J. Vacik, P.E.M. Verheggen, Th., and J. Zuska, *J. Chromatogr.*, 1970, **49**, 262.
80. Z. Ryslavy, P. Bocek, M. Delm, and J. Janak, *J. Chromatogr.*, 1978, **147**, 369.
81. J.C. Reijenga, H.J.L.A. Slaats, and F.M. Everaerts, *J. Chromatogr.*, 1983, **267**, 85.
82. I. Sziklai, J. Grof, O. Ribari, J. Manyhart, and P. Piffko, *Acta Oto. Laryngol.*, 1985, **99**, 25.
83. J. Konecna, L. Krivankova, and P. Bocek, *Biochem. Clin. Bohemoslov.*, 1987, **16**, 67.
84. B. Gustavsson, O. Almersjo, M. Berne, and J. Waldenstrom, *J. Chromatogr.*, 1983, **276**, 395.
85. G. Schmitz, U. Borgmann, and G. Assmann, *J. Chromatogr.*, 1985, **320**, 253.

86. W. Tschope, R. Brenner and E. Ritz, 'Analytical and Preparative Isotachophoresis', ed. C.J. Holloway, Walter de Gruyter, Berlín, 1984, pp. 207.
87. H. Klein and R. Teichmann, *J. Chromatogr.*, 1982, **250**, 152.
88. R. Jannasch, *Pharmazie*, 1983, **38**, 379.
89. S. Fanali, F. Foret, and P. Bocek, *J. Chromatogr.*, 1985, **330**, 436.
90. K. Rubach and P. Offizors, *Dtsch. Lebensm. Rundsch.*, 1983, **79**, 88.
91. L. Hansen, J. Sollenberg, and K. Wiberg, *J. Chromatogr.*, 1984, **312**, 489.
92. P.A. Pfeifer, G.K. Bonn, and O. Bobleter, *Fresenius' Z. Anal. Chem.*, 1983, **315**, 205.
93. Z. Stransky, *J. Chromatogr.*, 1985, **320**, 219.
94. L. Krivankova and P. Bocek, *Electrophoresis*, 1985, **6**, 143.
95. V. Madajova, D. Kaniansky, E. Cizmarova, and M. Hudec, *J. Chromatogr.*, 1985, **320**, 131.
96. M. Valcárcel and M.D. Luque de Castro, *J. Chromatogr.*, 1987, **3**, 393.
97. M. Valcárcel and M.D. Luque de Castro, 'Continuous Separation Techniques in FIA', Chapter VII of 'Selective Sample Handling and Detection in HPLC', ed. K. Zech and R.W. Frei, Elsevier, Amsterdam, 1989.
98. W.B. Furtman 'Continuous Flow Analysis. Theory and Practice', M. Dekker, New York, 1976.
99. M. Valcárcel and M.D. Luque de Castro 'Flow Injection Analysis. Principles and Applications', Ellis Horwood, Chichester, 1987.
100. J. Ruzicka and E. Hansen 'Flow Injection Analysis', 2nd edn., Wiley, New York, 1988.
101. B. Karlberg and G.E. Pacey 'Flow Injection Analysis. A Practical Guide', Elsevier, Amsterdam, 1989.
102. M. Valcárcel 'Flow Injection Analysis', Chapter 11 of 'Sample Introduction in Atomic Spectroscopy', ed. J. Snedon, Elsevier, Amsterdam, 1990.
103 M. Valcárcel and M. Gallego, Chapter 5 of 'Flow Injection Atomic Spectroscopy', ed. J.L. Burguera, M. Dekker, New York, 1989.
104. M. Valcárcel, *Analyst (London)*, 1987, **112**, 729.
105. M. Gallego, M. Silva, and M. Valcárcel, *Anal. Chem.*, 1986, **58**, 2265.
106. G.E. Pacey, D.A. Hollowell, K.G. Miller, M.R. Straka, and G. Gordon, *Anal. Chim. Acta*, 1986, **179**, 259.
107. P.Martínez Jiménez, M. Gallego, and M. Valcárcel, *J. Anal. At. Spectrom.*, 1987, **2**, 211.

Subject Index